FUNDAMENTAL CONCEPTS IN ENVIRONMENTAL STUDIES

For Undergraduate Students

Dr. D.D. MISHRA
M.Sc., Ph.D., F.I.C.S., M.N.A.Sc.
Former Principal
Government College, Udaipura, Raisen (M.P.)

S Chand And Company Limited
(ISO 9001 Certified Company)

S Chand And Company Limited

(ISO 9001 Certified Company)

Head Office: D-92, Sector–2, Noida – 201301, U.P. (India), Ph. 91-120-4682700

Registered Office: A-27, 2nd Floor, Mohan Co-operative Industrial Estate, New Delhi – 110 044, Phone: 011-49731800

www.**schandpublishing.com**; e-mail: **info@schandpublishing.com**

Marketing Offices:

Chennai : Ph: 23632120; chennai@schandpublishing.com
Guwahati : Ph: 2738811, 2735640; guwahati@schandpublishing.com
Hyderabad : Ph: 40186018; hyderabad@schandpublishing.com
Jalandhar : Ph: 4645630; jalandhar@schandpublishing.com
Kolkata : Ph: 23357458, 23353914; kolkata@schandpublishing.com
Lucknow : Ph: 4003633; lucknow@schandpublishing.com
Mumbai : Ph: 25000297; mumbai@schandpublishing.com
Patna : Ph: 2260011; patna@schandpublishing.com

First Edition 2008
Revised Edition & Reprints 2009, 2010, 2014 (Twice), 2016, 2017 (Twice), 2018 (Twice), 2019 (Thrice), 2020 (Twice), 2022 (Twice), 2024 (Twice)

Reprint 2025 (Twice)

ISBN: 978-81-219-2937-0 **Product Code:** H6EVS64ENVS10ENZX0XR

PRINTED IN INDIA

By Vikas Publishing House Private Limited, Plot 20/4, Site-IV, Industrial Area Sahibabad, Ghaziabad – 201 010 and Published by S Chand And Company Limited, A-27, 2nd Floor, Mohan Co-operative Industrial Estate, New Delhi – 110 044.

Dedicated to

My Father

Late Shri Ram Babu Mishra

Whose Inspiration is Always with Me

PREFACE TO THE REVISED EDITION

Environmental science is an interdisciplinary subject that integrates the Physical and Biological Sciences with other academic fields of study. Basically when Environmental Science as a subject was introduced in the undergraduate syllabus of all Indian Universities, this book "Fundamental Concepts in Environmental Studies" had been designed to meet the requirements. The book was liked by all including students and faculty members. It covers complete syllabus as per UGC curriculum for undergraduate students of all disciplines of various Universities.

Keeping in view the present syllabus, suggestions and questions papers of different Universities, the book is thoroughly revised, so that it will be more effective and useful for competitive exams also in view of the present scenario.

In addition to certain corrections, topics like Hydrologic Cycle, Air Pollution, Solar and Wind Energies are modified in the light of present requirement. Some new topics like Dissolved Oxygen, Biological Oxygen Demand, Chemical Oxygen Demand, Natural Geysers, Environmental Club, Green Accounting, Honey and Bee Keeping, Social Forestry are also introduced. With additional data, new topics and necessary diagrammes, the book will be of immense use and more popular among students and readers.

Criticism and suggestions from readers will be greatly acknowledged.

AUTHOR

SYLLABUS

Core Module Syllabus of Environmental Studies for Undergraduate Courses of All Branches of Higher Education Foundation Course Environmental Studies

(Questions will be set from each Unit/Section)

(For B.A., B.Com., B.Sc. (Home Science), B.A. (Mang.), B.B.A., B.C.A. and Other Undergraduate Courses)

Theory 40 **Internal 10** **Total 70 marks**

UNIT-I: The Multidisciplinary Nature of Environmental Studies

Definition, scope and importance

Need for public awarencess (2 lectures)

UNIT-II : Natural Resources, Renewable and Non-renewable Resources

Natural resources and associated problems.

(*a*) **Forest Resource :** Use and over-exploitation, deforestation, case studies. Timber extraction, mining, dams and their effects on forests and tribal people.

(*b*) **Water Resources :** Use and over-utilization of surface and ground water, floods, drought, conflicts over water, dams-benefits and problems.

(*c*) **Mineral Resources :** Use and exploitation, environmental effects of extracting and using mineral resources, cases studies.

(*d*) **Food Resources :** World food problems, changes caused by agriculture and overgrazing, effects of modern agriculture, fertilizer-pesticide problem, waterlogging, salinity, case studies.

(*e*) **Energy Resources :** Growing energy needs, renewable and non-renewable energy sources, use of alternate energy sources. Case studies.

(*f*) **Land Resources :** Land as a resource, land degradation, man-induced landslides, soil erosion and desertification.

— Role of an individual in conservation of natural resources. (8 lectures)

— Equitable use of resources for sustainable lifestyles.

UNIT-III: Ecosystems

- Concept of an ecosystem.
- Structure and function of an ecosystem.
- Producers, consumers and decomposers.
- Energy flow in the ecosystem.
- Ecological succession.
- Food chains, food webs and ecological pyramids.
- Introduction, types, characteristic features, structure and function of the following ecosystems :
  - (*a*) Forest ecosystem
  - (*b*) Grassland ecosystem (6 lectures)
  - (*c*) Desert ecosystem
  - (*d*) Aquatic ecosystems (ponds, streams, lakes, rivers, oceans, estuaries)

UNIT-IV: Biodiversity and Its Conservation

- Introduction-definition : genetic, species and ecosystem diversity.
- Biogeographical classification of India.
- Value of biodiversity : consumptive use, productive use, social, ethical, aesthetic and other values.

- Biodiversity at global, national and local levels.
- India as a mega-diversity nation. (8 lectures)
- Hot-spots of biodiversity.
- Threats to biodiversity : habitat loss, poaching of wildlife, man, wildlife conflicts.
- Endangered and endemic species of India.
- Conservation of biodiversity : in-situ and ex-situ conservation of biodiversity.

UNIT-V : Environmental Pollution

Definition

- Causes, effects and control measures of :
  - (*a*) Air pollution
  - (*b*) Water pollution
  - (*c*) Soil pollution
  - (*d*) Marine pollution
  - (*e*) Noise pollution
  - (*f*) Thermal pollution
  - (*g*) Nuclear hazards
- Solid waste management : causes, effects and control measure of urban and industrial wastes.
- Role of an individual in prevention of pollution.
- Pollution case studies.
- Disaster management : floods, earthquake, cyclone and landslides. (8 lectures)

UNIT-VI : Social Issues and the Environment

- From unsustainable to sustainable development
- Urban problem related to energy.
- Water conservation, rain water harvesting, watershed management.
- Resettlement and rehabilitation of people; its problems and concerns. Case studies.
- Environmental ethics : issues and possible solutions.
- Climte change, global warning, acid rain, ozone layer depletion, nuclear accidents and holocaust. Case studies.
- Wasteland reclamation.
- Consumerism and waste products.
- Environment Protection Act.
- Air (Prevention and Control of Pollution) Act.
- Wildlife Protection Act.
- Forest Conservation Act.
- Issues involved in enforcement of environmental legislation.
- Public awareness. (7 lectures)

UNIT-VII : Human Population and the Environment

- Population growth, variation among nations.
- Population explosion-Family Welfare Programme.
- Environment and human health.
- Human rights.
- Value education.
- HIV/AIDS.
- Women and child welfare.
- Role of information technology in environment and human health.
- Case studies. (6 lectures)

CONTENTS

UNIT 1

Multidisciplinary Nature of Environmental Studies

DEFINITION

The word 'environment' is derived from the French word **'environner'** means "to encircle or surround." It is a composite word for the conditions/surroundings in which organism or group of organisms live. The environment is a very wide term. It includes total physical and biotic word, in which biological beings live, grow, get nourished and develop their natural characteristics. In otherwords, it concerns with the "**Biosphere**" which include all biotic parts of hydrosphere, lithosphere and atmosphere. The environment consists of both biotic and abiotic substances, i.e., consists of air water, food, sunlight, temperature, electricity, etc. Thus environment can be defined in a number of ways, but common definition is —

"Environment is the sum of all social, economical biological, physical or chemical factors which constitute the surroundings of men/living organism, who is both creator and moulder of this environment."

The environment for any living organism is always been changing never constant or static. This change is sometimes slow, some times rapid or so. Some of these changes are irreversible (e.g. eutrophication of lake) while others are cyclic (e.g. the annual climatic cycle) or transient (i.e., droughts). Now because natural biogeographical environment fluctuate with time, it is not easy to distinguish change brought about by man.

Like other organisms, man is also affected by environment. These changes in environment may benefit or harm the man or other organisms living in it. Environmental science is multidisciplinary branch of science involving, chemistry, physics, life science, agriculture public health, botany, medical sciences, geography, and many other fields. Environmental science or studies is the study of the characteristics, composition functions and systematic study of different components of the natural environmental systems. The environment includes both physical or non living (abiotic) and living (biotic) environment. Economics, sociology, education and mass-communication do help in understanding the socio-economic aspects of environment. Mathematics, statistics and computer science also help in modeling and management of environment.

With increasing scientific knowledge, man is able to modify the environment to suit his immediate needs much more than any other organism. Since the very beginning of human civilization man started interfering with the environment. He devastated forests for the use of tree as wood, land under cultivation. He had polluted the rivers and other water resources. The traditional concept that, natural resources are abundant for man to use or abuse has been responsible for massive degeneration of nature, natural systems, environment and wild life. The natural systems in which man exists along with all other species must maintained in a healthy and functional state.

The environmental science is, therefore, a mutidisciplinary science, which may require attention of experts from different branches of science when dicisions ragarding environmental matters have to be taken. In industrialized developing countries of the world, India occupies 7th place. India has good industrial infrastructure in several industries like chemical, power, nuclear energy, food, petroleum, pesticides, insecticides, plastic etc. A number of industrial effluents and emissions, especially toxic gases and spewed in to the air daily. A rapid increase in Atomic and nuclear energy has added a huge amount of radioactive substances in the atmosphere. Thus the environment is detoriated to such an extent that it has crossed the critical limit and has become lethal to all organisms, including men.

The craze of progress in agriculture, industry, transportation and technology is taken as the general criterian of development of any nation. Such activities of man has created adverse effects on all living organisms in the biosphere. Today environment has become foul, contaminated, undesirable, therefore, harmful for the health of living organisms (including man). So for as pollution is concerned, environment includes the air, the water, the soil, the noise, the building, the landscapes, the oceans, the lakes, the rivers, the parks, the vehicles and many other things. Not only addition of constituents in these adversely alter the natural quality of the environment but also removal of constituents caused pollution.

SCOPE AND IMPORTANCE

Environment consists of all living and non living things which surround us. Therefore the basic components of the environment are :–

1. Atmosphere or the air
2. Hydrosphere or the water
3. Lithosphere or the rocks and soil
4. The biosphere

Environment influence and shaped our life. It is from the environment that we get food to eat, water to drink, air to breathe and all necessities of day to day life are available from our environment. Thus it is the life support system. Hence the scope & importance of the environment can be well understand.

The basic concepts of environmental studies are interesting and important too not only to the scientists engagged in various fields of science and technology but also to the personnel involved in resource planning and material management. It is now universally realised that any future developmental activities have to be viewed in the light of its ultimate environmental impact. The tremendous increase in industrial activity during the last few decades and the release of obnoxious industrial wastes in to the environment, have been considerable concern in recent years from the point of view of the environmental pollution. Environmental pollution on one hand and deforestation, soil erosion, population explosion, global warming interence in ecosystem and biosphere on the other are threatening the very existence of life on the earth. According to Jantsch 1970 environmental science attempts to solve the major environmental problems with the help of interdisciplinary and transdisciplinary approaches i.e., the entire knowledge of all the disciplines of science such as Maths, Physics, Chemistry, Biology, Geology, Geography, Computer Science, Medical Science and Biotechnology, as well as of social sciences such as Economics Sociology and Psychology. Because, environmental science is a subject which draws heavily from environmental biology, but depends more on transdisciplinary approach.

Since, ecology, environmental biology and environmental science can not be defined without reference to environment, they show too much overlapping which is inevitable and unavoidable. Some major problems and disciplines which provide input to environmental science are mentioned in fig.

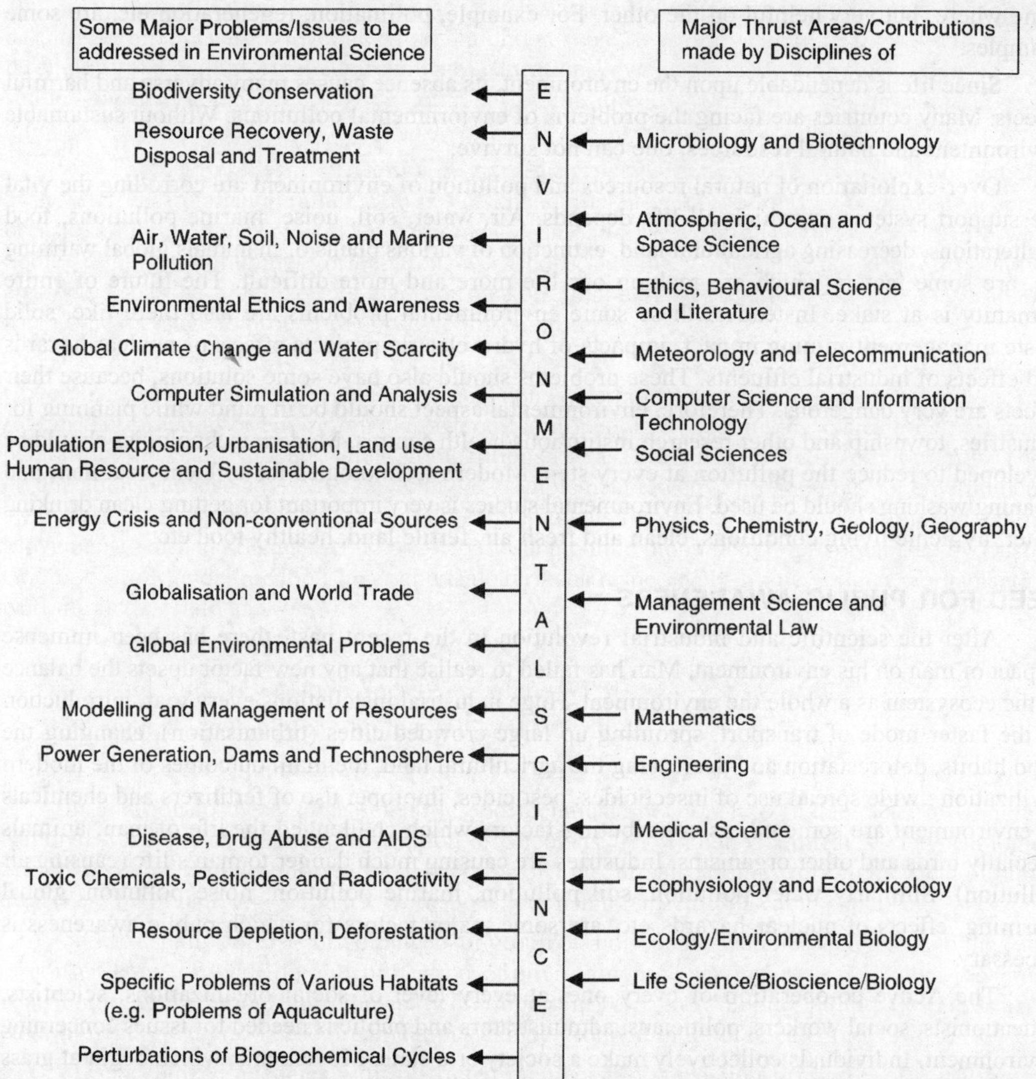

Fig. 1.1 Some major problems and issues and disciplines which provide input to solving these problems in environmental science.

Environment is responsible for creating conditions suitable for the existance of a healthy biosphere on this planet. The load of pollutant dicharged is also diluted and chemically modified. It regulates the temperature on earth where life activities are possible. Harmful ultraviolet radiations are absorbed in the stratosphere by the vital ozone layer. These rays can severely damage the terrestrial life. The atmosphere of our environment is a quick and effective media for transfer, transport and dissemination of gaseous wastes.

There are some natural and spontaneous process in our environments, by which we get plentiful supply of clean, air, fresh water and fertile soils, which regrate endlessly by biogeochemical cycles. Our planet which is rich in biodiversity have millions of species including beautiful, intriguing, towering trees and other coral nets where all types of animals live together. All animals depend on plants for their food. Insects, bacteria, virus and other organisms are one hand harmful some where, but very helpful on the other. For example, pollination, regeneration etc. are some examples.

Since life is dependable upon the environment, its absence causes many adverse and harmful effects. Many countries are facing the problems of enviornmental pollutions. Without sustainable environment and natural resources, one can not survive.

Over-exploitation of natural resources and pollution of environment are corroding the vital life support systems on which all life depends. Air, water, soil, noise, marine pollutions, food adulterations, decreasing agricultural land, extinction of various plants of mammals global warming etc. are some factors which are making our life more and more difficult. The future of entire humanity is at stake. Instead of these some environmental problems are also there like, solid waste management mining impact, impacts of hydro-electric projects effects of nuclear hazards and effects of industrial effluents. These problems should also have some solutions, because their effects are very dangerous. Therefore, environmental aspect should be in mind while planning for industries, township and other research institutions/health centres. Modern technologies should be developed to reduce the pollution at every step. Modern technologies for effluent treatment and cleaning/washing should be used. Environmental studies is very important for getting clean drinking water, hygienic living conditions, clean and fresh air, fertile land, healthy food etc.

NEED FOR PUBLIC AWARENESS

After the scientific and industrial revolution in the recent past, there has been immense impact of man on his environment. Man has failed to realise that any new factor upsets the balance of the ecosystem as a whole/the environment. Huge industrial installations every year, introduction of the faster mode of transport, sprouting up large crowded cities (urbanisation), changing the food habits, deforestation and decreasing the agricultural land, the main outcomes of the modern civilization : wide spread use of insecticides, pesticides, improper use of fertilizers and chemicals in environment are some others contributing factors which challenged the life of man, animals specially birds and other organisms. Industries are causing much danger to man's life (causing air pollution). Similarly water pollution, soil pollution, marine pollution, noise pollution, global warming, effects of nuclear hazards etc. are some major factors for which public awareness is necessary.

The Active co-operation of every one, at every level of social organizations, scientists, educationists, social workers, politicians, administrators and public is needed for issues concerning environment. Individuals collectively make a society or a state. Movements, which begin at grass root levels, effects the ideologies and policies of a country or the nation as a whole more effectively, than the policies introduced from top to downwards. When the opinion of the public will change, it will effect the govt. policies, which transform in to actions. Therefore little efforts on the part of each individual shall add up to introduce significant improvements of the environment.

Over exploitation of natural resources is a basic concern for everybody. Food shortage will increase in frequency and severity if population growth, soil erosion and nutrient depletions will continue at the existing rate. Therefore, it is our duty and we can accept the family planning schemes. This will not only reduce the population but also solve the problems of food and rehabilitation. By burning fossil fuels (oil, coal and natural gas), we release carbon-dioxide and other heat absorbing gases, that cause global warming and may bring about sea level rise and catastrophic climatic changes. Acid rain is the result of it. Chlorinated compounds such as chlorofluorocarbons used in refrigerator and air conditioner also contribute to global warming as well as damaging the stratospheric ozone that protect us from cancer causing ultraviolet radiations in sunlight.

Now a days every body talks about environment but how many of us are serious about it. How many of us (from all walks of life) have clear concepts of environment. There must be planning about the effects and control measures of environmental pollution. Govt. should initiate and help by awareness compaigns to save environment. There should not be the political propaganda but should be the integral part of our educational programmes. By writing on walls the word "save water", "save oil" is not enough for Govt. or people. We should opt some programmes relating to it. We should discourage to use fuel vehicles, until it is not necessary. For short routes, we should use bicycle; on foot. We should accompany the four seater or so with others over use of water, for cleaning and other purposes should be decreased. Rain water harvesting is another example for using the rain water instead flowing out. Any government at its own level can not achieve the goals of sustainable development until the public has a participatory role in it. It is only possible only when public aware about the ecological and environmental issues. For example ban the littering of polythene can not be successful until the public understands the environmental implications of the same. Public should understand about the fact that if we degrading our environment, we are harming ourselves. This is the duty of we educated people to educate the others about the adverse effect of environment.

For the first time, the attention of general public was attracted at global level when "Earth Summit" in 1992 was held in Rio de Janerio on environment and development. Later on another world summit on "Sustainable Development" at Johannesburg in 2002 was also held to discuss the environment and aware the public to save the environment. In this directions, United Nations has organised several conferences in different parts of the world (Stockholm 1972, Vienna 1985, Montreal 1987, Brazil 1992 etc) to work out the action plan from time to time for fighting with menace of environmental pollution. We should keep the earth green and alive as it provides shelter, food and protective cover. The soil degradation, soil erosion, deforestation, losing wetlands, land conversion etc. are the measure issues which force ourselves to think and aware the public in this regard. Because human himself is responsible for these environmental deterioration. Therefore, it is necessary to check all these destructive processes. Govt. also doing some efforts on national level but still much more has to be done.

The marine ecosystem includes, the oceans, seas, sea shores, bays and esmaries of the world. The physical factors like wares, tides, currents, salimities, temperature, pressures and sunlight dominate life in the ocean and determine the make up of biological communities. These communities have significant effect on biomass, leakage from oil tankers, oil drilling, catchment area (coastline) and rivers polluted the sea water, which effects sensitive flora and fauna, various species of invertibrate, mammals, coral reps, fishes and other organisms.

Diesel vehicles emits particles in their exhaust which have a diameter less than 10 microns (PH-10). It is easily inhaled. Any amount of these particles in the air are dangerous for health (particularly effects lungs). In India about 20 million people are asthmatics. Mine waste and effluents from mining and metallergical industries give a number of physical and chemical problems to human beings. Certain other industries like paper and pulp industries, fertilizer industries, explosive industries, soap and detergent industries, chemical industries, food processing industries, textile, tannery, leather, petroleum industries release/discharge undesirable and harmful constituents which are responsible for air and water pollution, causes great public concern.

Sewage begins to cause nuisance as it starts to become stale. It is therefore necessary to dispose it off as soon as possible. Proper methods of disposal and its treatment should be applied otherwise causes the chronic diseases. When sewage is applied continuously on a part of land, the pores or voids of the soil are clogged and free circulation of air is prevented. As a result anaerobic conditions are developed in place of aerobic conditions and the land is not capable of taking further sewage load. At this stage, decomposition of sewage takes place and offensive gases are produced. This is called the sewage sickness of land. People should aware of it.

The noise which is increasing pollution is one of the important factor of environment due to populations explosion, rapid industrializations and urbanisations. We should know the consequences of noise pollutions. Ear drum can be damage when exposed to very loud and sudden noise. Noise pollution affects human health, comfort and efficiency. It causes contraction of blood vessels, high blood pressure, mental distress, high cholesterol, heart attacks, neurological problems, birth defects, abortion etc. The department of environment realised the importance of creating a sound research base for scientific studies relating to environmental problems. Environmental protection act was introduced in 1976 as the 42nd amendment act in the constitution.

Only by celebrating "World Environmental Day" we can not get rid of this concern. Govt. along can not do any thing until unless every citizen is aware of the environmental pollution & their effects. This is the time to make aware and motivate each and every individual for environmental consciousness.

QUESTIONS

Short answer questions :–

1. Define environment.
2. Explain, "Environmental science is a multidisciplinary science."
3. Discuss some measure issues, which help in solving the problems in environmental science.
4. Define ecology.
5. What is synecology ?

Long answer questions :–

1. Discuss in details the scope and importance of environment.
2. Write an essay on "public awareness to protect our environment".
3. Give an account of role of public in Environmental protection.

UNIT

2

Natural Resources

Man is the most highly evolved animal. Since its appearance on earth man and his associates has been dependent on nature. The development of new ideas enabled him to identify materials of nature and to put them in to different uses for his comfort and development. In this series water, air, sunlight, land, minerals, forest, coal, oil, metals etc. were available and were identified by man. As the development was in progress, he started to make use of these natural gifts. He converted most of the natural sources in to natural resources for his own interest. This was possible later, when development and research technology was developed to transform these natural sources or materials in to more valuable goods. Thus, natural resources can be defined as—**Things/materials of the nature, that can be put to some use by human beings for their growth, development, comfort and other necessities are called as "Natural Sources".**

For example, air, water, soil, forests, animals, minerals, metals energy and other substances are some examples of natural resources that are utilised by human beings. At present, total global production is nearly enough to match the human demand for energy, materials etc., if we judiciously destribute the resources available to us. But for future, the situation appears pretty grim.

The resources are not equally distributed throughout the world. We can realise the value of resource only when it is rather scarce. Overexploitation threatens most of our natural resources now.

The primitive man was a part of environment. His basic requirement were limited i.e. food and shelter. Food was collected from the surroundings while caves, bushes, trees were enough to provide shelter. Those days natural materials, fields and living beings only were considered useful. But exploitation, destruction, development of natural resources began right from the time when man learned to use fire, metal copper for arms, tools, pans, domestic animals and grow plants for food. Now the boundaries of environment is from polar ice to equitorial mountain tops and from deep down ocean to ionosphere of sky. Everything present in this environment is regarded as natural resource.

TYPES OF NATURAL RESOURCES

All the natural resources can be divided into two categories—

(*i*) Exhaustible natural resources

(*ii*) Inexhaustible natural resources

Exhaustible natural resources are soils, forests, water, coal, petroleum, natural gas, minerals etc. These are consumed or exhausted through continuous use or misuse. Exhaustible natural resources can be further divided into two—

(*a*) Renewable natural resources

(*b*) Non-renewable natural resources

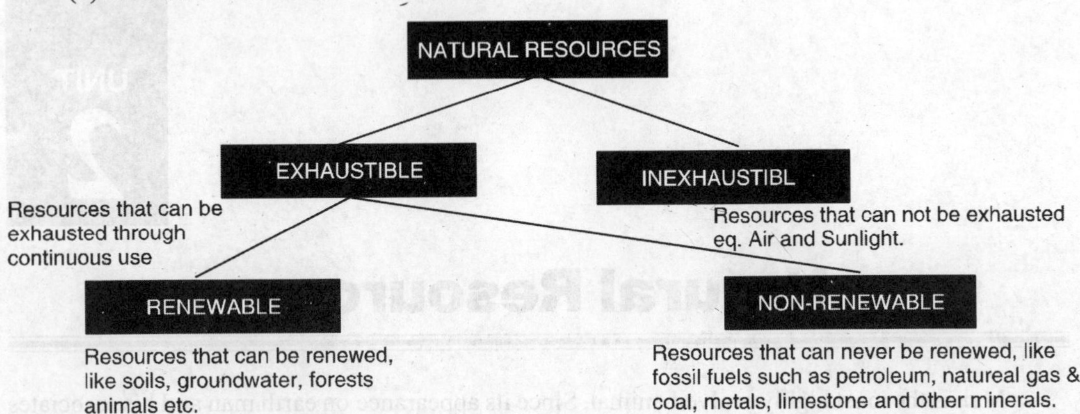

Fig. 2.1. Natural Resources

Inexhaustible natural resources are those which can not be exhausted through continuous use or misuse eq. air and sunlight etc.

RENEWABLE RESOURCES

The natural resources which are consumed/exhausted/depleted through continuous use and can be recovered by very hard efforts taken up for long periods are called **Renewable Resources**, for eg. soils, forests, groundwater etc. In other words we can say that all renewable resources are replenished through natural cycles or manually. For example oxygen in air is replenished through photosynthesis. Forest is maintained themselves and manually. Similarly fresh water is available through cycles & manually too.

Most of the removable resources are interdependent to each other. Forests maintained the environment/climate, plants need to check soil erosion & soil is needed for plants. Air and insects are needed for pollination. Wood, fibre, fodder, fruits, vegetables, milk etc. are developed directly or indirectly by recent photosynthetic activity. Thus the renewal of these resources will continue at as long as photosynthesis continues on this planet. These resources are the life support system which can fulfill all human needs. But its productivity/renewability is limited or depend upon availability of water, nutrients and environmental conditions.

The natural resources are useful to human society in one way or other. Hence we should ensure a continuous yield of useful plants, animals and materials by establishing a balanced cycle of harvest and renewal (**Odum - 1971**).

NON-RENEWABLE RESOURCES

Non renewable resources are not replenishable or we can not get back our coal and petroleum reserves in our life time, if ones they are consumed/exhausted completely. Non-renewable resources are metals (iron, copper, zinc etc.), coal, oil deposits, minerals, stone, mineral, salts (Phosphate, nitrates, carbonates etc.) etc.

Minerals are often called the '*STOCK*' resources, because their new materials can only be extracted from the earth's crust once. But even in the transformed state in which they are used, they are not lost to the planet and so are ideally available for reuse. Metals/minerals come from a very slow process of geo-chemical concentration, which took millions of years to form. Therefore, these deposits which occur today can disappear at some point of time in future.

Coal, petroleum and natural gas are called as '*FOSSIL FUELS*' because they are formed from dead remains of plants and animals buried in the earth long long ago. They are called fuels because they are burnt to give off energy. Coal have a high heating value, hence it is a useful fuel. Since fossil fuels are non-renewable sources of energy, therefore it is essential to explore more and more alternatives. Today we are utilizing solar energy, wind energy, ocean, geothermal & atomic energy as an alternative sources of energy.

Minerals, rocks, salts and chemicals etc. are termed as '*abiotic resources*', as biological activity is not involved in their formation. Continuous over exploitation of these resources shall exhaust many of our valuable deposits, which took millions of years to form. They can not be duplicated within the human scale of time.

NATURAL RESOURCES AND ASSOCIATED PROBLEMS

Being most highly developed/evolved animal, man possesses certain special characteristics. Man apply all their power and intelligence for food and development. They adopt new ways to fulfil their needs and often make improvements in old ways to derive resources and to fulfil their desire more efficiently. This is how they develop new technologies for utilization of natural resources. As we have seen, natural resources are exhaustible and non-exhaustible, the exhaustible resources are renewable and non renewable, therefore proper utilization of our natural resources is the need of today. Human beings utilize most of resources like air, water, land, minerals, flora, fauna, fuels, energy etc. for their growth and development.

Now the problem is, how and up to what extent human beings should utilize various resources. Resources are valuable gift of nature. Hence use of natural resources should depend on knowledge, availability, type, quantity, value, necessity etc. The use of resources should be in limit not to exhaust them so that ecological balance within the nature should also remain undisturbed.

In the past, the man was not so advanced and was satisfied with what he received from the nature due to his limited needs. Thus, there was a complete balance among all the components of the natural environment in the past. At present, the story has become very different due to human activities of consumption and misuse of natural resources and the explosion of population. Therefore, nature has lost its capacity of tolerance. Fast growth in human population has led to the fast rate of consumption of natural resources and creation of wastes. Human activities of cutting trees, killing animals, damaging and mixing poisonous wastes in air and water are creating severe crisis in the environment and natural resources. To some extent nature tries to recycle some of these resources. But, because of spurt in population these are used up faster than these are replenished, leading to their depletion.

Directly or indirectly there are some problems from the over exploitation of natural resources. They are, lowering of water lable, extinction of wild animals, soil erosion, siltation of rivers, floods, climatic change, interruption of water cycle, loss of aquatic plants/animals, various diseases, ozone layer depletion, global warming, acid rains, green house effects, shortage of food, problem of housing, fossil fuels etc. Using wood as fuel not only cause severe damage to earth but also produce pollution.

Oceans are also provide different types of food materials and minerals. Large quantities of petroleum and natural gas are also obtained from oceans. Thus they are valuable contributors in the development and prosperity of human beings. But due to human activities, oceans have been put under heavy conditions of serious stress.

Now the conservation of natural resources should be the priority of everyone citizen. Methods of conservation management should be followed.

By making rules only, we can not protect natural resources. Awareness programmes, social forestry, joint forest management, van panchayat, organising the seminars, public programmes are some means by which we can protect and conserve natural resources. During the recent years many societies and NGO's have come forward in this direction.

Human activities should be refined using high technology which adversely affect the natural resources and environment under environmental resource management. Population should be controlled because it is the root cause of resource exploitation. We should plan for conservation of natural resources for their long term and maximum usage. Non-renewable resources should be used in as sustainable manner i.e. they should remain available for future generations.

FOREST RESOURCES

India is one of the 12 mega diversity countries, commanding 7% of the world's biodiversity and supporting 16% of the major forest types, varying from alpine pastures in the Himalayas to temperate sub tropical, tropical forests and mangroves in the coastal areas. But nearly half of the country's area is degraded, affected by problems of soil degradation and erosion. Sixty percent forests are located in ecologically sensitive zones. These forests need to be managed in a way to ensure that they are ecologically protected and maintained, as well as sustained at the highest productivity level to meet the growing population's burgeoning demands for fuel, food, fodder, and timber.

Country data

Total land area (thousand ha)	297,319
Total forest area 1995 (thousand ha)/% of total land area	65,005/21.9
Natural forest 1995 (thousand ha)	50,385
Total change in forest cover 1990-95 (thousand ha)/annual change (%)	36/0
Population total 1997 (million)/Annual rate of change 1995-2000 (%)	960.2/1.6
Rural population 1997 (%)	72.6
GNP per person 1995 in US$	340

Source of data: FAO - State of the World's Forest 1999

India has a large and diverse forest resource. Its forest types vary from tropical rainforest in the north-east, to desert and thorn forests in Gujarat and Rajasthan; mangrove forest in West Bengal, Orissa and other coastal areas; and dry alpine forests in the western Himalaya. The most common forest types are tropical moist deciduous forests, tropical dry deciduous forests, and wet tropical evergreen forests.

According to the Forest Survey of 1997, the country has 76.5 million ha of forest. The degraded area was 26.13 million ha and there was another 5.72 million ha of scrub; thus, in total 31.85 million ha of forests were degraded or open.

The land use outside for habitations (rural and urban), industries and infrastructure, such as roads, rivers, canals, railway lines, under permanent snow, rocks, desert, or not available for other reasons amounted to 264 million ha. It consists of cultivated land of 142 million ha, forestland of 67 million ha, fallows of 24 million ha, pastures of 12 million ha, tree groves of 3 million ha, and cultural waste of 16 million ha. Thus, in order to achieve the national goal of one third of the country under forest/tree cover, an area of 29.7 million ha has to be brought under plantations.

It was reported that the country's achievement in raising forest plantations, in terms of area, has been impressive. Up to 1998, the total area of tree plantations was 28.38 million ha, of which about 17 million ha were planted before 1990's. The current annual rate of plantation is 1.2 million. The quality of these plantations varies considerably. It should be noted that forest plantations are a means to meet the increasing demand for industrial raw material or for direct consumption, i.e. fuel wood, but not to justify deforestation or claim restoration of biodiversity and other environmental services.

There are other woodlands established in small blocks on non-forestry lands, which are not included in the forest survey because of limitations of interpretation of satellite data.

The performance of forest plantations, in terms of survival, growth and yield, has been poor caused by several factors, including inadequacies in site selection and site-species matching, poor planting stock, lack of proper maintenance and protection (from fire, grazing, pests and diseases), lack of timely tending/thinnings, delays in fund allocation, and inadequately trained staff. In this regard, some people are of the opinion that a master plan for tree plantations should be developed specifying categories, management regimes, utilisation and investment needs; and emphasis should be given to enhancing qualitative and quantitative productions.

Involvement of the private sector in plantation development has not been substantial and are not adequately supported by the government through relevant research, extension, technological packages, input delivery, market information or credit facilities. This sector is dominant in the area of harvesting and processing. It was noted that the needs and problems relating to this area are different from those producers of wood in rural areas.

According to the latest State of Forest Report, 1999, the total forest cover was 633.73 million ha or 19.39% of the geographical area, with dense forest accounting for 11%. The Report stated that the forest cover has increased by 4,000 km2 since the last survey in 1997. Thus, the overall decline in the forest cover has been halted. In this regard, the Minister of Environment and Forests stated that a major constraint facing the afforestation programmes is funding, which requires Rs 66.95 billion per year (1US$ approximately equal to Rs 43.5 in September 2000) in order to achieve one-third forest cover within the next 20 years. Rs 16 billion per year is available from both the central and state budget together to be allocated for afforestation. Involvement and investment from various NGOs, corporate, public and private sectors to fund this sector is being approached. In this connection, consultation with several donor agencies, including international and bilateral banks, for possible support have also been carried out. In this regard, WB and EU have provided substantial support to several forestry programmes in some states.

In regard to national parks, sanctuaries and other reserves, the country's achievement in terms of area is substantial. The Protected Areas (PA) cover about 14.8 million ha, or about 14% of the forest area, consisting of 80 national parks, 441 wildlife sanctuaries and 8 biosphere reserves. However, the condition of several PAs is poor because of fire, grazing and inadequate management. The Management plans of some PAs are not comprehensive. Some are below the minimum size required to be effective.

Non-wood forest products (NWFPs) have a great potential to support the socio-economic development of the country and also the principles of sustainable forest management. These products are essential to local communities. Some products have great potential for export. Some products have also provided employment and income earning.

Deforestation

The conversion of forested areas to non-forest is deforestation. Historically, this meant conversion to grassland or to its artificial counterpart, grainfields; however, the Industrial Revolution added urbanization and technological uses. Generally this removal or destruction of significant areas of forest cover has resulted in a degraded environment with reduced biodiversity. In developing countries, massive deforestation has accompanied mankind's progress since the Neolithic, and has shaped climate and geography.

Deforestation (whether deliberate or unintended) is the result of the removal of trees without sufficient reforestation. There are many causes, ranging from extremely slow forest degradation to sudden and catastrophic wildfires. Deforestation can be the result of the deliberate removal of forest cover for agriculture or urban development, or it can be an unintentional consequence of uncontrolled grazing (which can prevent the natural regeneration of young trees). The combined effect of grazing and fires can be a major cause of deforestation in dry areas. In addition to the direct effects brought about by forest removal, indirect effects caused by edge effects and habitat fragmentation can greatly magnify the effects of deforestation.

While tropical rainforest deforestation has attracted most attention, tropical dry forests are being lost at a substantially higher rate, primarily as an outcome of slash-and-burn techniques used by shifting cultivators.

DEFINITION OF DEFORESTATION

Deforestation is the loss or continual degradation of forest habitat due to either natural or human related causes. Agriculture, urban sprawl, unsustainable forestry practices, mining, and petroleum exploration all contribute to human caused deforestation. Natural deforestation can be linked to tsunamis, forest fires, volcanic eruptions, glaciation and desertification. The effects of human related deforestation can be mitigated through environmentally sustainable practices that reduce permanent destruction of forests or even act to preserve and rehabilitate disrupted forestland. (See Reforestation and Treeplanting)

The term deforestation is often the source of disagreement between various interest groups. Conservation groups often use broad definition while groups seeking to maintain the status quo often use a narrow definition.

Deforestation defined broadly can include not only conversion to non-forest, but also degradation that reduces forest quality - the density and structure of the trees, the ecological services supplied, the biomass of plants and animals, the species diversity and the genetic diversity. Narrow definition of deforestation is: the removal of forest cover to an extent that allows for alternative land use. The United Nations Research Institute for Social Development (UNRISD) uses a broad definition of deforestation, while the Food and Agriculture Organization of the UN (FAO) uses a narrow definition.

Definitions can also be grouped as those which refer to changes in land cover and those which refer to changes in land use.

Land cover measurements often use a percent of cover to determine deforestation. This type of definition has the advantage in that large areas can be easily measured, for example from satellite photos. A forest cover removal of 90% may still be considered forest in some cases. Under this definition areas that may have few values of a natural forest such as plantations and even urban or suburban areas may be considered forest.

Land use definitions measure deforestation by a change in land use. This definition may consider areas to be forest that are not commonly considered as such. An area can be lacking trees but still considered a forest. It may be a land designated for afforestation or an area designated administratively as forest.

USE OF THE TERM DEFORESTATION

The term deforestation has been used to refer to fuel wood cutting, commercial logging and the slash and burn technique, a component of some shifting cultivation agricultural systems. It is also used to describe forest clearing for annual crops, for grazing, and establishment of industrial forest plantations. The meaning of the term is ambiguous enough and so charged with emotion that the use of a more precise term might be better suited in specific cases. Related terms are forest decline, forest fragmentation and forest degradation, loss of forest cover and land use conversion.

The term also has a traditional legal sense of the conversion of Royal forest land into purlieu or other non-forest land.

CAUSES OF DEFORESTATION

Present causes

While short-sighted, market-driven forestry practices are often one of the leading cause of forest degradation, the principal human-related causes of deforestation are agricultural and livestock grazing, urban sprawl mining and petroleum extraction. Causes include demand for farm land and fuel wood. Underlining causes include poverty, lack of reform. The causes of deforestation are complex and often differ in each forest and country. Government policies, such as ones in Brazil, make it a priority to resettle some of the country's numerous landless people. The largest cause as of 2006 is slash-and-burn activity in tropical forests. Slash-and-burn is a method sometimes used by shifting cultivators to create short term yields from marginal soils. When practiced repeatedly, or without intervening fallow periods, the nutrient poor soils may be exhausted or eroded to an unproductive state. Slash-and-burn techniques are used by native populations of over 200 million people worldwide.

Prehistory

Deforestation has been practiced by humans for thousands of years. Fire was first tool that allowed humans to modify the landscape. The first evidence of deforestation shows up in the Mesolithic. Fire was probably used to drive game into more accessible areas. With the advent of agriculture fire became the prime tool to clear land for crops. In Europe there is little solid evidence before 5000 BP. Mesolithic foragers used fire to create openings for red deer and wild boar. In Great Britain shade tolerant species like oak and ash are replaced in the pollen record by hazels, brambles, grasses and nettles. Removal of the forests led to decreased transpiration resulting in the formation of upland peat bogs. Widespread decreased in elm pollen across Europe between 6400–6300 BP and 5200–5000 BP, starting in southern Europe and gradually moving north to Great Britain, may represent and clearing by fire at the onset of Neolithic agriculture.

Pre-industrial history

The historic silting of ports along the southern coasts of Asia Minor (*e.g.* Clarus and the examples of Ephesus, Priene and Miletus, where harbors had to be abandoned because of the silt deposited by the Meander) and in coastal Syria during the last centuries BC, and the famous silting up of the harbor for Bruges, which moved port commerce to Antwerp, all follow perious of increased settlement growth (and apparently of deforestation) in the river basins of their hinterlands. In early medieval Riez in upper Province, alluvial silt from two small rivers raised the riverbeds and widened the floodplain, which slowly buried the Roman settlement in alluvium and gradually moved new construction to higher ground; concurrently the headwater valleys above Riez were being opened to pasturage.

A typical progress trap is that cities are built in a woody area providing wood for some industry (e.g. shipbuilding, pottery) which starts consuming it so fast – and without proper replanting – that it becomes impossible to obtain it close enough to remain competitive, leading to the city's abandonment, as happened repeatedly in Ancient Asia Minor. Especially the combination of mining and metallurgy went along this self-destructive path.

Meanwhile most of the population remaining active in (or indirectly dependend on) the agricultural sector, the main pressure in most areas remained land clearing for crop and cattle farming; fortunately enough wild green was usually left standing (and partially used, e.g. to collect firewood, timber and fruits, or to graze pigs) for wildlife to remain viable, and the hunting privileges of the elite (nobility and higher clergy) often protected significant woodlands.

Industrial Pressure : The massive use of Charcoal on an industrial scale was a new acceleration of the onslaught on western forests.

ENVIRONMENTAL EFFECTS

Atmospheric pollution

Deforestation is often cited as one of the major causes of the enhanced greenhouse effect. Trees and other plants remove carbon (in the form of carbon dioxide) from the atmosphere during the process of photosynthesis. Both the decay and burning of wood releases much of this stored carbon back to the atmosphere. A.J. Yeomans asserts in Priority One (http://www.yeomansplow.com.au/priority-one.htm) that overnight a stable forest releases exactly the same quantity of carbon dioxide back into the atmosphere. Others state that mature forests are net sinks of CO_2 (see Carbon dioxide sink and Carbon cycle).

Wildlife

Some forests are rich in biological diversity. Deforestation can cause the destruction of the habitats that support this biological diversity - thus causing population shifts and extinctions.

Hydrologic cycle and water resources

Trees, and plants in general, affect the hydrological cycle in a number of significant ways:

- their canopies intercept precipitation, some of which evaporates back to the atmosphere (canopy interception);
- their litter, stems and trunks slow down surface runoff;
- their roots create macropores - large conduits - in the soil that increase infiltration of water;
- they reduce soil moisture via transpiration;
- their litter and other organic residue change soil properties that affect the capacity of soil to store water.

As a result, the presence or absence of trees can change the quantity of water on the surface, in the soil or groundwater, or in the atmosphere. This in turn changes erosion rates and the availability of water for either ecosystem functions or human services.

The forest may have little impact on flooding in the case of large rainfall events, which overwhelm the storage capacity of forest soil if the soils are at or close to saturation.

Soil erosion

Deforestation generally increases rates of soil erosion, by increasing the amount of runoff and reducing the protection of the soil from the tree litter. This can be an advantage in excessively leached tropical rain forest soils. Forestry operations themselves also increase erosion through the development of roads and the use of mechanized equipment.

China's Loess Plateau was cleared of forest millennia ago. Since then it has been eroding, creating dramatic incised valleys, and providing the sediment that gives the Yellow River its yellow color and that causes the flooding of the river in the lower reaches (hence the river's nick-name 'China's sorrow').

Removal of trees does not always increase erosion rates. In certain regions of southwest US, shrubs and trees have been encroaching on grassland. The trees themselves enhance the loss of grass between tree canopies. The bare intercanopy areas become highly erodible. The US Forest Service, in Bandelier National Monument for example, is studying how to restore the former ecosystem, and reduce erosion, by removing the trees.

Landslides

Tree roots bind soil together, and if the soil is sufficiently shallow they act to keep the soil in place by also binding with underlying bedrock. Tree removal on steep slopes with shallow soil thus increases the risk of landslides, which can threaten people living nearby.

CONTROLLING DEFORESTATION

Farming

New methods are being developed to farm more food crops on less farm land, such as high-yield hybrid crops, greenhouse, autonomous building gardens, and hydroponics. The reduced farm land is then dependent on massive chemical inputs to maintain necessary yields. In cyclic agriculture, cattle are grazed on farm land that is resting and rejuvenating. Cyclic agriculture actually increases the fertility of the soil. Selective over farming can also increase the nutrients by releasing such nutrients from the previously inert subsoil. The constant release of nutrients from the constant exposure of subsoil by slow and gentle erosion is a process that has been ongoing for billions of years.

Forest management

Efforts to stop or slow deforestation have been attempted for many centuries because it has long been known that deforestation can cause environmental damage sufficient in some cases to cause societies to collapse. In Tonga, paramount rulers developed policies designed to prevent conflicts between short-term gains from converting forest to farmland and long-term problems forest loss would cause, whilst during the seventeenth and eighteenth centuries in Tokugawa Japan the shoguns developed a highly sophisticated system of long-term planning to stop and even reverse deforestation of the preceding centuries through substituting timber by other products and more efficient use of land that had been farmed for many centuries. In sixteenth century Germany landowners also developed silviculture to deal with the problem of deforestation. However, these policies tend to be limited to environments with good rainfall, no dry season and very young soils (through volcanism or glaciation). This is because on older and less fertile soils trees grow too slowly & silviculture to be economic, whilst in areas with a strong dry season there is always a risk of forest fires destroying a tree crop before it matures.

Afforestation

Today, in China, where large scale destruction of forests has occurred, the government has required that every able-bodied citizen between the age of 11 and 60 plant three to five trees per year or do the equivalent amount of work in other forest services. The government claims that at least 1 billion trees have been planted in China every year since 1982. In western countries, increasing consumer demand for wood products that have been produced and harvested in a sustainable manner are causing forest landowners and forest industries to become increasingly accountable for their forest management and timber harvesting practices.

The Arbor Day Foundation's (http://www.arborday.org/) Rain Forest Rescue Program is a charity that helps to prevent deforestation. The charity uses donated money to buy up and preserve rainforest land before the lumber companies can buy it. The Arbor Day Foundation then protects the land from deforestation. This also locks in the way of life of the primitive tribes living on the forest land.

CASE STUDIES

Brazil

In Brazil the rate of deforestation is apparently driven by commodity prices. Recent development of a new variety of soyabean has lead to displacement of beef ranches and slash and burn farmers which in turn move further into the forest.

Indonesia

There are large areas of forest in Indonesia that are being lost as native forest is cleared by large multi-national pulp companies and being replaced by plantations.

United States

Upon arrival European-Americans began clearing large areas of forest for wood and agriculture. Beginning in about 1850 farm land began to be abandoned because of soil exhaustion and competition from the mid-west. Also, mechanization allowed land formerly used as pastures for horses to revert to forest. From 1850 to about 1920 the amount of forest land in the United States actually increased. Today the trend in forest cover increase has reversed as urban sprawl causes conversion of forest as the forest is transformed to suburbs.

Timber Extraction

Once world bank study in 1989 has argued that tree crop estates are a better employment generating option than even forest plantations, notwithstanding the high density of useable timber in the plantations. In Malaysia, timber using industry i.e. timber extraction is on large scale. In Indonesia timber manufacturing employment is as high as 3.7 million. Newspaper Kompas (24 Oct., to 1989) reported that small regional sawmills alone were employ 200,000 people in Indonesia. Even in Sarawak, employment in the timber industry is about 55000 indirect timber extraction, plus a large number in associated tasks (ITTC 1990).

In 1980, there was a serious attempt to developed woodbased industry in Sabah. Sawmills,veneer and plywood mills all increased sharply in number. Although sawn timber production increased by 40% during the same period, the total volume of exports fell by almost one third. The situation in Peninsula is less serious than that in Sabah only because of the much greater diversity of the economy. For the timber and timber using industries, it is worse.

With good prices, production expanded, and at the end of 1980s the Peninsula had 681 sawmills, 43 Veneer and plywood mills and more than 1200 small woodworking plants, furniture factories. There has been an increasingly severe shortage of timber for mills, for the building industry, for exporters of sawn timber and plywood and for the downstream factories. The future of the big plans for timber manufacture is not withstanding optimism, uncertain and is potentially in conflict with the conservationist gcals now formally adopted by the Govt.

Unlike Sabah and Peninsula, Kalimantan and Sarawak have shortage of timber due to exploitation of forest. They have banned export for time being. If it is, then on very high rate. In Indonesia has been warned not to exceed an annual capacity of 10 million m^3, otherwise real shortage of raw material could arise.

In India also, millions of people are engaged in timber industries. Not only they are earning but solving the problem of employment to some an extent. Plywood industries are also increased. There is gap between demand and supply, this results import of loggs from other countries like Malaysia, Indonesia etc. Forest area is also receding. Day by day, the demand of loggs is increasing due to population growth. Timber shortage are clearly either real or in prospect, posing a severe threat to both entrepreneurship and employment. Given a low rate of success with enrichment planting, growing emphasis is now being placed on plantation forestry to ensure future supply. Indonesia has already stepped up its reforestation efforts using mainly Acacia mangium, with a view to the future establishment of pulp & paper mills.

A major conflict is emerging between those whose interest lies from the production of roundwood, sawn timber, plywood and those whose depend upon the sustainable supply of timber to the sawmills, plywood mills and factories. Unauthorised invasion by cultivators too is now a days a problem. In 1991 the International Tropical Timber Organisation issued a document that how a sustainable tropical timber production could be achieved.

Mining

Mining is the extraction of valuable minerals or other geological materials from the earth, usually (but not always) from an ore body, vein, or (coal) seam. Materials recovered by mining include bauxite, coal, diamonds, iron, precious metals, lead, limestone, nickel, phosphate, rock salt, tin, uranium, and molybdenum. Any material that cannot be grown from agricultural processes must be mined. Mining in a wider sense can also include extraction of petroleum, natural gas, and even water.

History

The oldest known mine in the archaeological record is the "Lion Cave" in **Swaziland.** At this site, which by radiocarbon dating is 43,000 years old, paleolithic humans mined for the iron containing mineral hematite, which they ground to produce the red pigment ochre. Sites of a similar age where Neanderthals may have mined flint for weapons and tools have been found in Hungary.

Another early mining operation was the turquoise mine operated by the ancient Egyptians at Wady Magharch on the Sinai Peninsula. Turquoise was also mined in pre-Columbian America in the Cerillos Mining District in New Mexico, where a mass of rock 200 feet (60 m) in depth and 300 feet (90 m) in width was removed with stone tools; the mine dump covers 20 acres (81,000 m^2.) Black gun powder in mining was first used in a mineshaft under Banska Stiavnica, Slovakia in 1627, in the same town in 1762 the first Mining Academy in the world was established.

Mining in the United States became prevalent in the 19th century. Mining for minerals and precious metals, such as in the California Gold Rush in the mid 1800s, was very important in westward expansion to the Pacific coast along with ranching and exploration of oil and gas fields. During this time period many white Americans and post-slavery African Americans, with the aid of railroads, travelled west for work opportunities in mining. Many western cities such as Denver and Sacramento originated as mining towns.

Steps in the mining process

1. Prospecting to locate ore
2. Exploration to defining the extent and value of ore where it was located
3. Conduct resource estimate to mathematically estimate the extent and grade of the deposit
4. Conduct mine planning to evaluate the economically recoverable portion of the deposit
5. Conduct a feasibility study to evaluate the total project and make a decision as whether to develop or walk away from a proposed mine project. This includes a cradle to grave analysis of the possible mine, from the initial excavation all the way through to reclamation.
6. Development to create acess to an ore body
7. Exploitation to extract ore on a large scale
8. Reclamation to make land where a mine had been suitable for future use

Environmental effects and mitigation

Environmental issues can include erosion, formation of sinkholes, loss of biodiversity and contamination of groundwaters by chemicals from the mining process and products.

Modern mining companies in many countries are required to follow strict environmental and rehabilitation codes, ensuring the area mined is returned to close to its original state, or an even better environmental state than before mining took place. In some countries with pristine environments, such as large parts of Australia, this is impossible despite the best intentions. Past mining methods have had, and methods used in countries with lack of environmental regulations can continue to have, devastating environmental and public health effects.

Mining can have adverse effect on surrounding surface and ground water if protection measures are not exercised. The result can be unnaturally high concentrations of some chemical elements over a significantly large area of surface or subsurface. Coal mining releases approximately twenty toxic release chemicals, of which 85% is said to be managed on site. Combined with the effects of water and the new 'channels' created for water to travel through, collect in, and contact with these chemicals, a situation is created where mass-scale contamination can occur. In well-regulated mines hydrologists and geologists are implementing careful measures to mitigate any type of water contamination that could be caused by mines. In modern American mining, operations must, under federal and state law, meet standards for protecting surface and ground waters from contamination, including acid mine drainage (AMD). To mitigate these problems water is continuously monitored at coal mines. The five principal technologies used to control water flow at mine sites are : diversion systems, containment ponds, groundwater pumping systems, subsurface drainage systems, and subsurface barriers. In the case of AMD, contaminated water is generally pumped to a treatment facility that neutralizes the contaminants.

Case Studies:

Some examples of environmental problems associated with mining operations are :

Ashio Copper Mine, Ashio, Japan was the site of substantial pollution at the end of nineteenth century

Berkeley Lake, an abandoned pit mine in Butte, Montana that has filled with water which is now acidic and poisonous. In 2003, a water treatment plant came on-line, initially treating "new" water entering the pit and thereby reducing the rate of rise of pit water. Treated water is currently used in the concentrator of the nearby Montana Resources Continental Pit, but it is clean enough to return to Silver Bow Creek. Eventually, water in the pit itself will be treated.

Britannia Mines a former copper mine near Vancouver, British Columbia. Copper from the abandoned mine washes into Howe Sound, polluting the water. No animal life remains there now - Latest reports are that after a water treatment plant was put in, fish are returning to Britannia Bay - may be for the first time ever. The name used by the First Nations tribes of Britannia Beach, even before mining started, means "The Place of No Fish".

Scouriotissa, a copper mine in Cyprus that has been abandoned. Contaminated dust blows off this site.

Tar Creek, an abandoned mining area in Picher, Oklahoma that is now an Environmental Protection Agency superfund site. Water in the mine has leaked through into local groundwater, contaminating it with metals such as lead and cadmium. [1]

(http://www.health.state.ok.us/PROGRAM/envhlth/sites/ottawa.html)

Benefits of Large Dams

Water is essential for sustenance of all forms of life on earth. It is not evenly distributed all over the world and even its availability at the same locations is not uniform over the year. While the parts of the world, which are scarce in water, are prone to drought, other parts of the world, which are abundant in water, face a challenging job of optimally managing the available water resources. No doubt the rivers are a great gift of nature and have been playing a significant role in evolution of various civilizations, nevertheless on many occasions, rivers, at the time of floods, have been playing havoc with the life and property of the people. Management of river waters has been, therefore, one of the most prime issues under consideration. Optimal management of river water resources demands that specific plans should be evolved for various river basins which are found to be technically feasible and economically viable river basins which are found to be technically feasible and economically viable after carrying out extensive surveys. Since the advent of civilization, man has been constructing dams and reservoirs for storing surplus river waters available during wet periods and for utilization of the same during lean periods. The dams and reservoirs world over have been playing dual role of harnessing the river waters for accelerating socio-economic growth and mitigating the miseries of a large population of the world suffering from the vagaries of floods and droughts. Dams and reservoirs contribute significantly in fulfilling the following basic human needs :-

- WATER FOR DRINKING AND INDUSTRIAL USE
- IRRIGATION
- FLOOD CONTROL
- HYDRO POWER GENERATION
- INLAND NAVIGATION
- RECREATION

Water for drinking and industrial use :

- Due to large variations in hydrological cycle, dams and reservoirs are required to be constructed to store water during periods of surplus water availability and conserve the same for utilization during lean periods when the water availability is scarce.
- Properly designed and well-constructed dams play a great role in optimally meeting the drinking water requirements of the people.
- Water stored in reservoirs is also used vastly for meeting industrial needs.
- Regulated flow of water from reservoirs help in diluting harmful dissolved substances in river waters during lean periods by supplementing low inflows and thus in maintaining and preserving quality of water within safe limits.

Irrigation :

- Dams and reservoirs are constructed to store surplus waters during wet periods, which can be used for irrigating arid lands. One of the major benefits of dams and reservoirs is that water flows can be regulated as per agricultural requirements of the various regions over the year.
- Dams and reservoirs render unforgettable services to the mankind for meeting irrigation requirements on a gigantic scale.
- It is estimated that 80% of additional food production by the year 2025 would be available from the irrigation made possible by dams and reservoirs.
- Dams and reservoirs are most needed for meeting irrigation requirements of developing countries, large parts of which are arid zones.
- There is a need for construction of more reservoir based projects despite widespread measures developed to conserve water through other improvements in irrigation technology.

Flood Control :

- Floods in the rivers have been many a time playing havoc with the life and property of the people. Dams and reservoirs can be effectively used to control floods by regulating river water flows downstream the dam.

- The dams are designed, constructed and operated as per a specific plan for routing floods through the basin without any damage to life and property of the people.
- The water conserved by means of dams and reservoirs at the time of floods can be utilized for meeting irrigation and drinking water requirements and hydro power generation.

Hydro power generation :

- Energy plays a key role for socio-economic development of a country. Hydro power provides a cheap, clean and renewable source of energy.
- Hydro power is the most advanced and economically viable resource of renewable energy.
- Reservoir based hydroelectric projects provide much needed peaking power to the grid.
- Unlike thermal power stations, hydro power stations have fewer technical constraints and the hydro machines are capable of quick start and taking instantaneous load variations.
- While large hydro potentials can be exploited through mega hydroelectric projects for meeting power needs on regional or national basis, small hydro potentials can be exploited through mini/micro hydel projects for meeting local power needs of small areas. Besides hydro power generation, mutli purpose hydroelectric projects have the benefit of meeting irrigation and drinking water requirements and controlling floods etc.

Inland navigation :

- Enhanced inland navigation is a result of comprehensive basic planning and development, utilizing dams, locks and reservoirs that are regulated to play a vital role in realizing large economic benefits of national importance.

Recreation :

- The reservoir made possible by constructing a dam presents a beautiful view of a lake. In the areas where natural surface water is scarce or non-existent, the reservoirs are a great source of recreation.
- Alongwith other objectives, recreational benefits such as boating, swimming, fishing etc. linked with lakes are also given due consideration at the planning stage to achieve all the benefits of an ideal multipurpose project.

While dams provide a yeoman service to the mankind, the following impacts of the construction of dams are required to be handled carefully :-

- Resettlement and Rehabilitation
- Environment and forests
- Sedimentary issues
- Socio economic issues
- Safety aspects

The above problems related to the construction of dams may be resolved successfully in case the approach of management is objective, dynamic, progressive and responsive to the needs of the hour.

Problems from Dams

We have seen the benefits of dams, but the otherside of coin is also not good when such projects are undertaken and hundreds crores of public money is spent, individual or organisations in the grab of PIL can not be permitted to challenge the policy decision taken. For such developmental projects thousands of acre land is acquired, results the public become landless. As far as relief and rehabilitation are concerned, people are not given properly. Therefore, if the interests of any affected persons are not protected in terms of the approved mechanism, they certainly have the right to approach appropriate authorities or even the supreme court in case its directions are not complied with. Later on people go on strike, & other demonstrations. "Narmada Bachao Andolan", "Chipko movements" etc. are such movements, which work for the relief & rehabilitation of affected people.

Due to these dams, most affected people are tribals, poor, labour class people etc. They can not oppose of their own. As regards reliefs & rehabilitation detailed and exhaustive plans have been drawn by the states. The conditions of the rehabilitation are such that the affected families are to be offered choices that should, in fact, make the quality of their lives better than the conditions they encountered in their original inhabitation. Eleborate mechanisms have also been established to monitor and ensure compliance and even a grievances redressal mechanism has been put in place. Govt. also take more than adequate care to protect the interests of the oustees.

CASE STUDY

The Chipko Movement (1987)

(*India*)

"... for its dedication to the conservation, restoration and ecologically-sound use of India's natural resources."

The forests of India are a critical resource for the subsistence of rural people throughout the country, especially in hill and mountain areas, both because of their direct provision of food, fuel, fodder and because of their role in stabilising soil and water resources. As these forests have been increasingly felled for commerce and industry, Indian villagers have sought to protect their livelihoods through the Gandhian method of *satyagraha* or non-violence resistance. In the 1970s and 1980s this resistance to the destruction of forests spread throughout India and became organised and known as the Chipko Movement.

The first Chipko action took place spontaneously in 1973 and over the next five years spread to many districts of the Himalaya in Uttar Pradesh. The name of the movement came from a word meaning 'embrace': the villagers hugged the trees and thus saved them by putting their bodies in the way of the contracots' axes. The Chipko protests in Uttar Pradesh achieved a major victory in 1980 with a 15-year ban on green felling in the Himalayan forests of that State by order of India's the then Prime Minister, Indira Gandhi. The movement later spread to Himachal Pradesh in the north, Karnataka in the south, Rajasthan in the west, Bihar in the east and to the Vindhyans in central India. In addition to the ban in Uttar Pradesh, the movement succeeded in halting clear felling in the Western Ghats and the Vindhyas, as well as generating pressure for a natural resources. Policy is more sensitive to people's needs and environmental factors.

The Chipko Movement was the result of hundreds of decentralised and locally autonomous initiatives. Its leaders and activists have primarily been village women, acting to save their means of subsistence and their communities. Men have been involved, too, however, and some of them have given wider leadership to the movement. One of the most prominent leaders have been Sunderlal Bahuguna, a Gandhian activist and philosopher, whose appeal to Mrs Gandhi resulted in the green-felling ban and whole 5,000-kilometre trans-Himalayan footmarch in 1981-83 was crucial in spreading the Chipko message.

In the late 1980s, Bahuguna joined the campaign which already for many years had been opposing construction of a proposed Himalayan dam on the river near his birthplace of Tehri. In 1989 he began the first of a series of hunger strikes to draw political attention to the dangers posed by the dam and in due course the Chipko Movement gave birth to the Save Himalaya Movement.

Bahuguna ended a 45-day fast in 1995 when the Indian government promised a review of the Tehri dam project. But the promise was not kept and the following year he committed himself to another fast, only broken after 74 days when the Prime Minister gave a personal undertaking to conduct a thorough review, largely on Bahuguna's terms. The veteran environmentalist, then in his

70th year, told the Prime Minister that the Himalayan glaciers were receding at an alarming rate. If this was not checked, the glacier feeding the Ganges would disappear within 100 years.

"We in Himalaya are facing a crisis of survival due to the suicidal activities being carried out in the name of development... The monstrous Tehri dam is a symbol of this... There is need for a new and long-term policy to protect the dying Himalaya. I do not want to see the death of the most sacred river of the world - the Ganga - for short-term economic gains."

– Sunderlal Bahuguna

WATER-RESOURCES

Water is a crucial natural resource, its availability greatly influences the health of the people and development potential of the area. Water as a resource in relation to its needs is becoming increasingly scarce. Proper assessment of the availability of this resource from surface and sub surface sources is crucial for its proper planning, development and efficient management.

About 70% of the global surface is covered with water in the form of oceans, seas, river, lakes, ponds. Total quantity of water available on the earth is 1386 million cubic kilometers. 97.3% of the water available on earth is saline and only 2.7 is available as fresh water. Most of which lies frozen in polar regions or in deep aquifers, not available for use.

The mean annual rainfall, taking the country as a whole is 1170 mm. This gives an annual precipitation of about 4000 km^3. A significant part of this precipitation returns back to the atmosphere as evaporation. A large part of the remaining precipitation seeps into the ground and the balance flows through streams, rivers and collects in water bodies adding to the surface flow. A part of the water which seeps in to the ground remains as soil moisture in the upper layers and the rest adds to the ground water resource.

India's Water Resources

India is a country of vast biological, geographic, and climatic diversity. It has total geographic area of 329 Mha; excluding bodies of water, India's total land area is estimated at 297 Mha.

India is bordered in the north by the 2,500-kilometer long Himalayan Mountains. Melting snow and glaciers provide a continuous flow for numerous rivers running south from the Himalayas into the vast Indo-Gangetic Plain, which is dominated by the Ganges River and its tributaries. Heavy rains are typical in the Himalayas during the monsoon months between June and October, causing frequent floods. Southern India consists largely of the Deccan Plateau, which is flanked by the Western Ghats running along the west coast and the smaller Eastern Ghats on the east coast. The Deccan rivers are rainfed and fluctuate in volume; many of these rivers are not perennial.

India receives average annual precipitation of 4000 km^3, out of which 700 km^3 is immediately lost to the atmosphere, 2150 km^3 soaks into the ground, and 1150 km^3 flows as surface runoff (CGWB 1996, NCA12). See Figure 2.2.

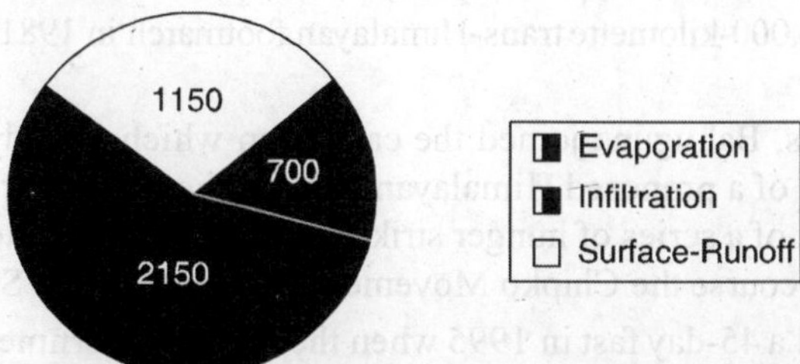

Fig. 2.2: Distribution of Precipitation
(Source : Central Ground Water Board, 1965)

The basinwise details of water resources are given in Table 2.2. The total water resources in the country has been estimated by the National Commission as 1,953 km^3.

Nearly 62 per cent or 1,202 km^3 or total water resources lie in Ganga-Brahmaputra-Meghna basin. The remaining 23 basins have 751 km^3 of total water resources.

River Basin

River basin is the basic hydrologic unit for planning and development of water resources. It is useful and convenient to make the assessment of water resources basin wise.

Earlier, the entire country was divided into twenty river basins by the Central Water Commission. These comprised of 12 major basins (having a drainage area exceeding 20,000 sq. km.) and 8 other river basins each combining a number of medium and minor rivers. However, it was seen that some areas such as the area of North Ladakh not draining into Indus and areas of Andaman, Nicobar and Lakshadweep islands are not covered in the 20 river basins. The Commission in consultation with CWC reviewed the grouping of the river basins. Based on this review, the drainage area of the country has now been divided into 24 basins, which includes 2 basins, namely Area of North Ladakh not draining into Indus and drainage area of Andaman, Nicobar & Lakshadweep Islands. The minor rivers of east coast from Mahanadi to Kanyakumari have been divided into five basins in place of the earlier two basins and the rivers draining into Myanmar and Bangladesh have been grouped under two separate basins, in place of the earlier one basin. The rivers of west coast south of Tapi are grouped into one basin in place of the earlier two basins. Area of inland drainage of Rajasthan which was one of the twenty basins, has been deleted. The list of basins alongwith their catchment areas and locations are given in Fig. 2.3.

During the review undertaken, it was noticed that there were some discrepancies in the reported drainage area of the Indus basin. Both the reports of the second Irrigation Commission (1972) and the Ravi & Beas Waters Tribunal (1987) had adopted the area of Indus Basin as 3,21,289 sq. km. on a watershed basis. However, as per the delineation of the basin boundary shown in the map of the basin in the CWC Publication (1989) "Major River Basins of India – An Overview" the area of the basin appears to be more than 3,21,289 sq. km. The Working Group has adopted the area of 3,21,289 sq. km. for the basin, since the same has been used by the earlier reports.

Based on the physiography, the river systems of India can be classified into four groups, viz. (i) Himalayan rivers, (ii) Deccan rivers, (iii) Coastal rivers, and (iv) Rivers of the inland drainage basin. The main Himalayan River Systems are those of Indus and Ganga-Brahmaputra-Meghna system. The Himalayan rivers receive very heavy rainfall in monsoon months and the rivers swell, causing frequent floods. The flows in the summer months are due to melting of snow and glaciers and therefore, have continuous flow throughout the year. The important river systems in the Deccan are West flowing rivers of Narmada and Tapi and the East flowing rivers of Brahmani-Baitarani, Mahanadi, Godavari, Krishna, Pennar and Cauvery. The Deccan rivers are rainfed and some of them are non-perennial. There are numerous coastal rivers which are comparatively small. Most of them are non-perennial. While only handful of such rivers drain into the sea near the deltas of east coast, there are as many as 600 such rivers on the west coast. The west coast rivers are short in length and have limited catchment areas. A few rivers in Rajasthan do not drain into the sea. They drain into salt lakes or get lost in sands with no outlet to sea.

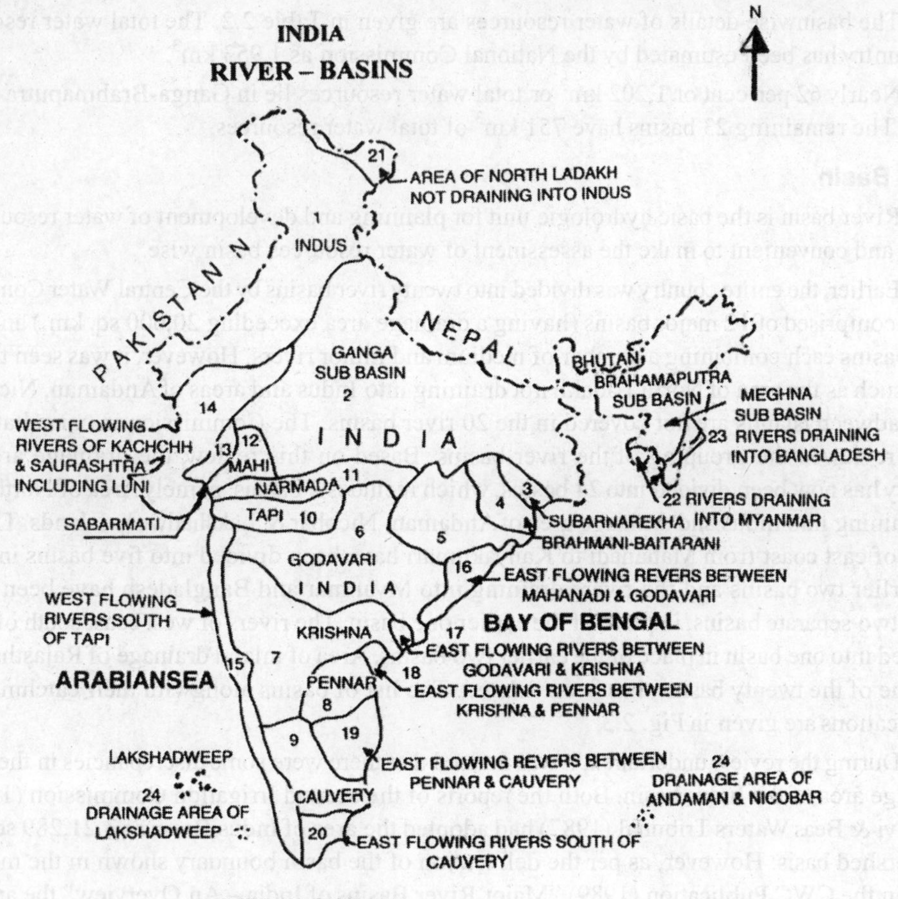

S.NO.	*Name of Basin*	*S.NO.*	*Name of Basin*
1.	Indus	15.	West Flowing Rivers South of Tapi.
2.	Ganga-Brahmaputra-Meghna Basin	16.	East Flowing Rivers between Mahanadi and Godavari
3.	Subarnarekha	17.	East Flowing Rivers between Godavari and Krishna
4.	Brahmani-Baitarani	18.	East Flowing Rivers between Krishna and Pennar
5.	Mahanadi	19.	East Flowing Rivers between Pennar and Cauvery
6.	Godavari	20.	East Flowing Rivers South of Cauvery
7.	Krishna	21.	Area of North Ladakh not draining into Indus
8.	Pennar	22.	Rivers draining into Bangladesh
9.	Cauvery	23.	Rivers draining into Myanmar
10.	Tapi	24.	Drainage Area of Andaman and Nicobar and Lakshadweep
11.	Narmada		
12.	Mahi		
13.	Sabarmati		
14.	West Flowing Rivers of Kachchh and Saurashtra including Luni.		

Fig. 2.3 : INDIA - RIVER BASINS

AVAILABILITY OF WATER RESOURCES

India is one of the few countries in the world endowed with abundant land and water resources. Average annual precipitation including snowfall over the country is 4000 billion cubic meters (BCM). In addition, it receives another 200 BCM from rivers flowing in from other countries. Average annual water resources in various river basins are estimated to be 1869 BCM, of which the utilizable volume of water has been estimated to be 1086 BCM including 690 BCM of surface water and 396 BCM of ground water. The rest of the water is lost by evaporation or flows into the sea and goes waste. The utilization of water is expected to be 784 to 843 BCM by the year 2025. Though the present utilization level is only about 50%, the availability of water is highly irregular. It is not available in places of need, at times of need and in required quantities.

In the major part of the country, rainfall is the only sources for water which is available mainly during the monsoon season lasting for less than 3 months. Due to tropical climate and its geographical, location, the country experiences vast spatial and temporal variation in precipitation. About one-third of the country's area is drought prone. The south and western parts comprising the states of Rajasthan, Gujarat, Andhra Pradesh, Madhya Pradesh, Maharashtra, Tamil Nadu and Karnataka are the drought prone states. On the other hand, north and north eastern regions including states of Uttar Pradesh, Bihar, West Bengal and Assam are subjected to periodic floodings. The recent drought that the country is facing has not only created water scarcity but also led to migration of people and cattle from water scarcity areas to adjoining areas causing hardships to them but also social and economic problems in the areas where they migrate. In such areas, women have to fetch water from wells, ponds, lakes etc. from long distances which consume lot of time, energy and affect the health of the women folk. Therefore, concentrated and dedicated efforts are needed not only to harness the rain water but also conserve it so that it can be utilised in the time of need. The demand is maximum in summer when the available water resources dwindle. In the time of crisis, ground water is the most dependable source. The problem is more serious in urban areas, where natural recharge has been reduced due to reduction in permeable areas as a result of increased urbanisation.

Surface Water: There are twenty river basins of which 12 are major basins having drainage areas of 20,000 km^2 (Fig. 2.3). The annual average runoff estimated to be 1952.87 BCM. The storage built up in various river basins through major and medium irrigation projects is about 173.73 BCM. The major and medium irrigation projects under construction and identified would account for 75.42 BCM and 132.3 BCM respectively taking the total to Rs. 381.45 BCM. If minor irrigation structures are included the total storage goes upto 420 BCM. However it is estimated that about 49.5 BCM and 75.2 BCM capacities of the existing reservoirs will be lost through siltation by the year 2020 and 2025 AD respectively. Total live storage capacity of all reservoirs would be about 462 km^3.

The average annual utilisable water resources through conventional schemes of all the 20 river basins taking into account the uneven nature of distribution of water resources and the topographic constraints is estimated to be 690.3 BCM which is about 35% of the total surface water resources. This indicates that vast amount of runoff is going as waste to sea.

Table 2.1 Rainwater availability in India

Monsoon Activity	*Mham*	*Percentage Share*
South West	296	74
North East	12	3
Pre monsoon	52	13
Post monsoon	40	10
Total	400	100%

Source : Fertiliser Statistics, FAI, New Delhi (1994)

The interbasin transfer proposals investigated by the National Water Development Agency (NWDA) envisage utilization of additional 200-250 km^3 of surface water. According to another study, carried out by the Central Ground Water Board (CGWB), an additional quantity of 160.5 km^3 water can be artifically recharged by utilizing surplus monsoon runoff. Many of these proposals are at conceptual stage only and would require further investigations and studies – and broad agreements between concerned State Governments – before a reasonable estimate of additional utilisable surface water can be made. In view of this any substantial use of surface water through interbasin transfer and/or artificial recharge by utilising surplus monsoon runoff is not for seen till the middle of the 21st century.

Ground Water Resources: Replenishable ground water resources assessed is of the order of 431.32 BCM and utilisable ground water resources as 395.6 BCM of which 325.6 BCM is available for irrigation and the rest for domestic and industrial uses (Central ground water Board 1995). Thus the total utilisable ground water resource amounts to about 92% of potential replenishment.

The gross available and utilisable water resources of the country, are, therefore 2384.5 BCM and 1086.0 BCM respectively. The available and utilisable water resources per capita based on 1991 population are 2830 m^3 and 1288 m^3/capita/year respectively. However, there is a wider variation in its availability in basins. Out of the 20 basins, 4 basins had more than 1700 m^3/capita/year utilisable water resources while 9 basins had between 1000 and 1700 m^3/capita/year, in 5 basins it is between 500 m^3 to 1000 m^3/capital/year and 2 basins it is less than 500 m^3/capita/year. With increase in population which has crossed one billion the availability of land and water resources gets further reduced.

According to a standard definition for water availability if it falls below 1000 m^3/capita/year water supply begins to hamper health, economic development and human beings resulting in scarcity conditions requiring to take up immediate remedial measures. At less than 500 m^3/capita/year, water supply is primary constraint to life. So many river basins are already facing water stress requiring remedial measures.

According to the initial estimates of the Central Ground Water Board an additional quantity of about 10,081 km^3 of static ground water can be exploited. Generally static water is not regular replenished on annual basis, its one time use is only possible as a short time strategy. Further experience indicates that large scale use of static water or its mining is usually associated with surface settlement which may cause heavy damages to properties of land and soils and may, possibly, trigger seismic activities. Therefore, any appreciable use of static ground water on regular basis, is not foreseen till the middle of the 21st century.

Contribution of rainfall to world's water supply

Rain is the main source of fresh water. Its contribution is as

1. Water in sea & oceans 97.2%
2. Water in glaciers and ice caps 2.14%
3. Underground water 0.16%
4. Surface water 0.009%
5. Soil moisture 0.005%

Rainfall is seasonal and is not evenly distributed. On an average, annually about 112 cm rain is received over the plains of India. Highest rainfall in India is about 250 cm while lowest is 15-20%. In Cherrapunji (Meghalaya) rainfall is about 1087 cm annually.

WATER DEMAND

During 1990 the total utilisation of water for all uses, was about 518 BCM or 609 m^3/capita/year.

The water demand to meet the requirement of domestic, industrial and irrigation is given in Table - 2.2.

Table 2.2. Requirement of water for various uses in BCM

Sl. No.	*Category*	*Year 2010*			*Year 2025*			*Year 2050*		
		Low	*Medium*	*High*	*Low*	*Medium*	*High*	*Low*	*Medium*	*High*
1	Irrigation	489	536	556	619	688	734	830	1008	1191
2	Domestic	39.4	41.6	61	47	52	78	59	67	104
3	Industries	37	37	37	61	67	79	69	81	116
		555.4	614.6	654	727	807	881	958	1156	1411

The total medium water requirement of the country would be for the above purposes is about 614.6, 807 and 1156 BCM by the years 2010, 2025 and 2050 respectively.

Water demand for irrigation to maximise agricultural production with the maximum possible level of irrigation to achieve self sufficiency in food grains is given in Table 2.3.

Table 2.3. Water demand for irrigation

Year	*Low demand scenario million tons*	*Water requirement BCM*	*Middle demand scenario million ton*	*Water requirement BCM*	*High demand scenario million ton*	*Water demand BCM*
2010	249	489	265	536	271	576
2025	322	619	349	688	365	734
2050	469	830	539	1088	605	1191 (1451 km^3)

Other than three requirements, water also required for energy development, inland Navigation, Flood Moderation, ecological considerations, to compensate evaporation losses from reservoirs, environment afforestation, etc. If the country desires to irrigate all cultivable land at 100% intensity of irrigation by the year 2050, the water requirement would be 1451 km^3 and the total requirement will be 1824 km^3. This is a hypothetical and rather unrealistic scenario.

HYDROLOGICAL CYCLE (WATER CYCLE)

In this solar-driven water cycle, (see figure on next page) water evaporates from the earth's surface into the atmosphere and is returned back as rain or snow. Part of this precipitation evaporates back into the atmosphere. Another part flows into streams, rivers, and lakes, commencing its journey back to the sea. Still another part percolates into the soil and becomes soil moisture or groundwater. Plants extract soil moisture into their tissues and release it into the atmosphere in the process of evapotranspiration. Much of the groundwater eventually works its way back into the flow of surface waters.

The National Research Council (1991) report defines the *hydrologic cycle* as "the pathway of water as it moves in its various phases through the atmosphere, to the Earth, over and through the land, to the ocean, and back to the atmosphere" as shown in **Fig. 2.4.** During the cycle, which has no beginning or end, a single water molecule may assume various states, returning to the hydrologic pathway as new chemical compounds are mixed with various solid and liquid substances. As shown in the figure, water evaporates from the oceans and the land surface to become part of the atmosphere; water vapor is transported and lifted in the atmosphere until it condenses and precipitates on the land or oceans; precipitated water may be intercepted by vegetation, become overland flow over the

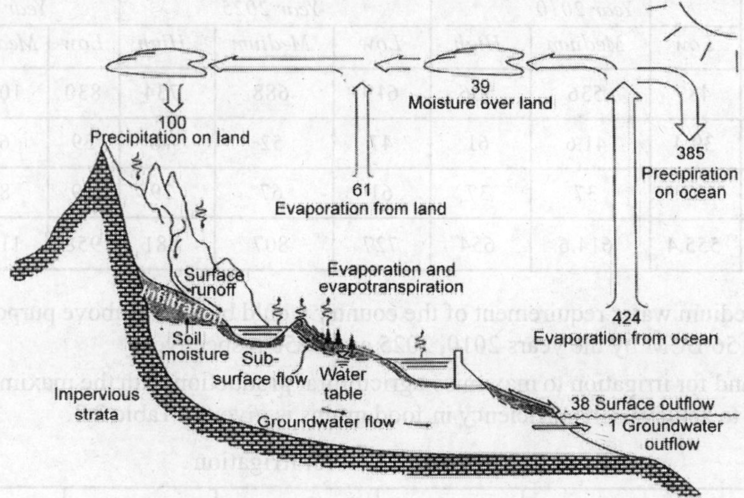

Fig. 2.4. Hydrologic cycle with global annual average water balance given in units relative to a value of 100 for the rate of precipitation on land.

ground surface infiltrate into the ground, flow through the soil as subsurface flow, and discharge into streams as surface runoff. Large amounts of the intercepted water and surface runoff returns to the atmosphere through evaporation. Infiltrated water may percolate deeper to recharge groundwater, and later emerge in springs, or seepage into streams, to form surface runoff. Finally, this water may flow out to the sea or evaporate into the atmosphere. Throughout this cycle, water may take on many quality aspects.

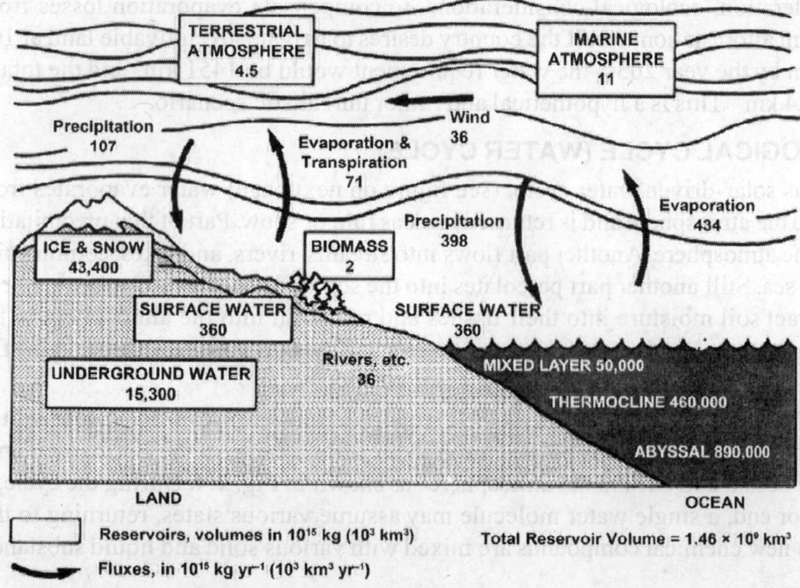

Fig. 2.5. The hydrologic cycle at global scale (*National Research Council 1986*)

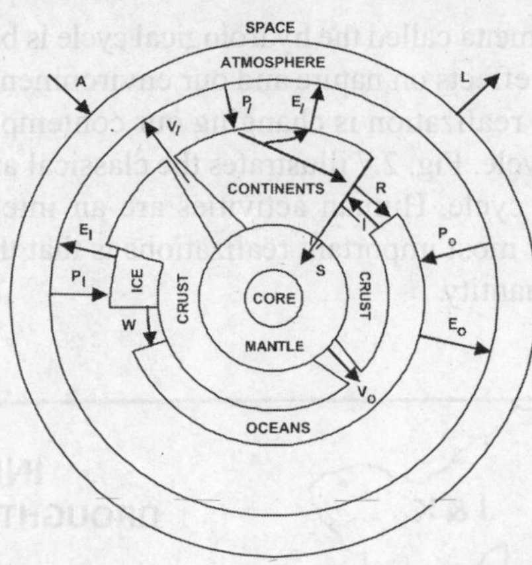

A = Additions of water from space
E_o = Evaporation from oceans
E_I = Evaporation (i.e. sublimation) from ice
E_l = Evapotranspiration from land
I = Intrusion of seawater into continental aquifers
L = Loss of water to space
P_o = Precipitation on oceans
P_I = Precipitation on ice
P_l = Precipitation on land
R = Runoff from continents
S = Subduction of water containing crut
V_o = Volcanic venting to oceans
V_l = Volcanic venting to atmosphere
W = Wastage of ice sheets to ocean

Fig. 2.6. The hydrologic cycle as a global geophysical process. Enclosed areas represent storage reservoirs for the earth's water, and the arrows designate the transfer fluxes between them. (*National Research Council 1991*)

The hydrologic cycle can also be viewed on a global scale, as shown in Fig. 2.5. Our knowledge of the amount of water in space and in the earth's mantle is very limited. There is evidence that space and the earth's mantle both exchange water with the primary crustal, ice, the atmosphere, and the ocean. The hydrologic cycle can also be viewed as a global geophysical process, as shown in Fig. 2.6. Water vapor and methane molecules are diffused into space, causing loss of the hydrogen in water. These hydrogen atoms subsequently escape by photochemistry. The addition of water from space is a controversial tissue. Volcanic activity vents water vapor to the atmosphere and liquid vapor to the ocean. Water recirculates on a geological time scale by the subduction of water-contaning crustal material.

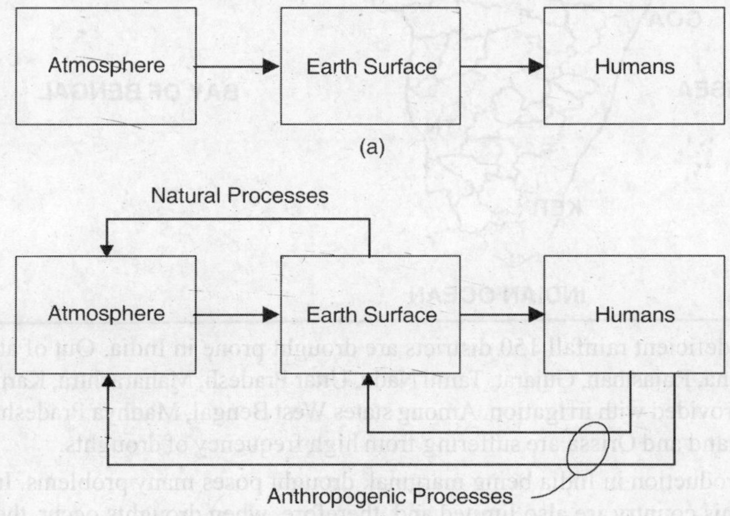

Fig. 2.7. The role of humans in the hydrologic cycle, (a) classical viewpoint, (b) modern viewpoint. (*National Research council, 1982*)

This fascinating phenomena called the hydrological cycle is being changed by human activities. We have begun to realize the effects on nature and our environment brought about the these changes in the hydrologic cycle. This realization is changing our contemporary views of the ineractive role of people in the hydrologic cycle. Fig. 2.7 illustrates the classical and modern view points of the role of people in the hydrologic cycle. Human activities are an integral and inseparable part of the hydrologic cycle. One of our most important realizations is that the quality of water in this cycle is of as much concern as the quantity.

DROUGHT

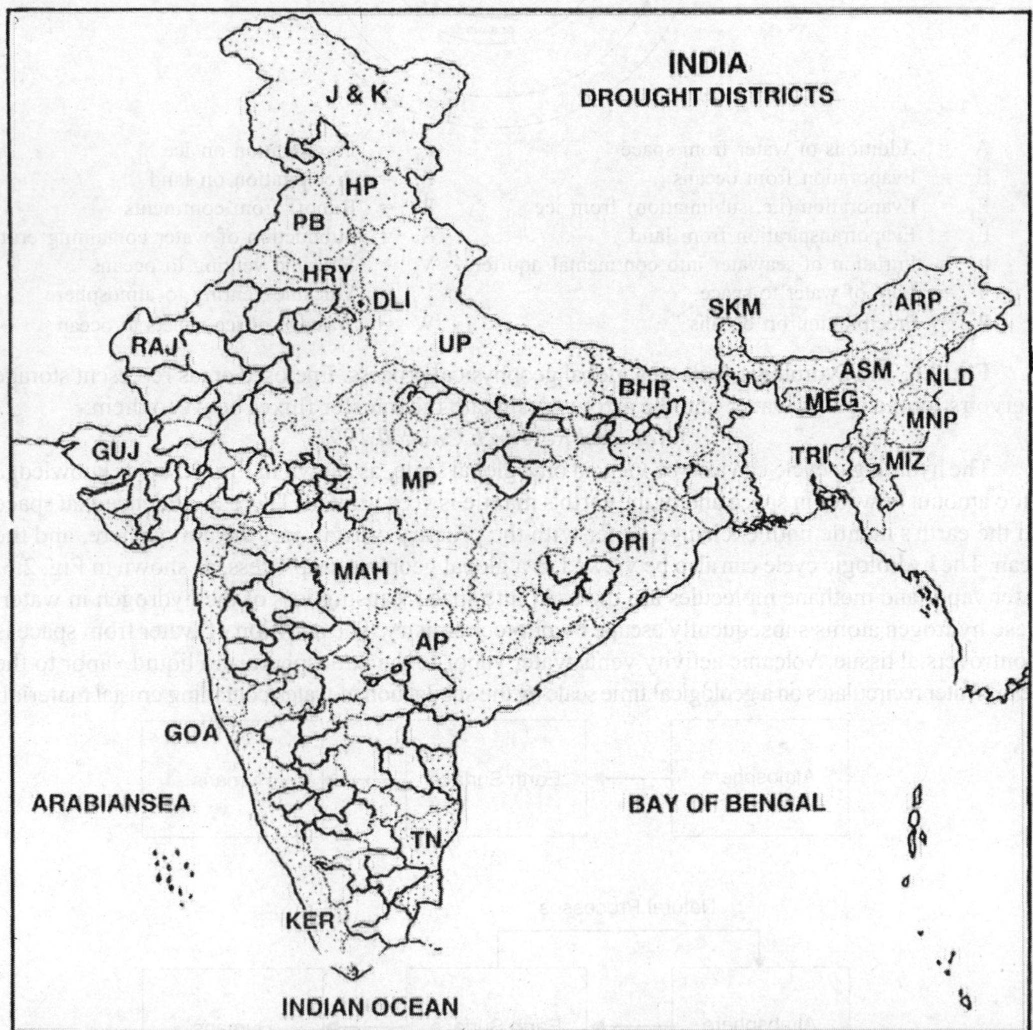

Due to deficient rainfall 150 districts are drought prone in India. Out of above 7 districts of Punjab, Haryana, Rajasthan, Gujarat, Tamil Nadu, Uttar Pradesh, Maharashtra, Karnataka and Andhra Pradesh are provided with irrigation. Among states West Bengal, Madhya Pradesh, Andhra Pradesh, Bihar, Jharkhand and Orissa are suffering from high frequency of droughts.

Food production in India being marginal, drought poses many problems. Irrigation facilities available in this country are also limited and, therefore, when droughts occur, they cause partial or complete crop failure. If failures occur in consecutive years, it becomes a national calamity, putting great strain on the economy of the country.

According to the practice followed by the Indian Meteorological Department, drought is taken to have occurred over an area where the annual rainfall is less than 75 per cent of the normal. When the annual rainfall is less than 50 per cent of the normal, it is called severe drought. Drought can occur anywhere in India, depending upon the distribution of rainfall.

Attempts to quantify drought in the form of an index have been made by using various techniques. Pilot studies have been conducted in India, with special reference to Bihar State, in order to arrive at a rational definition. The Palmer drought index, which takes into account rainfall, evapo-transpiration and soil moisture is considered a comprehensive approach to this problem. The computation of the Palmer drought index have been carried out for the different sub-divisions of the country. The computations show that, on the average, drought is experienced on 20-25 per cent of the days in each of the months of the *kharif* season over large areas of the country.

Another aspect of great interest in respect of drought is whether there is any periodicity in its occurrence. The power-spectrum analysis of the rainfall series and that of the Palmer drought index series show some relation to the quasi-biennial and the eleven-year sunspot cycles in some areas. The amplitude of the cycle is, however, too small to exert a significant influence.

Connected with the above points is also the problem whether the rainfall series over an area exhibits any trend, cyclic variation or persistence. A number of studies have been carried out on this aspect over different regions of the country. By and large, over the areas studied, it has been found that the rainfall series is found to be of a random nature with no significant trend. Hence it would not be feasible statistically to predict the amounts of rainfall during any particular year from a knowledge of the past rainfall occurrences alone. The probabilities of the rainfall occurrences can, however, be used to estimate the risk of poor rainfall and the attendant drought.

OVER EXPLOITATION OF WATER

The exploitation of groundwater resources more than its annual replenishment has caused the continuous declining of water levels, declining of well yield, drying of shallow wells, deterioration of ground water quality, sea water intrusion into coastal aquifers and high cost of energy required to lift the water from great depths which becomes uneconomical for poor farmers to continue agriculture.

Though India is blessed with a good water resources, but its distribution over the country is not uniform/proper. Even in the high rainfall areas like Meghalaya and Kerala water scarcity is felt in summer months due to over exploitation of water and mismanagement. There is large amount of rainfall annually flowing out as runoff to sea. The annual replenishment to groundwater resources is getting reduced due to reduction in natural recharge mechanism by man's interaction desides recurring droughts.

There is uneven distribution of water resources coupled with over utilization of groundwater resources has resulted in imbalance between recharge and development. There exists gap between available utilizable water resource and future need of water for the country. Due to over utilization of water, Punjab, Haryana, Tamilnadu and Gujarat and the states where the water tables have declined steeply. In Gujarat more than 90% wells water table dropped by 0.5 meters to 9.5 meters (CGWB). In Haryana, the average depth of ground water is fallen by 1 to 33 cm annually in different parts of the state.

CONFLICTS OVER WATER

Water being the basic requirement for life and necessary for almost all socio-economic activities is facing ever greater demand. Its relative demand increases with degree of scarcity. As we have noted above, a large part of the country already faces water scarcity conditions and it is expected that by the middle of the next century most regions of the country would face some degree of scarcity. These conditions have already created a number of inter state water dispute. If the such conditions continue, we can also expect the next world war will be on water. Bitterness over these disputes is increasing with passage of time. These water disputes have multiple facets, examples of two such instances are as :

1. Urban water demands are concentrated in space, therefore, pose serious problems at local levels. Water demands in mega cities are growing much faster than envisaged and are putting heavy strain on water resources. It is creating difficult problem for the surrounding rural areas, leading to serious conflicts.

2. Since the urban water supply are met from surface (river) flows, there will be conflict with upstream users, specially farmers, over the quantum of withdrawls, while the down stream users will be affected by the polluted waste waters released by urban areas. Such conflicts already exists : between Delhi and Haryana, and between Chennai and the farmers in drought prone districts of Andhra Pradesh.

In future such conflicts are likely to increase in number and escalate in magnitude unless an effective mechanism is evolved to resolve than expeditiously and judiciously.

MINERAL RESOURCES

Minerals, being the vital raw material for many basic industries, play an important role in the industrialisation and overall development of nation. Minerals are generally called the "stock" as they are the non-renewable resources. Minerals are the definite chemically bonded substances, created through chemical processes between organic and inorganic matters present in the earth's crust. They may be solid or liquid. Since the prosperity of a nation depend upon the proper use of minerals, hence they should be conserved and should not be misused. Govt. should promote the research in this field of mining minerals.

The history is hundreds year old. Iron, steel, copper, zinc, lead, gold, silver, cobalt etc. metals were extracted from minerals in India. But now, building materials coal, iron ore, mangenese ore, gold, petroleum, natural gas, copper ore, ilmenite, glass sand, lead and zinc ores, chromite, raynite, ilmenite, magnesite, gypsom, monazite, beryl, dolomite, bauxite etc. are produced from minerals in India. The minerals from metals like bismuth, cadmium, graphite, platinum, tungsten tin, silver, gold are extracted, are in least quantity.

Types of minerals

Minerals available in earth crust can be divided into three types

1. Metallic minerals
2. Non-metallic minerals
3. Mineral fuels

Some other classifications of minerals are also given by scientists. They are classified as strategic and critical depending on the use and importance.

1. Metallic Minerals—We cannot extract metal directly from minerals. There is difference between minerals and ores. Therefore, for extracting metals, minerals are treated by different processes before extraction. Metallic minerals are generally found in combined state. According to availability of metals, metallic minerals are further divided into following :—

(*a*) Ferrous alloys : Most common metal (which is used largely) is iron. Other than iron are aluminium, lead, zinc, copper etc. All are found in rich quantities, found in native as well as in combined state. Iron pyrite, Lynonite, Haematite, Magnetite are examples of ferro alloys. Certain other metals, non-metals are contaminated with these as impurities.

(*b*) Non-ferrous alloys : The minerals/alloys of this type contain the metals like titanium, antimony, arsenic, beryllium, copper, zirconium, cerium, lithium etc. These metals are costlier than proceeding metals. Here the iron found as an impurities.

(*c*) The minerals/alloys containing very least quantity of metals whose extraction is costlier. These metals are generally used in jewellary eg. gold, platinum, silver, irridium etc.

2. Non-metallic minerals : Minerals, whose yield products are other than metals comes in this head. They are called the non-metals. They are further divided on the basis of physical and chemical properties. Graphite, pyrolusite, dolomite quartz, kaoline, fire clay, felspar, mica, asbestos, gypsom fluorite, chrome/red ochre, lime stone, borax, phosphorite, ilmanite, flint, dymond, calcite, sand stone, stones like phylite, cyanite, lime stone, ruby, sapphire. Emarald, amber, spodumene etc. are the examples of non-metallic minerals.

3. Mineral fuels : These include the materials used to provide energy, for example coal, natural gas, fossil fuels and petroleum etc. These are the important source of energy, hence they have tremendous importance for mankind.

Coal is the most commonly available fuel which is used as domestic as well as industrial fuel. It is of different type i.e. Anthracite, Bituminous, Lignite etc. The type and quality of the coal depend upon the percentage of carbon present in them. It is the principal source of energy in world. It is used in various ways in different industries like cement, glass, railways, textile, sugar, paper, steel etc. It is also largely used in domestic way. USA, China, Britain, Germany, South Africa, Australia are richest coal containing countries in world.

Petroleum is used in the manufacture of large number of petro-chemicals. It is drilled out from the sources as crude oil. Crude oil is refined before use as petrol, diesel, kerosine etc.

Minerals in nature : The man is using minerals since long. From lacs of year back primitive man was using flint, quartiz etc. for preparation of their tools. This was called "stone age". Later they use metals therefore, the period was named after as "copper age", "bronze age", and "iron age". Now present age is "machine age" because machines are prepared from minerals and they run by mineral fuels.

The formation of mineral deposits is a very slow geo-chemical or biological process, which takes millions of years to develop mineral deposits. Most of the minerals are widely distributed in earth's crust. Studies shows that, there are number of ways by which mineral deposits are formed. They are

1. Molten rock materials, which is a complex collection of a number of substances, when cooled, the crystallization of different minerals takes at different temperatures. These are settled in different bands, giving the mineral deposits.

2. Sodium chloride, gypsom, salt peter etc. Water soluble minerals are obtained by evaporation of lake/sea water. The compounds of iron and manganese as chemical sediments are also formed by precipitation from lake or sea water.

3. Deposits of minerals like asbestos, talc, graphite etc. are formed by intense heat and pressures inside earth's crust.

4. When the pH, temperature, solubilities are changes, the rock materials in solution/suspension are deposited in sufficient amounts to form mineral deposits as water current slow down.

5. Mineral deposits are also formed by oxidation and reduction reactions.

6. Formation of mineral deposits are also take place by micro-organisms. It is mainly autotrophic bacteria which are involved in mineralization reactions.

There are also other views for formation of mineral deposits. When the plants, dead animals, wild life & other ecosystems are accumulated below in earth. Biological process convert them in to mineral deposits.

Mineral resources of India : India has sufficient quantities of iron, aluminium, titanium, copper, lead, zinc ores. India is fairly rich in mineral resources. We possess good deposits of most of mineral elements which we needed in large quantities. However, other economically important minerals are not present in sufficient quantities. Iron minerals, which are most important ingredient of today's economy are found in sufficient quantity in our country. We are presently exporting it to other countries. The similar case is for aluminium also. Our country ranks Vth among the aluminium rich countries of the world. At present we have sufficient aluminium stock for the domestic market. We are exporting to other countries as well.

Zinc, lead ore reserves in India are estimated to be about 390 million tons. Good quality of these elements minerals/ores are being depleted at a fast rate.

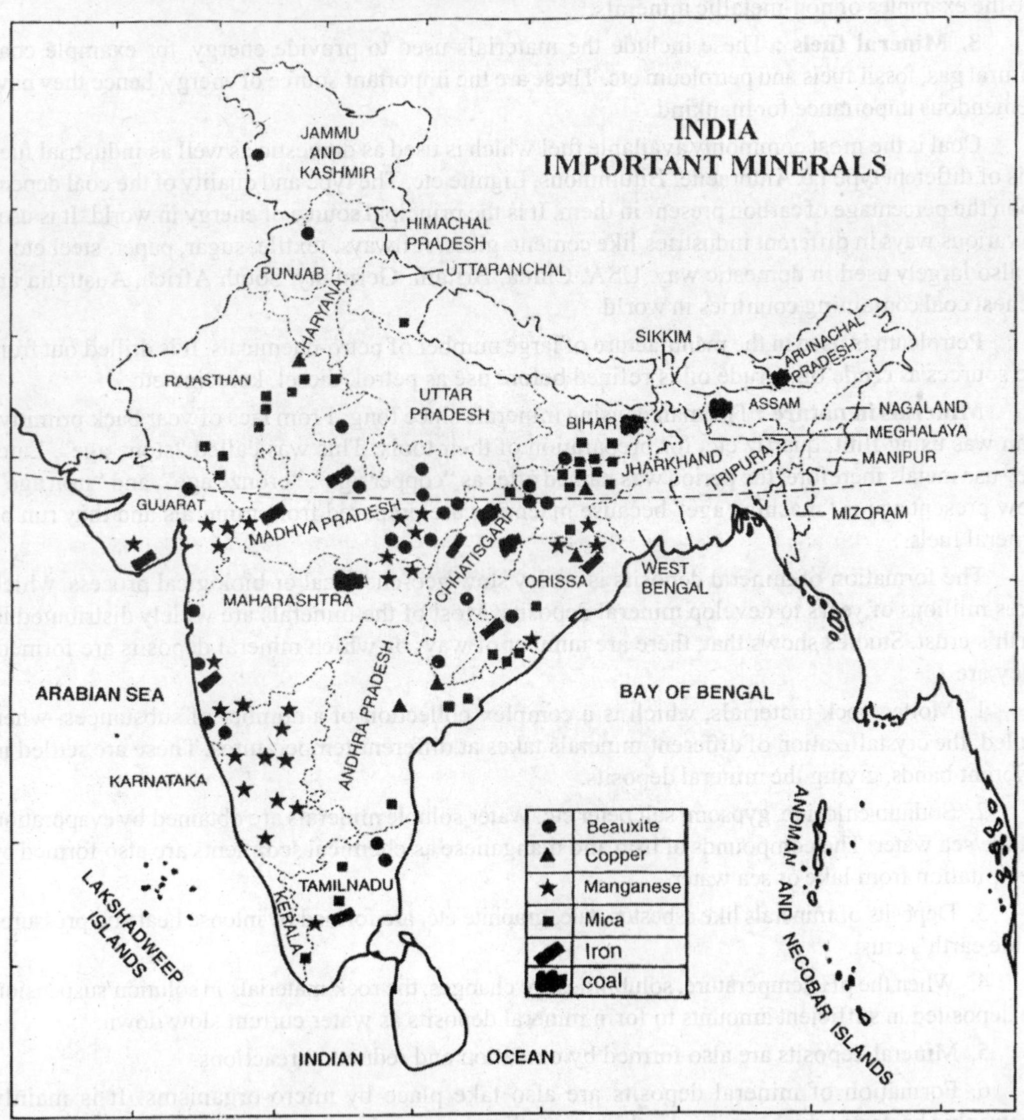

India has a large number of economically useful minerals and they constitute one-quarter of the world's known mineral resources. About two-thirds of its **Iron deposits** lies in a belt along Orissa and Bihar border. Other haemaite deposits are found in Madhya Pradesh, Karnataka, Maharashtra and Goa. Magnetite Iron-ore is found in Tamilnadu, Bihar and Himachal.

India has the world's largest **deposits of coal.** Bituminous coal is found in Jharia and Bokaro in Bihar and Raniganj in West Bengal. Lignite coals are found in Neyveli in Tamilnadu.

Next to Russia, India has the largest supply of **Manganese.** The manganese mining areas are Madhya Pradesh, Maharashtra and Bihar-Orissa area. **Chromite deposits** are found in Bihar, Cuttack district in Orissa, Krishna district in Andhra and Mysore and Hassan in Karnataka. **Bauxite deposits** are found in western Bihar, southwest Kashmir, Central Tamilnadu, and parts of Kerala, U.P. Maharashtra and Karnataka.

India also produces third quarters of the world's **mica.** Belts of high quality mica are, Bihar, Andhra and Rajasthan. **Gypsum** reserves are in Tamilnadu and Rajasthan. **Nickel ore** is found in Cuttack in Bihar and Mayurbanj in Orissa. **Ilmenitc** reserves are in Kerala and along the east and the west coastal beaches.

Silimanite reserves are in Sonapahar of Meghalaya and in Pipra in M.P. **Copper ore** bearing areas are Agnigundala in Andhra, Singhbhum in Bihar, Khetri and Dartiba in Rajasthan and parts of Sikkim and Karnataka.

The Ramagiri field in Andhra, Kolar and Hutti in Karnataka are the important **gold mines.**

The Panna **diamond belt** is the only diamond producing area in the country, which covers the districts of Panna, Chhatarpur and Satna in Madhya Pradesh, as well as some parts of Banda in Uttar Pradesh.

Petroleum deposits are found in Assam and Gujarat. Fresh reserves were located off Bombay. The potential oil bearing areas are Assam, Tripura, Manipur, West Bengal, Punjab, Himachal, Kutch and the Andamans.

India also possesses the all-too valuable nuclear **uranium** as well as some varieties of **rare earths.**

A quarter of all mining is carried out in the southern part of Orissa. Gold, silver and diamonds make up a small part of other natural resources available in India. The gemstones are-found in Rajasthan. Major portion of the energy in India is generated from coal. It is estimated that India has around 120 billion tons of coal in reserve, enough to last for around 120 years. Huge reserves of the petroleum have been found off the coast of Maharashtra and Gujarat and M.P. Electrical energy generated by hydroelectric power, coal and nuclear energy. Half of the hydroelectric power is generaied by snow field reservoirs high up in the Himalayas. In Madhya Pradesh important minerals like diamond, tin ore, coal, copper ore, alexandrite, iron ore, dolomite, rock phosphate, manganese ore, lime stone, granite, marble, corumdum, pyrophylite, diaspore, Bauxite etc. are found in different quantities. Chhatisgarh (new state of M.P.) is rich in minerals and forest products.

Environmental effects of extracting and using mineral resources

Mining, minerals and mineral based industry indeed play an extremely important role in the development of mankind. The total geographical area of India i.e. 329 million hectares constitute 2.4% of the world land area. Out of this about 82500 hectares is sustaining mining activities of some kind or the other. As the mining activity grows, the per capita availability of land is declining at a very high rate. The extra emphasis on mining and minerals is directly related to growing population and better standard of living.

The environment means the surroundings. The components of environment include soil, water, air, land, landscape, and living creatures. The environment is more damaging by open cast mining than underground mining. Not only environment, mining also effects human health. Over exploitation causes the wastage of mineral wealth and derelict of land. Mineral deposits should not be over exploited because they are non-renewable. Derelict land is that land which has been abandoned as useless. Dereliction is the result of thoughtless, uncontrolled ruthless exploitation of natural resources.

This land is the permanent damage not usable for agriculture. There are following environmental effects of mining—

i. Land degradation due to lowering of the surface levels at some places and creation of large mounds at other places;

ii. Deforestation in the mining areas, i.e., the loss of valuable soil cover, resulting in the possibility of enhancement of soil-erosion;

iii. The loss of top and sub-soil;

iv. Adverse effect on ground water table. The local water table is lowered as a result of opencast mining. The replenishment of acquifers is adversely affected as the mined out terrains are completely dismembered of acquiducts leading to acquifers. As a result the affluent discharge of rain water is increased leaving the water-table completely or partly unrecharged. This also increases the salinity of remaining ground water;

v. Due to increased discharge of rain water passing through the terrains, disturbed by surface mining, the local drainage system is polluted, which on joining the main drainage feature, affects it also;

vi. The frequency of land slides increases substantially as a combined result of factors as stated above;

vii. The erosion of soil is enhanced;

viii. The agricultural lands are affected by silt and the fine material mined but not recovered. It also clogs the surface water channels;

ix. The disturbance caused adversely affects the well-balanced pH and diminishes the regenerative qualities of soil, etc.;

x. The disturbance caused to the floral and faunal population is immense;

xi. The heavy earth-moving machinery and blasting cause problems of noise, vibration and the release of noxious gases in the atmosphere;

xii. The aesthetic damage caused to the landscape reduces its recreational value.

xiii. Mine drainage has polluted streams, rivers, lakes even seas.

xiv. Fumes from smelters damage forests and spread pollution over large area (air pollution)

xv. Mining and mineral based industries with their effluents create pollution problems. Asbestos, cement and other chemical industries are very hazardous. People are not supposed to live in surrounding areas.

xvi. Mining causes the reduction of forests i.e. deforestation. Thus flora and fauna are also destroyed. Wild life also effected. Land becomes barren and this results in increased incidents of land slides.

xvii. The people related with mining and extraction effected by polluted environment (Dust & poisonous gases) lead to skin and lung diseases.

xviii. Mining affects the sub segments of the environment like forests, vegetation, soil cover, humus and ground water. Dust and toxic gases indirectly affects air, humidity, temperature.

xix. Deforestation and climatic change results poor rainfall and affects flora & fauna.

WORLD FOOD PROBLEM

Before the 21st century, it was felt that world food production is not sufficient for the present population. Food production was less because people were using the old techniques, seed etc. Later on when population pressure starts, the new ways of food production, using fertilizers, pesticides, insecticides etc. are discovered to increase the yield. In 1999 International Food Policy Research Institute (IFPRI) reported the increase in world food consumption by 2020, discussing the impact of this on both developed and developing countries. The report considers the six emerging issues, nutrition, grain prices, WTO, agroecological approaches to small scale farming, biotechnology, information technology and precision farming. In world food submit 1996 in Rome the following points were discussed—

(*i*) Reduce world hunger

(*ii*) Agricultural supply and demand

(*iii*) Population growth

With respect to crop production and agricultural growth the similar discussion on "goals, solutions and actions necessary to end hunger" were also held in IFPRI conference held in Bonn, Germany 2001. All this shows the awareness of we people to increase the food production in view of increasing population growth. World leaders also agreed that the contribution of irrigation to incremental food production should be substantial.

Different scenarios have been examined to explore a number of issues such as the expansion of irrigated agriculture, the increase in food production in rainfed areas, and the public acceptance of genetically modified crops. Some opinion were that the world may face urgent food and agriculture problems. Analysis believe that what is needed is a new and greener revolution to once again increase productivity and boost production.

Severe droughts and sharply rising food prices spurred national governments and international agencies to address the food crisis of the 1960s and 1970s. The 'Green Revolution', consisting of crop variety improvements, increased use of fertilizers and expansion of irrigation, averted the projected shortages in food production. According to some experts, another food crisis predicted by advocates of a new boom in investment for irrigation is not yet in view. Food grain prices have remained stable for the last 15 years. There is hunger in the world, but that is because the hungry cannot translate their needs into demand or civil disorders disrupt food flows. However, according to the authoritative Consultative Group on International Agricultural Research (CGIAR), the world is entering the 21st century on the brink of a new world food crisis that is as dangerous, but far more complicated than the threats it faced in the 1960s (Shah and Strong, 2000).

Much could be said on the role of demographic and economic factors, such as world trade, price commodities and agricultural subsidies to farmers in meeting the challenge. However, the purpose of this paper is not to contribute more to the dabate between experts on food security. It is to examine the probable consequences of the business-as-usual scenario that has been the prevailing model for the development of irrigated agriculture, particularly of the large-scale irrigation systems, in many countries. It also projects the likely benefits of increased investment in irrigation and advocates a new approach to design and management of irrigation systems in association with institutional and policy reforms.

CHANGES CAUSED BY AGRICULTURE

Agriculture in US and India was food raising activities, and over half of the current crops comes from plants such as corn, cotton, potatoes, tobacco etc. The agricultural methods in most parts of the world were primitive. Fields were dug by oxen pulling wooden plows, seeds were broadcast by lands and grains were harvested with scythes. From the Indians the first American settlers learned how to clear land till the fields and grow the corn, that was crucial to their initial survival. The mid-1800s began an era of great change which brought advances in cultivation method, breeding of improved crop varieties and use of fertilizers and crop rotations to maintain soil productivity. A vigorous market soon developed for soil amendments such as guano, manure, crushed bone and lime. By 1860 seven factories had been established in US to manufacture mixed chemical fertilizers. The use of pesticides also began. By 1893 there were 42 patented insecticides offered by several manufacturers. In 1840 the benefits of irrigation were discovered. Govt. passed several laws to assist western states in developing extensive and costly irrigation systems. Farm labour requirements diminished with the introduction of mechanised invention of machines for tilling, planting, reaping and threshing vastly increased farm efficiency. In 1892 the first successful gasoline powered tractor was introduced in Iowa. Later on tractors gradually became popular.

Ever since colonial days, agricultural leaders have been interested in increasing the productivity. As land became less available, people became more interested in maintaining soil fertility and increasing crop yields. In 1914 Govt. responded to this need by providing funds for state agricultural extension programme assist farmers in adopting improved farming methods. In 1930s national attention was focussed on the need for soil and water conservation measures to maintain farm productivity. The rich soils produced bountiful crops and between 1870 to 1910 the population of plains states increased by a factor of 10. In response to soil and water conservation programmes, the soil conservation service was established in 1935. In America untill world war II, productivity increased in view of additional demand for food. Conversion of animal power to mechanical power resulted in increased output. Use of fertilizers and pesticides including DDT increased by 50% in between 1940–1944. But in India the productivity was not so, because England ruled India till 1947. The production was not sufficient. Some time we import the food grains and other food materials. After the independance we progressed by giving the facilities to farmers. Farmers didn't have high quality mechanical power, seeds, fertilizers, pesticides etc. For all people were dependent on Govt.

FOOD RESOURCES

We know that only plants can produce food, hence they are called producers. Food, which is necessary for all living organisms consists of proteins, enzymes, carbohydrates, minerals etc. There are various types of animals depend upon the types of food. Herbivores take plants directly as food. Meat eaters are called Carnivores. Animals eat every thing i.e. vegetables or meat called Omnivores. Some animals eat flesh of dead animals are called Scavengers. Thus agriculture and domestic animals are the principal sources of food.

Cereals, pulses, grains, vegetables, fruits etc. we get from agriculture. Domestification of cattle and poultry are necessary for food production from animals. Fish is the another animal source of food. For the last so many years, our food production increased by 50%, at the same time population growth is increased. Still we are not getting production at par with other countries like U.S.A. The 1996 world food summit warmed the countries to reduce the number of undernourished and malnutrition.

The land and water resources are limited, therefore, it is needed to increase the productivity, urgent steps are also needed to reduce the population. Fishes also play an important role to fulfil the world food demand to some an extent. In India, open ocean is under utilised and coastal areas are overexploited for fish industry. Thus fish production from aquatic resources goes high, solving the food problem.

OVERGRAZING

Grazing management is the foundation of grassland based live-stock production since it affects both animal and plant health and productivity. Overgrazing can occur under continuous or rotational grazing. It can be caused by having too many animals on the farm or by not properly controlling their grazing activity. Overgrazing reduces plant leaf areas which reduces interception of sunlight and plant growth. Plants become weakened and have reduced root length and pasture sod weakens. The reduced root length makes the plants more susceptible to death during dry weather. The weakened sod allows weed seeds to germinate and grow. If the weeds are unpalatable or poisonous, major problems can result.

One indicator of overgrazing is that the animals run short of pasture. Under continuous grazing, overgrazed pastures are predominated by short grass species such as blue grass and will be less than 2-3 inches tall in the grazed areas. Soil may be visible between plants in the stand, allowing erosion to occur. Under rational grazing overgrazed plants do not have enough time to grow to the proper height between grazing events. The animals are turned in to a paddock before the plants have restored carbohydrate reserves and grown back roots lost after the last defoliation. As the sod thins, weeds encroach in to the pasture. Another indicator is that the livestock run out of pasture, and hay needs to be fed early in the fall. Overgrazing is also indicated in livestock performance and condition.

Overgrazing can increase soil erosion. Reduced soil depth, soil organic matter and soil fertility hurt the land's future productivity. Soil fertility can be corrected by applying the appropriate lime and fertilizers. However the loss of soil depth and organic matter takes years to correct. Their loss is critical in determining the soil's water holding capacity and how well pasture plants do during dry weather.

To prevent overgrazing, match the forage supplement to the herd's requirement. This means that a buffer needs to be in the system to adjust for the last spring growth of cool season forages. One buffer may state producers use is to harvest hay in May and June and allow the cattle to graze the after math in August and September.

Another potential buffer is to plant warm season perennial grasses such as switchgrass, which do not grow early in the session. This reduces the acreage that the live stock can use early in the session, making it easier for them to keep up with the cool season grasses. The animals then use the warm-season grasses during the heat of the summer, and the cool season grasses recover for all grazing.

EFFECTS OF MODERN AGRICULTURE

Between 1950 and 1975 agricultural productivity in American history changed more rapidly. Total farm output increased more than half. This change was due to technological innovations, development of hybrid strains and other genetic improvements and a fourfold increase in the use of pesticides and fertilizers. Not only in America, all over world productivity and means of farming were changed. Thus the agriculture has become more intensive producing higher yields per acre. It also has become more expensive, relying on purchase of machinery and chemicals to replace heavy labour requirements of the past. To remain competitive farmers have been forced to become more efficient.

Although the intensification of agriculture has vastly increased productivity, it also has had a number of potentially detrimental environmental consequences, ranging from rapid erosion of fertile topsoils to contamination of drinking water suppliers by the chemicals used to enhance farmland productivity.

IMPACTS

Damage to Soil. Soil erosion from farmland threatens the productivity of agriculture fields and causes a number of problems elsewhere in the environment. An average of 10 times as much soil erodes from American agricultural fields as is replaced by natural soil formation processes. Because it takes up to 300 years for 1 inch of agricultural topsoil to form, soil that is lost is essentially irreplaceable. The consequences for long-term crop yields have not been adequately quantified. The amount of erosion varies considerably from one field to another, depending on soil type, slope of the field, drainage patterns, and crop management practices; and the effects of the erosion vary also. Areas with deep organic loams are better able to sustain erosion without loss of productivity than are areas where topsoils are shallower.

Erosion affects productivity because it removes the surface soils, containing most of the organic matter, plant nutrients, and fine soil particles, which help to retain water and nutrients in the root zone where they are available to plants. The subsoils that remain tend to be less fertile, less absorbent, and less able to retain pesticides, fertilizers, and other plant nutrients. Why then is erosion allowed to continue at excessive levels on many U.S. farms? Often the short-term costs of implementing erosion control measures far exceed the immediate economic benefit to the farmer, but such cost-benefit analyses fail to take into account the long-term losses of fertility and water-holding capacity of the soil. Up to a certain point, increased fertilization and irrigation will compensate for the lower soil fertility. Long-term loss of farmland productivity and damage to the environment from eroded sediments, therefore, often are overlooked in the need for short-term economic gains.

Over the past 50 years, the negative effects of soil erosion on farm productivity have been masked by improved technology and increasing use of fertilizers and pesticides. Ironically, many of these measures used to increase the short-term productivity of American farms are also causing excessive erosion, which threatens productivity over the long term. For example, diminished use of cover crops leaves soils unprotected from wind and rain during much of the year, and increased mechanization has led to use of larger fields without windbreaks or drainage contours.

The effects of erosion are also felt elsewhere in the environment. A recent study estimated the off-site cost of cropland erosion in the United States to be in the range of a billion dollars per year (Clark, Haverkamp, and Chapman 1985). Eroded soil clogs streams, rivers, lakes, and reservoirs, resulting in increased flooding, decreased reservoir capacity, and destruction of habitats for many species of fish and other aquatic life. The eroded soils contain nutrients and other chemicals that are beneficial on farm fields, but can impair water quality when carried away by erosion. As a result, drinking water supplies may contain nitrate or organic chemicals in concentrations that exceed public health standards, or surface waters may become clogged with excessive plant growth from the added nutrients.

Even when soil erosion is not excessive, intensive agriculture can impair soil quality by depleting the natural supplies of trace elements and organic matter. In natural ecosystems, soil fertility is maintained by the diverse contributions and recycling of nutrients by a wide range of plant and animal species. When this diversity is replaced by a single species grown year after year, some trace elements are depleted if not replaced by fertilization. The organic content of the soil also diminishes unless crop residues or other organic materials are supplied in sufficient quantities to replace that consumed over time.

Contamination of Water. In the Northeast water supplies are generally plentiful, but are increasingly becoming threatened by contamination. Farming is one potential source of such contamination. Surface runoff carries manure, fertilizers, and pesticides into streams, lakes, and reservoirs, in some cases causing unacceptable levels of bacteria, nutrients, or synthetic organic compounds. Similarly, water percolating downward through farm fields carries with it dissolved chemicals, which can include nitrate fertilizers and soluble pesticides. In sufficient quantities these can contaminate groundwater supplies.

Fertilizers. Nutrients are lost from agricultural fields through runoff, drainage, or attachment to eroded soil particles. The amounts lost depend on the soil type and organic matter content, the climate, slope of the land, and depth to groundwater, as well as on the amount and type of fertilizer and irrigation used.

The three major nutrients in fertilizers are nitrogen, phosphorus, and potassium. Of these, nitrogen is the most readily lost because of its high solubility in the nitrate form. Leaching of nitrate from agricultural fields can elevate concentrations in underlying groundwater to levels unacceptable for drinking water quality. In the Suffolk County area of Long Island, for example, almost 10 per cent of private wells tested for nitrate exceed the 10 mg/l drinking water standard.

Phosphorus does not leach as readily as nitrate because it is more tightly bound to soil particles. However, it is carried with eroded soils into surface water bodies, where it may cause excessive growth of aquatic plants. If this process proceeds far enough, lakes and reservoirs become choked with decaying mats of algae, which have offensive odour and can cause fish kills from the resulting lack of dissolved oxygen.

Potassium, the third major nutrient in fertilizers, does not cause water quality problems because it is not hazardous in drinking water and is not a limiting nutrient for growth of aquatic plants. It is tightly held by soil particles and so can be removed from fields by erosion, but generally not by leaching.

Pesticides. The trend toward intensive crop production in modern farming has led to increased potential for damage by pests and diseases. Predators that would be present in a mixed biological community are not supported by large fields of a single crop; so farmers, instead, rely on chemical measures for crop protection. Use of pesticides on U.S. farms has risen 10-fold over the past 40 years as agriculture has become more intensive. One drawback to this is that pesticides generally kill not only the pest of concern, but also a wide range of other organisms, including beneficial insects and other pest predators. Once the effect of the pesticide wears off, the pest species is likely to recover more rapidly than its predators because of differences in the available food supply. Previously unimportant species may also become significant crop pests when their natural predators are killed by pesticide applications.

Another drawback to the increasing pesticide use is the development of resistance in pest species. The individual pests that survive pesticide applications continue to breed, gradually producing a population with greater tolerance to the chemicals control pest populations.

Following World War II, DDT and related chlorinated hydrocarbons were introduced as potent new pesticides and were used throughout the world for protection of agricultural crops, as well as control of mosquitoes, lice and other human pests. In 1962 Rachel Carson's book *Silent Spring* brought public attention to the fact that these organic compounds are highly persistent in the environment and accumulate in animal tissues, causing water contamination, fish kills, and decline of some bird populations. DDT was banned for agricultural use in the United States in 1973, and since that time it and similar chlorinated hydrocarbons have been replaced by less persistent, but more acutely toxic, compounds.

In the past couple of decades, awareness has been growing of the many potential problems caused by the heavy use of chemicals in modern agriculture. This, combined with the rapid rise in the cost of fertilizers and pesticides, has led many farmers to seek ways of reducing their reliance on chemical intensive methods of farming. A small but growing percentage of farmers are farming with no synthetic chemicals, and many others are reducing their overall chemical use. Agriculture research has begun to focus on ways of maintaining environmental quality while producing acceptable crop yields. One example is integrated pest management, aimed at controlling pests through a combination of methods that minimize undesirable ecological effects. Continuing research and education need to be conducted on farming practices that produce profitable yields while maintaining environmental quality and the long-term productivity of the land.

WATERLOGGING

Another problem associated with excessive irrigation on poorly demand soils is waterlogging. This occurs (as is common for salinization) in poorly drained soils where water can't penetrate deeply. For example, there may be an impermeable clay layer below the soil. It also occurs on areas that are poorly drained topographically. What happens is that the irrigation water (and/or seepage from canals) eventually raises the water table in the ground – the upper level of the groundwater – from beneath. Growers don't generally realize that waterlogging is happening until it is too late – tests for water in soil are apparently very expensive.

The raised water table results in the soils becoming waterlogged. When soils are water logged, air spaces in the soil are filled with water, and plant roots essentially suffocate – lack oxygen. Waterlogging also damages soil structure.

Worldwide, about 10% of all irrigated land suffers from water logging. This is an area about the size of Idaho. As a result, productivity has fallen about 20% in this area of cropland.

Both waterlogging and salinization could be reduced if the efficiency of irrigation systems could be improved, and more appropriate crops (less water hungry) could be grown in arid and semi-arid regions. In addition, increasing the cost of water to more closely reflect its true value would encourage its conservation, rather than using incentives that essentially encourage wasting water, as some water rights laws in the US do (e.g., the "use it or lose it" approaches discussed above). We can consider excessive irrigation to be another example of humanities' attempt to increase "K" in ways that are unsustainable. For example, one could argue that the Central Valley of CA shouldn't be lush and green; it is one of the most agriculturally productive regions of the world only because of irrigation, which may not be sustainable there.

SALINITY

In many areas of India, crop production is limited because of salinity or alkalinity or both. It is estimated that about 7 million hectares in the country have either gone out of cultivation or this area produces low yield of crops. Three class of saline and alkali soils are recognized. They are—

1. SALINE SOILS. The soils containing toxic concentrations of soluble salts in the root-zone are called saline soils. Electrical conductivity in the saturation extract of such soils taken as a measure of salts is greater than 4.0 mmhos/cm. Exchangeable sodium percentage is less than 15 and the pH is less than 8.5. The soluble salts mainly consist of chlorides and sulphates of sodium, calcium and magnesium. Because of the white encrustation due to salts, the saline soil is also called white alkali.

2. NON-SALINE ALKALI OR SODIC SOILS. These soils do not contain any large amount of neutral salts and, as such, the electrical conductivity is less than 4 mmhos/cm. The detrimental effect of alkali soil on plants is largely due to toxicity of a high amount of exchangeable sodium and the pH. Alkali soils have an exchangeable sodium percentage of more than 15 and a pH greater than 8.5. Such soils have low infiltration rate and the physical condition is unfavourable. Because of high alkalinity, resulting from sodium carbonate, the surface soil is discoloured and black, and, hence the term black alkali is frequently used to designate the non-saline alkali soil.

3. SALINE-ALKALI SOILS. This group of soils is both saline and alkali. They have appreciable amounts of soluble salts, as indicated by the electrical conductivity values of more than 4 mmhos/cm. Also, the exchangeable sodium percentage is greater than 15. The pH, however, is likely to be less than 8.5.

The soil salinity or alkalinity or both have many adverse effects, which are summarized below :

1. Causing low yields of crops or crop failure in extreme cases.
2. The limiting of the choice of crops, because some crops are sensitive to salinity or alkalinity or to both.
3. Rendering the quality of fodder is poor, at times, the fodder grown on alkali soils may contain a high amount of molybdenum and a low amount of zinc, causing nutritional imbalance and diseases among live-stock.
4. Creating difficulties in the construction of buildings and roads and their maintenance.
5. Causing excessive run-off and floods owing to low infiltration, resulting in damage to crops in the adjoining areas.

Causes of salinity. In arid and semi-arid areas, salts formed during the weathering of soil minerals are not fully leached. During the periods of higher-than-average rain-fall, the soluble salts are leached from the more permeable high-lying areas to the low-lying areas, where, if the drainage is restricted, salts accumulate on the surface as water evaporates. The excessive irrigation of the uplands containing salts thus results in the accumulation of salts in the valleys. In areas having a salt layer at lower depths in the profile, seasonal irrigation may favour the upward movement of the salts. Salinization is also caused owing to the irrigation of soils with saline water. In all these cases, restricted drainage is usually the main reason. Rise in the water-table within 2 metres of the surface due to irrigation, the obstruction of natural drainage because of developmental activities, e.g. roads and canals, and the siltation of natural drainage may also cause soil salinity. In the coastal areas, the ingress of sea-water induces salinity in the soil. When sodium ions predominate in the soil solution, and carbonates are present, alkali soils are formed.

Reclamation. The salinity or alkalinity depend upon their mode of formation and physiographic position. Since the degree of salinity or alkalinity may vary as such methods of reclamation also differ.

(*i*) If the problem is only of salinity, the salts need to be leached below the root-zone and not allowed to come up. In practice, however, this might be difficult to accomplish, especially in deep and fine-textured soils containing more salts in the lower layers. Under these conditions, a provision of some kind of subsurface drains becomes important. If the soil contains a sandy layer at a lower depth, the leaching of the salts below this layer will check the rise of salts.

(*ii*) The reclamation of alkali soils needs the addition of a soil amendment, containing soluble calcium salts. The commonly used amendment is gypsum. In the course of reclamation, sodium on the exchange complex is replaced by calcium. The sodium salts, thus formed, are leached down.

(*iii*) The number and frequency of leaching, the quantity of gypsum to be added and the techniques involved vary from region to region, depending upon the clay mineralogy of the soils, the intensity of the problem, the subsequent use of the soils, the availability and quality of irrigation water and the economics of these operations. Hence, the state authorities engaged in this work should be consulted to draw a schedule of operations for reclamation.

ENERGY RESOURCES

Energy is needed by all living organisms and vegetations for biochemical reactions of their cells. It is a power which is needed in one form or other for work done. Long before most of the power available to human society was limited to solar energy trapped by green plants which produced organic matter. Biological oxidation of the organic matter provided fuel to muscle power. The fire was the first form of known energy used for cooking, heating purposes. The formation of fossil fuels (coal, oil & natural gas) is also due to photosynthesis carried on by plants which occurred millions of years ago. Now the things are changed drastically. For the developmental activities, energy sources have their own importance. Energy consumption of a nation is usually considered as an index of its development.

GROWING ENERGY NEEDS

Energy is the prime input of a country. It is converted into heat & electricity. For every activity to be performed required energy in the form of heat, light, electricity and even food (in the form of energy) for our body. Food energy is measured in calories. India has fast growing developing economy, with the GDP growth rate exceeding 6% in recent years. Xth plan projected 8% growth rate economy. This growth has been accompanied by a steady increase in energy consumption. Primary commercial energy demand grew at annual rate of 6% upto 2001. It will go more rapidly than in the past as country's reforms process accelerates.

As the economy grows, energy intensity rises following corresponding increase in energy consumption. However, beyond a certain level of per capita income, energy intensity begins to decline. These linkages between energy and economic factors, manifested in energy elasticity and energy intensity are broadly related to—

(*i*) Demographic changes, including a relatively faster growth in urban areas, higher per capita GDP and per Capita gross saving.

(*ii*) Efficient end-use devices.

(*iii*) Technological improvements in conversion equipments.

(*iv*) Inter fuel substitution with more efficient alternatives.

As the world population is growing (@ 6 billion) we need energy much faster rate to meet the industrial, food, residental, transport, agriculture etc. requirements. So it was realised that, energy sources will not meet the requirement last long. In early 70s interest was shown world over to look for other alternatives sources of energy. Therefore, thrust was given for renewable sources of energy like solar, wind, hydro, biomass, tidal, hydrothermal, hydrogen energy etc.

ENERGY CONSUMPTION

India's commercial energy consumption is about one-fourth that of the world average. Therefore, India's consumption of coal, oil, and natural gas is only a small part of total world consumption. The consumption of commercial fuels is steadily rising throughout the country, with coal continuing to be the most prominent energy source. The fuel composition of commercial energy consumption varies significantly from sector to sector. Coal continues to meet over 70 per cent of the industry's energy needs, but petroleum products, especially natural gas and power, are steadily replacing coal. In the field of agriculture, draught animal power continues to be predominantly used in various farming activities, but this use is also not well documented. In transportation, the direct use of coal is virtually nil, following the substitution of steam traction by electric traction. Consequently, the use of electric power has gone up in the Railways sector.

The consumption of petroleum products has significantly increased following a seven-fold increase in registered motor vehicles between 1981 and 1997 (MoST, 1999). While a large part of the household sector continues to depend on traditional fuels for cooking, this data has not been documented formally. There appears to be some substitution of petroleum products by power in the household sector, probably resulting from extensive electrification, whereby kerosene gets replaced by electricity for lighting purposes. But in overall terms, petroleum consumption has grown manyfold and the oil intensity of the Indian economy has been sharply increasing.

ENERGY SOURCES (RENEWABLE & NON-RENEWABLE SOURCES OF ENERGY)

There are following two types of energy sources to meet the requirement.

1. Renewable or non-conventional or inexhaustible energy sources.—These sources are continuously replenished by natural processes. For example, solar energy, wind energy, bio energy, hydropower etc. These energy systems convert such energy in to a form which we can use. Renewable energy sources are essentially flows of energy. At present total potential of 126000 MW assessed by different non-conventional energy sources.

At present, most important non-conventional energy source is wind energy, for which a capacity of 1800MW has been set up in the country. There is also a large potential for tapping of ocean energy, geo thermal energy and tidal power, but the techno-economic viability for power generation from these sources has still to be established.

2. Non-renewable or conventional or exhaustible energy sources.—Examples of this are coal, petroleum, natural gas and nuclear power. These are traditional sources available to us. All these sources are limited and takes millions of years for formation. As a result of unlimited use, they will exhaust one day. Therefore we should conserve these for longer period.

In addition to commercial fuels, coal, oil—natural gas, and power. India consumes large quantities of traditional fuels. The traditional energy sector is not well documented. Firewood, dung cake and agricultural wastes constitute the primary source of energy for cooking in over 90 per cent of rural households. These fuels are also used for water heating and space heating. Traditional fuels also comprise an important source of energy in over 30 per cent of urban households and their use is most prevalent among the lower income categories. Draught animal power continues to be used in farming and transportation in villages and in small towns. Traditional fuels are often classified as non-commercial sources because they are mostly gathered or collected from the neighbourhood. While this is true in rural areas, the urban poor who use these fuels pay heavily for fuelwood (average of over Rs 5/kg). The price varies significantly across regions, being particularly high in large cities. The transition from non-commercial energy to commercial has to be planned as part of the integrated sustainable energy policy framework. Energy sources have a key role in the preparation and implementation of an integrated sustainable energy policy.

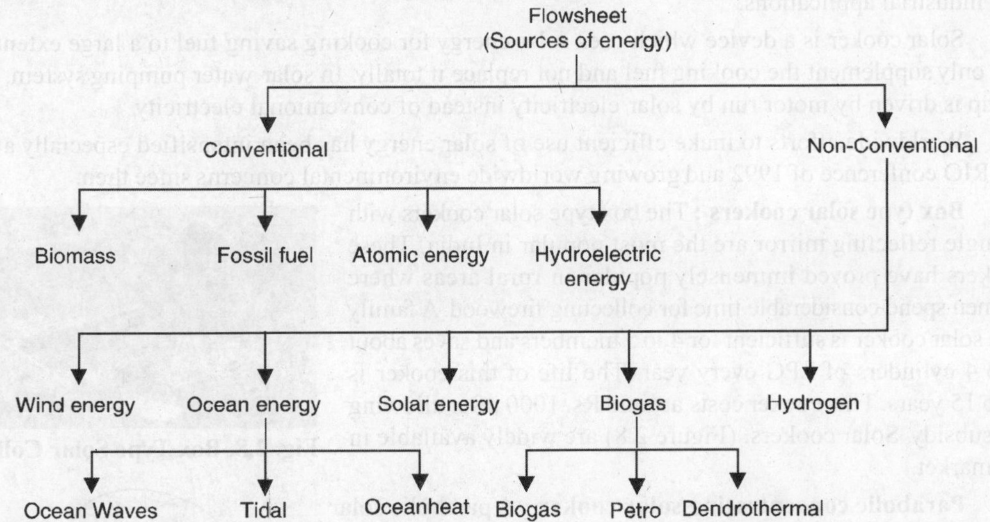

Various forms of renewable energy sources

Here we will discuss the application potential of commercially viable renewable energy sources. For Renewable energy potential in India see table.

Table 2.4. Renewable Energy Potential in India

Energy source	*Estimated potential*
Hydropower	84,000 MW
Small hydro	15,000 MW (latest assessment by MNES)
Ocean thermal power	50,000 MW
Wave energy	20,000 MW
Tidal power	10,000 MW
Wind energy	45,000 MW (as per MNES)
Solar energy	5×10^{15} kWh/year
Bioenergy	16,000 MW Biomass
Cogeneration	3,500 MW Cogeneration in

Source : MNES 2002

(*i*) Solar Energy : Solar energy is the most readily available and free source of energy since prehistoric times in a most primitive manner. It is estimated that solar energy equivalent to over 15000 times the world's annual commerical energy consumption reaches the earth every year. It is non-polluting helps in lessening green house effect. Solar energy can be utilised through two different routes, as solar thermal route and solar electric route (Solar photovoltaic). Solar thermal energy is used for cooking, heating, drying, timber seasoning, distillation, electricity generation, cooling, refrigeration, cold storage etc. Solar photovoltaic uses sun's heat to produce electricity for lighting home & building, running motors, pumps, electric appliances, lighting streets, village electrification, powering of remote telecommunication and railway signals etc. Solar cells convert solar energy into DC electricity (through silicon solar cells) but in solar thermal route, solar energy can be converted in to thermal energy with the help of solar collectors and receivers known as solar thermal devices. These devices are used in solar water heaters, air heaters, solar cookers and solar dryers for domestic and industrial applications.

Solar cooker is a device which uses solar energy for cooking saving fuel to a large extent. It can only supplement the cooking fuel and not replace it totally. In solar water pumping system, the pump is driven by motor run by solar electricity instead of conventional electricity.

Worldwide efforts to make efficient use of solar energy has been intensified especially after the RIO conference of 1992 and growing worldwide environmental concerns since then.

Box type solar cookers : The box type solar cookers with a single reflecting mirror are the most popular in India. These cookers have proved immensely popular in rural areas where women spend considerable time for collecting firewood. A family size solar cooker is sufficient for 4 to 5 members and saves about 3 to 4 cylinders of LPG every year. The life of this cooker is upto 15 years. This cooker costs around Rs. 1000 after allowing for subsidy. Solar cookers. (Figure 2.8) are widely available in the market.

Fig. 2.8. Box Type Solar Collector

Parabolic concentrating solar cooker : A parabolic solar concentrator comprises of sturdy Fibre Reinforced Plastic (FRP) shell lined with Stainless Steel (SS) reflector foil or aluminised polyester film. It can accommodate a cooking vessel at its focal point. This cooker is designed to direct the solar heart to a secondary reflector inside the kitchen, which focuses the heat to the bottom of a cooking pot. It is also possible to actually fry, bake and roast food. This system generates 500 kg of steam, which is enough to cook two meals for 500 people (see Figure 2.9). This cooker costs upward of Rs. 50,000.

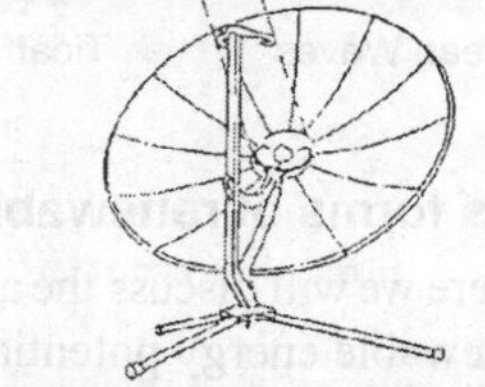

SUNBASKET

Fig. 2.9. Parabolic Collector

Positioning of solar panels or collectors can greatly influence the system output, efficiency and payback. Tilting mechanisms provided to the collectors need to be adjusted according to seasons (summer and winter) to maximise the collector efficiency.

The period four to five hours in late morning and early afternoon (between 9 am to 9pm) is commonly called the "Solar Window". During this time, 80% of the total collectable energy for the day falls on a solar collector. Therefore, the collector should be free from shade during this solar window throughout the year - Shading, may arise from buildings or trees to the south of the location.

Solar Electricity Generation

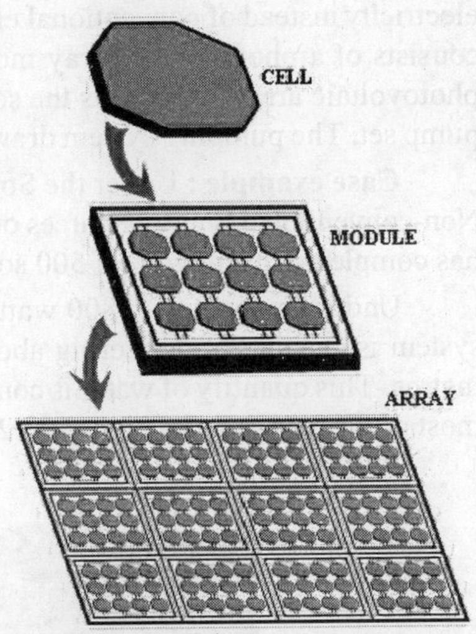

Fig. 2.10 Solar Photovoltaic Array

Solar Photovoltaic (PV) : Photovoltaic is the technical term for *solar electric*. Photo means "light" and voltaic means "electric". PV cells are usually made of silicon, an element that naturally releases electrons when exposed to light. Amount of electrons released from silicon cells depend upon intensity of light incident on it. The silicon cell is covered with a grid of metal that directs the electrons to flow in a path to crate an electric current. This current is guided into a wire that is connected to a battery or DC appliance. Typically, one cell produces about 1.5 watts of power. Individual cells are connected together to form a solar *panel* or *module*, capable of producing 3 to 110 Watts power. Panels can be connected together in series and parallel to make a solar *array* (see Figure 12.4). which an produce any amount of Wattage as space will allow. Modules are usually designed to supply electricity at 12 Volts. PV modules are rated by their peak Watt output at solar noon on a clear day.

Some applications for PV systems are lighting for commercial buildings, outdoor (street) lighting (see Figure 2.11), rural and village lighting etc. Solar electric power system can offer independence from the utility grid and offer protection during extened power failures. Solar PV system are found to be economical especially in the hilly and far flung areas where conventional grid power supply will be expensive to reach.

Fig. 2.11. Photovoltaic Domestic and Streelights

PV tracking systems is an alternative to the fixed, stationary PV panels. PV tracking systems are mounted and provided with tracking mechanisms to follow the sun as it moves through the sky. These tracking systems run entirely on their own power and can increase output by 40%.

Back-up system are necessary since PV systems only generate electricity when the sun is shining. The two most common methods of backing up solar electric systems are connecting the system to the utility grid or storing excess electricity in batteries for use at night or on cloudy days.

Performance. The performance of a solar cell is measured in terms of its efficiency at converting sunlight into electricity. Only sunlight of certain energy will work efficiently to create electricity, and much of it is reflected or absorbed by the material that make up the cell. Because of this, a typical commercial solar cell has an efficiency of 15%–only about one-sixth of the sunlight striking the cell generates electricity. Low efficiencies mean that larger arrays are needed, and higher investment costs. It should be noted that the first solar cell, built in the 1950s, had efficiencies of less than 4%.

Solar Water Pumps. In solar water pumping system, the pump is driven by motor run by solar electricity instead of conventional electricity drawn from utility grid. A SPV water pumping system consists of a photovoltaic array mounted on a stand and a motor-pump set compatible with the photovoltaic array. It converts the solar energy into electricity, which is used for running the motor pump set. The pumping system draws water from the open well, bore well, stream, pond, canal etc.

Case example : Under the Solar Photovolatic Water Pumping Programme of the Ministry of Non-conventional Energy Sources during 2000-01 the Punjab Energy Development Agency (PEDA) has completed installation of 500 solar pumps in Punjab for agricultural uses.

Under this project, 1800 watt PV array was coupled with a 2 HP DC motor pump set. The system is capable of delivering about 140,000 litres water every day from a depth of about 6–7 metres. This quantity of water is considered adequate for irrigating about 5–8 acres land holing for most of the crops. Refer Figure 2.12.

Fig. 2.12. Photovoltaic Water Pumping

***(ii)* Wind energy.** Wind energy is basically harnessing of wind power to produce electricity. The kinetic energy of the wind is converted to electrical energy. Since air flow from warmer to cooler regions, this causes, what we call winds and it is these airflows that are harnessed in windmills and wind turbines to produce power. Wind power is not a new development as this power in the form of traditional windmills – for grinding corn, pumping water, sailing ships have been used for centuries. The basic wind energy conversion device is the wind turbine. These turbines are generally of two types – vertical axis wind turbines and horizontal axis turbines depending upon the rotation of axis.

Wind electric generators (WEG) convert kinetic energy available in wind to electrical energy by using rotor, gearbox and generator. These are from 225 KW to 1000 KW are being installed in our country. In order for a wind energy system to be feasible, there must be an adequate wind supply. It requires an average annual wind speed of at least 75 km/h. Obstructions such as trees or hills can interfere with the wind supply to the rotors.

India has been rated as one of the most promising countries for wind power development with an estimated potential of 20,000 MW. India ranks fifth in the world in wind power generation. There are 208 wind potential stations in India. The financial, technical assistance are provided by Central Govt. to promote, support and accelerate the development of wind energy in India.

Wind Energy Technology. The basic wind energy conversion device is the wind turbine. Although various designs and configurations exist, these turbines are generally grouped into two types:

1. Vertical-axis wind turbines, in which the axis of rotation is vertical with respect to the ground (and roughly perpendicular to the wind stream),

2. Horizontal-axis turbines, in which the axis of rotation is horizontal with respect to the ground (and roughly parallel to the wind stream).

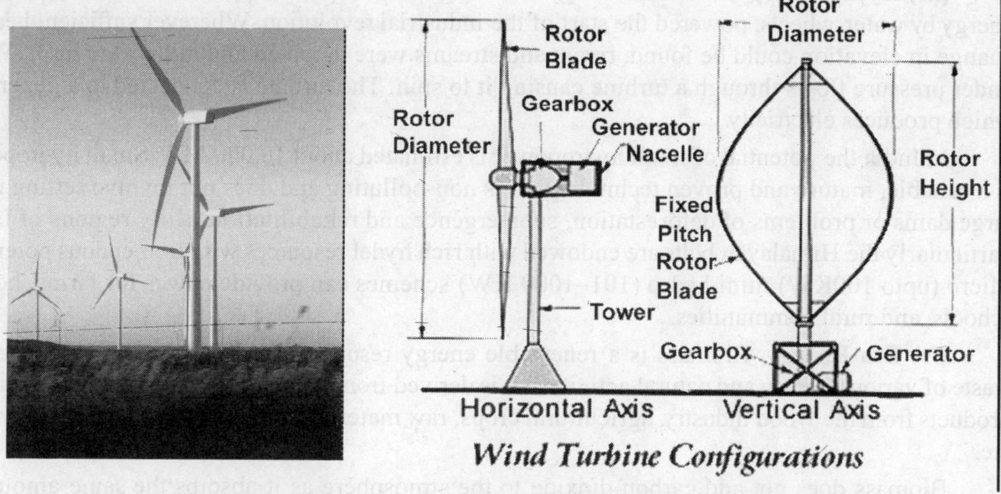

Fig. 2.13. Wind Turbine Configuration

The Figure 2.13 illustrates the two types of turbines and typical subsystems for an electricity generation application. The subsystems include a blade or rotor, which converts the energy in the wind to rotational shaft energy; a drive train, usually including a gearbox and a generator, a tower that supports the rotor and drive train, and other equipment, including controls, electrical cables, ground support equipment, and interconnection equipment.

Wind Potential : In order for a wind energy system to be feasible there must be an adequate wind supply. A wind energy system usually requires an average annual with speed of at least 15 km/h. The following table represents a guideline of different wind speeds and their potential in producing electricity.

Average Wind Speed km/h (mph)	Suitability
Up to 15 (9.5)	No good
18 (11.25)	Poor
22 (13.75)	Moderate
25 (15.5)	Good
29 (18)	Excellent

A wind generator will produce lesser power in summer tan in winter at the same wind speed as air has lower density in summer than in winter.

Similarly, a wind generator will produce lesser power in higher altitudes - as air pressure as well as density is lower than at lower altitudes.

The wind speed is the most important factor influencing the amount of energy a wind turbine can produce. Increasing wind velocity increases the amount of air passing the rotor, which increase the output of the wind system.

In order for a wind system to be effective, a relatively consistent wind flow is required. Obstructions such as trees or hills can interfere with the wind supply to the rotors. To avoid this, rotors are placed on top of towers to take advantage of the strong winds available high above the ground. The towers are generally placed 100 metres away from the nearest obstacle. The middle of the rotor is placed 10 metres above any obstacle that is within 100 metres.

***(iii)* Hydro Energy.** Potential energy of falling water, captured and converted to mechanical energy by water-wheels, powered the start of the industrial revolution. Wherever sufficient head or change in elevation could be found, rivers and streams were dammed and mills were built. Water under pressure flows through a turbine causing it to spin. The turbine is connected to a generator, which produces electricity.

In India the potential of small hydropower is estimated about 10,000 MW. Small hydropower is a reliable, mature and proven technology. It is non-polluting and does not involve setting up of large dams or problems of deforestation, submergence and rehabilitation. Hilly regions of India, particularly the Himalayan belts are endowed with rich hydel resources with tremendous potential. Micro (upto 100KW) mini hydro (101–1000 KW) schemes can provide power for farms, hotels, schools, and rural communities.

***(iv)* Bio Energy.** Biomass is a renewable energy resource divided from the carbonaceous waste of various human and natural activities. It is derived from numerous sources, including the by-products from the wood industry, agricultural crops, raw material from the forest, household wastes etc.

Biomass does not add carbon dioxide to the atmosphere as it absorbs the same amount of carbon in growing as it releases when consumed as a fuel. Its advantage is that it can be used to generate electricity with the same equipment that is now being used for burning fossil fuels. Biomass is an important source of energy and the most important fuel worldwide after coal, oil and natural gas. Bio-energy, in the form of biogas, which is derived from biomass, is expected to become one of the key energy resources for global sustainable development. Biomass offers higher energy efficiency through form of Biogas than by direct burning (see chart below).

Application. Bio energy is being used for Cooking mechanical applications, pumping power generation.

Some of the devices : Biogas plant/gasifier/ burner gasifier engine pump sets stirling engine pump sets produce gas/biogas based engine generator sets.

Biogas Plants. Biogas is a clean and efficient fuel, generated from cow-dung human waste or any kind of biological materials derived through anaerobic fermentation process. The biogas consists of 60% methane with rest mainly carbon-di-oxide. Biogas is a safe fuel for cooking and lighting. By-product is usable as high-grade manure.

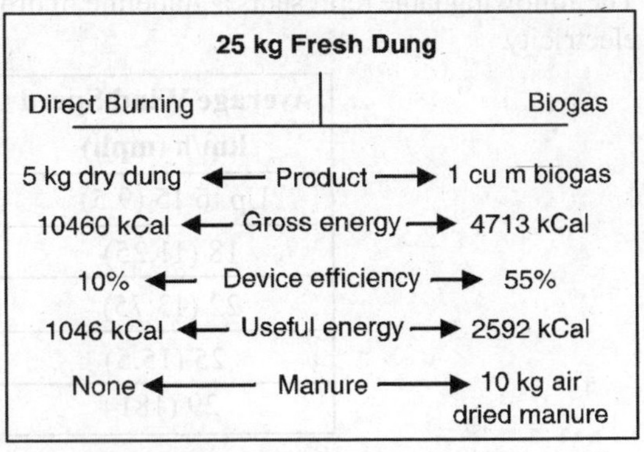

The Gobar (Hindi for "cow dung") Gas Research Station–established in 1960 as the latest of along series of Indian experimental projects dating back to the 1930's–has concentrated its efforts, as the name suggests, on generating methane gas for cow manure. At the station, Ram Bux Singh and his coworkers have designed and put into operation bio-gas plants ranging in output from 100 to 9,000 cubic feet of methane a day. They've installed heating coils, mechanical agitators and filters in some of the generators and experimented with different mixes for manure and vegetables wastes. Results of the projects have been meticulously documented and recorded. It's been a wild, exciting ride ... but our blindly wasteful squandering of the planet's fossil fuels will soon be a thing of the past. In the United States alone (the worst example, perhaps, but not really unusual among "modern" nations), every man, woman and child consumes an average of three gallons of oil each day. That's well over hundred billion gallons a year.

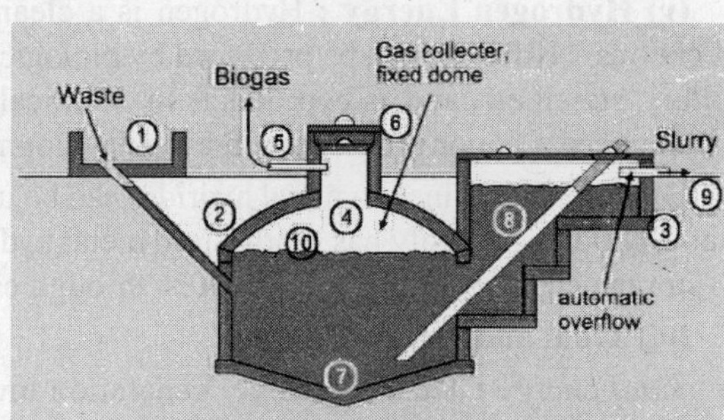

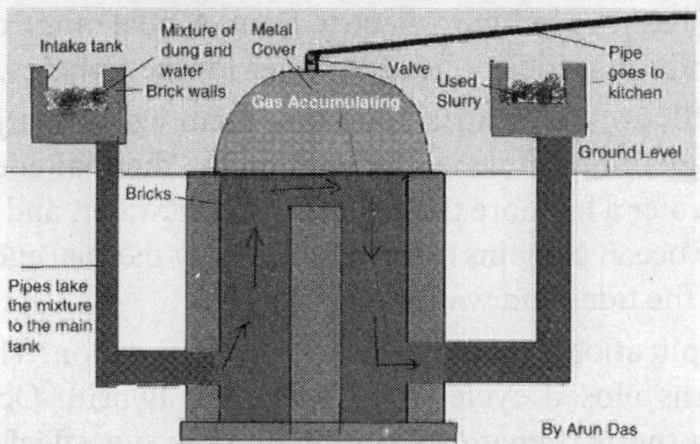

A typical biogas plant has the following components: A digester in which the slurry (dung mixed with water) is fermented in an inlet tank - for mixing the feed and letting it into the digester, gas holder/dome in which the generated gas is collected,outlet tank to remove the spent slurry, distribution pipe lives to transport the gas in to the kitchen and a manure pit, where the spent slurry is stored. Biomass fuels account for about 1/3 of the total fuel used in the country. It is used in 90% of rural and 15% of urban households. These are the cheap sources of energy in rural areas.

The process of densifying loose agro waste into a solidified biomass of high density, which can be conveniently used as a fuel is called Biomass Biquetting or Bio-coal. It is pollution free and ecofriendly. It can replace almost all conventional fuels like coal, firewood and lignite in heating, steam generation etc. It can be used directly in traditional Chulhas and furnaces or in the gasifier. It has high calorific value with low ash content without polluting gases.

Biomass gasifiers convert the solid biomass in to a combustible gas mixture normally called producer gas/mixture of CO, H_2, N_2 & CH_4 gas), has calorifical value 1000 – 1200 kCal/Nm^3.) Unlike other renewable energy sources, biomass can be converted directly into liquid fuels – Biofuels – for our transportation needs (cars, trucks, buses, airplanes and trains). The two most common types of biofuels are **Ethanol and Biodiesel.** Ethanol is an alcohol made by fermenting any biomass rich in carbohydrates while biodiesel is produced by plants such as rapeseeds, sunflowers and soyabeans, can be extracted and refined into fuel.

Biopower is the use of biomass to generate electricity. There are six major types of biopower systems - direct fired, cofiring, gasification, anaerobic digestion, pyrolysis and small modular. Most of the biopower plants in the world use direct fired systems. Through anaerobic digestion of biomass methane gas producéd which can be used as energy source. In addition to gas, liquid fuels can be produced from biomass through a process called pyrolysis. Several biopower technologies can be used in small modular systems, which generates electricity at a capacity of 5 megawatts or less.

Biomass cogeneration improves viability and profitability of sugar industries. Indian sugar mills are rapidly turning into biogas producer, the liftover of can after it is crushed and its juice extracted to generate electricity.

(v) Hydrogen Energy : Hydrogen is a clean fuel and energy storage medium for various applications. Hydrogen can be produced by biological conversions of various organic effluents like distillary, starch etc. and as byproducts in chemical processes. In January 2003 President Bush of US announced a major programme for the development of hydrogen energy as a freedom fuel.

Hydrogen contained in metal hydrides can be used in vehicles. The centre for Hydrogen energy Banaras Hindu University has synthesised metal hydrides, graphite nanofibres, which exhibit a very high storage capacity of the order of 40% through catalytic decomposition.

(vi) Tidal and Ocean Energy

Tidal energy : Tidal electricity generation involves the construction of a barrage across an estuary to block the incoming and outgoing tide. The head of the water is then used to drive turbines to generate electricity from elevated water in the basin as in hydro electric Dam. A tidal range of at least 7 m is required for economical operation and for sufficient head of water for the turbines.

Ocean Energy : Ocean covers more than 70% of earth surface making them world's largest solar collector. Ocean energy draws on the energy of ocean waves tides or on the thermal energy stored in the ocean. The seen warms the surface water a lot more than the deep ocean water, and this temperature difference stores thermal energy. The ocean contains two types of energy thermal energy from the sun's heat and mechanical energy from the tides and waves.

Ocean thermal energy is used for many applications, including electricity generation. There are three types of electricity conversion systems closed cycle, open cycle and hybrid. Ocean mechanical energy is quite different from ocean thermal energy. Even though the sun affects all ocean activity, tides are driven primarily by the gravitational pull of the moon and waves are driven primarily by the winds.

(vii) Geothermal Energy

The Earth's population lives between two great sources of energy the hot rocks beneath the surface of the earth and the sun in the sky. Geothermal energy is based on the core of the earth, which is very hot and it is possible to make use of this geothermal energy (in Greek, geothermal means heat from the earth). These are areas containing volcanoes, hot springs, geysers, and methane, under the water in the oceans and seas. In some countries, such as the USA, water is pumped from underground hot water deposits and used to heat houses.

Geothermal energy, which is derived from high temperature geothermal fluids, can be utilized tor power generation and thermal applications like greenhouse cultivation, space heating and cooking. Geothermal energy has been commercially exploited in as many as 20 countries with about 9000 MW installed capacity. The utilization of geothermal energy for the production of electricity dates back to the early part of the twentieth century. For 50 years the generation of electricity from geothermal energy was confined to Italy and interested in this technology was slow to spread elsewhere. In 1943, the use of geothermal hot water was pioneered in Iceland. Geothermal energy is now a commercial source of energy for electricity production and is widely used especially in USA which has over 7000 MW of geo-thermal energy.

Geothermal manifestations are widespread in India in the form of a large number of hot spring sites. This energy is being used for heating/power generation. Devices for harnessing geothermal energy includes: heat exchanger and steam turbines. In India, northwestern Himalayas and the western coast are considered to be potential geothermal areas. The geological survey of India has already identified more than 350 hot spring sites which can be explored to tap geothermal energy.

(viii) Chemical Sources of Energy

Fuel cells electrochemically produce direct current (DC) electricity through a reaction between hydrogen and oxygen. Such cells are electrochemical devices that convert the chemical energy of a fuel directly and very efficiently into electricity (DC) and heat, thus doing away with combustion.

The most suitable fuel for such cells is hydrogen or a mixture of compounds containing hydrogen. A fuel cell consists of an electrolyte sandwiched between two electrodes. Oxygen passes over one electrode and hydrogen over the other, and they react electrochemically to generate electricity, water and heat. Fuel cell technologies which are being commercialized including phosphoric acid, polymer electrolyte membrane fuel cells, solid oxide and molten carbonate fuel cells.

Though fuel cells have been used in space flights and combined supplies of heat and power, electric vehicles powered by this energy are considered the most promising option available to dramatically reduce urban air pollution. Compared to vehicles powered by the internal combustion engine, fuel-cell powered vehicles have very high energy near-zero pollution. Fuel-cell-powered EVs (electric vehicles) score over bettery-operated EVs in terms of increased efficiency and easier and faster refuelling. The Canadian company Ballard Power Systems and Germany's Dailmer-Benz are world leaders in the application of fuel cell technology to meet transportation needs. In India,diesel-run buses are the main means of transport, which emit significant quantities of SPM and SO_2. Fuel-cell powered buses can be introduced with relative ease and make a positive impact on urban air quality. Fuel cells are already in operation in several locations including Vancouver in Canada and Illinois and California in the USA, among others. Though rapid progress has been made, high initial cost is still the biggest hurdle in the widespread commercialization of fuel cells.

Fuel cells can also supply a combination of heat and power to commercial buildings, hospitals, airports and military installations at remote locations. Fuel cells have efficiency levels up to 55 per cent as compared to 35 per cent of conventional power plants. The emissions are very low as compared to conventional options. Moreover, fuel cell systems are modular (i.e. additional capacity can be added whenever required with relative ease) and can be set up wherever power is required.

India has world's largest programmes for renewable energy. Several renewable energy technologies have been developed and deployed in villages and cities. A Ministry of Non-conventional Energy sources (created in 1992) looking all these matters. Renewable Energy Development Agency Limited (IREDA) is also created to assist and provide financial assistance. IREDA covers a wide spectrum of financial activities including those that are connected to energy conservation and energy efficiency.

The estimated potential of various Renewable Energy technologies in India by IREDA are given below.

Energy source estimated potential

Solar Energy	20 MW/sq. km
Wind Energy	20,000 MW
Small Hydro	10,000 MW
Ocean Thermal Power	50,000 MW
Sea Wave Power	20,000 MW
Tidal Power	10,000 MW
Bio energy	17,000 MW
Draught Animal Power	30,000 MW
Energy from MSW	1,000 MW
Biogas Plants	12 Million Plants
Improved Wood Burning Stoves	120 Million Stoves
Bagasse-based cogeneration	3500 MW

Cumulative achievements in renewable energy sector (As on 31.03.2000)

Sources/Technologies	*Unit Upto 31.03.2000*
Wind Power	MW 1167
Small Hydro	MW 217
Biomass Power & Co-generation	MW 222
Solar PV Power	MW / Sq. km 42
Urban & MSW	MW 15.21
Solar Heater	m^2.Area 480000
Solar Cookers	No. 481112
Biogas Plants	Nos. in Million 2.95
Biomas Gasifier	MW 34
Improved Chulhas	Nos. in Million 31.9

Conventional or Non-Renewable Energy Sources : In early days wood was used as fuel in rural areas. Coal, oil and natural gas are also non renewable energy sources. But they are high priced. India is the third largest producer of coal. In the oil and natural gas sector, despite a continuous decline in balance, India continues to be one of the least explored regions. Sea below in table—

Table 2.5 : Energy Supply Indicators : 1999/2000

Reserves/ production	Coal (million tonnes)	Oil (million tonnes)	Gas (billion cubic metres)	Power
Reserves	211 593.6*	715	675	—
Installed capacity	—	—	—	97 837 MW
Production/generation	292.27	32.72**	27.43***	451 billion units

* coal reserves up to depth of 1200 M; **crude oil production only, petroleum products production 64.55 million tonnes; ***gross production includes flaring, etc.

Source : Planning Commission, Government of India (6)

Non-commercial energy sources in the country include firewood, animal dung and agricultural waste. Commercial as well as non-conventional energy sources, are converted to electricity which is the most convenient and favoured secondary energy source. The main sources of electricity production in the country are coal, hydro power, natural gas, nuclear energy and renewable energy sources.

(1) COAL

Coal reserves in the country comprise about 210 billion tonnes (1999). About 85 percent of the coal reserves are non-coking coal, the rest being coking coal. Coal reserves are unevenly distributed in the country, with the bulk reserves located in the eastern states of Bihar, West Bengal and Orissa. Central India, including Madhya Pradesh and Andhra Pradesh, also possess sizeable coal reserves of the order of 22 per cent of the total. Table 2.6 gives the distribution of coal reserves:

Table 2.6: Distribution of Coal Reserves (as on 1.1.1999)

S.No.	*State*	*Million tonnes*
1	Andhra Pradesh	13337
2	Arunachal Pradesh	90
3	Assam	320
4	Bihar	68756
5	Madhya Pradesh	42772
6	Maharashtra	6969

7	Meghalaya	459
8	Nagaland	20
9	Orissa	49061
10	Uttar Pradesh	1062
11	West Bengal	25903
	Total	**208749**

The total proven coal reserves are about 79 billion tonnes and the corresponding recoverable reserves of 26 billion tonnes at the current rate of production of about 300 million tonnes per annum, the reserves to production ratio is about 87. The estimated consumption of the total coal production is about 70 per cent. Electricity sector continues to depend mainly on coal. Coal requirements in the power sector are expected to substantially increase with demand far outstripping supply in the coming years.

(2) LIGNITE

Lignite reserves (Table 2.7) in the country are estimated to be about 29.72 billion tonnes. Eighty per cent is located in Tamil Nadu. Lignite is also mainly utilized to produce electricity.

Table 2.7: Lignite Reserves

(Million tonnes)

S.No.	*State/Location*	*Estimated reserves*
1.	Tamil Nadu and Pondicherry	
	Neyveli	3300
	Bahur	575
	Jayamkondacholapuram	1167
	Mannargudi	19500
	East of Veeranam	1420
	Kadikadu	133
	Michal Patti	23
2.	Gujarat	1541
3.	Rajasthan	1823
4.	Jammu and Kashmir	127
5.	Kerala	108

(3) PETROLEUM AND NATURAL GAS

India's oil and gas reserves are 732 million tonnes and 768 billion cubic metres, respectively (2001). The bulk of oil and natural gas resources are located in offshore areas where exploration, drilling and production are expensive, leading to much higher costs of domestic crude oil and natural gas as compared to imported oil and gas. Exploration for hydrocarbons has so far been limited to on-land and shallow water offshore areas.

Deep water exploration has now been included under the New Exploration Licensing Policy (NELP) and these areas were offered through competitive bidding routes. The success of the New Exploration Licensing Policy has been demonstrated by recent findings of natural gas in deep waters. The discoveries in deep water have opened up vast areas for exploration and exploitation but require specialized technologies and large scale investments.

(4) NUCLEAR POWER

The nuclear power programme in India is based on natural uranium and indigenous thorium reserves. Nuclear power generation started in India in 1969 with the commissioning of Tarapur Atomic Power Plant with a capacity of 2 × 200 and 20 MW based on Boiling Water Reactor (BWR) Technology with assistance from USA. This power station is currently under operation and supplies power to the western grid. Subsequent nuclear power projects in India have been based on Pressurized Heavy Water Reactor Technology (PHWR). The first such nuclear project was set up in Rajasthan with external assistance from Canada. Subsequently, this plant has been fully indigenized and further developed. At present Pressurized Heavy Water Reactors are in operation in Kalpakkam near Chennai, at Narora in Uttar Pradesh, and Kakrapara in Gujarat, all of them being totally indigenous.

Available reserves of natural uranium can support a nuclear programme of 10,000 MW based on Pressurized Heavy Water Technology. Through indigenous efforts and requisite support from industry, technology for production of fuel heavy water and all other necessary components have been developed for establishing 220-500 MW power units. The next phase of nuclear power developent in the country would be based on fast breeder reactors, which are fuelled by plutonium recovered and processed from the spent fuel of the first stage of the Pressurized Heavy Water Reactor Technology.

The Nuclear Power Programme in India is under the control of Department of Atomic Energy and its Nuclear Power Corporation. The current nuclear capacity is 2225 MW based on BWR/ PHWR technology and constitutes 2.65 per cent of the total installed capacity. An installed capacity of 11600 MW has been planned for the end of the Eleventh Plan which would rise to 20,000 MW by the year 2020. This would mean an addition of 900 MW capacity per year and this in turn, would require large scale investmens. A proposal to involve the private much needed investments in this highly capital intensive sector is under consideration. At present, however, the nuclear power sector is controlled totally by the government.

(5) HYDRO POWER

The total potential for hydro power in India, based on river systems, was assessed at 301117 MW at 60 per cent load factor, and economic potential of 84044 MW. Details of this potential are given in Table 2.8.

Table 2.8. Region-wise Distribution of Hydro-electric Potential

Region	*Energy (TWH)*	*Potential at 60% Load Factor (MW)*	*% Developed*	*% under development*
Northern	225.0	30155	14.30	8.03
Western	31.4	5679	31.94	26.97
Southern	61.8	10763	49.21	10.27
Eastern	42.6	5590	16.41	12.73
North-Eastern	239.3	31857	1.02	0.96
All India	**600.1**	**84044**	**15.07**	**7.23**

Source : Central Electricity Authority (1980)

Only 15% of this potential has been utilized so far while 7.23% are under development. This assessment is based on a CEA study of hydropower resources of the country conducted in 1980.

Case Study

Solar water-heating systems in Magarpatta Housing Complex, Pune

A view of housing complex in Magarpatta city with each house having a solar water heater

Solar water-heating systems are fast catching up in urban areas such as Bangalore, Pune, Hyderabad, and Mysore. The programme of the Ministry of Non-conventional Energy Sources of soft loans and other incentives has created a favourable environment for use of solar water-heating systems. A significant infrastructure for the manufacture, installation, financing, and servicing of solar water-heating systems has also been developed. The low pay-back period for urban households is an added attraction.

The result is that a number of builders and colony developers are now offering houses and apartments blocks fitted with solar water-heating systems. The latest to adopt this is Magarpatta city, a large housing complex, which is coming up on the outskirts of Pune. Magarpatta is one of the biggest housing complexes in India covering an area of over 550 acres. Magarpatta city is situated off Solapur Road at Hadapsar, Pune. Solar water-heating systems are fitted as an amenity to its flat holders, and as such its cost is included in the cost of the flat. The complex will eventually comprise about 10000 dwelling units in the form of individual houses and apartment blocks of various sizes including some buildings with as many as 11 floors. The complex will also have several buildings that will accommodate information technology-related business.

It has been reported that around 1500 flats have already been provided with such systems. An investment of about Rs. 2.25 crore has been made on these systems so far, and over 12000 units of electricity are estimated to be saved everyday through the use of these systems with the pay-back period (through savings in electricity bill) working out to be less than two years.

It is expected that the project will be completed by 2008. When completed, the township will probably have the biggest concentration of solar water heaters in any residential complex in India. It should serve as a model for future townships and residential colonies in the country.

LAND RESOURCES

In India, land is generally called as "*MOTHER LAND*". It is because of our life depend on it for food, fibre, fuel and other basic amenities. Therefore, it is the valuable gift of nature to human beings. Top layer of the land is called soil, which is renewable resource and essentail for survival of life. Though it is life support system, but it is overused which causes environmental problems. The existing land use pattern in different regions in India has been evolved as the result of the action and interaction of the various factors i.e. physical characteristics of land, the institutional frame work, the structure of other resources available and the location of the region in relation to other aspects of the economic development e.q. those relating to transport as well as to industry and trade.

Out of the total geographical area of 328 million hectares, the land use statistics are available for roughly 306 million hectares, constituting 93% of the total (1970-71) land available for cultivation is approximately 14 million hectares. But it is reducing day by day. It is due to mismanagement. The earth is made up of three principal layers cores, mantle and crust. Cores are inner most fluid layers. Mantle is the middle layer and crust is the upper most layer.

Till 1949-50, the land area in India was classified into five categories—(*i*) forests (*ii*) area not available for cultivation (*iii*) other uncultivated land (*iv*) fallow land (*v*) the net area sown. But after 1950-51 the new classification was adopted, where, classified in to 9 categories instead of 5. They are (*i*) forests, (*ii*) land put to non-agricultural use (*iii*) barren land unculturable land (*iv*) permanent pastures and other grazing lands (*v*) misc. tree crops and groves (*vi*) culturable waste (*vii*) fallow land (*viii*) current fallow (*ix*) net area sown.

Land-capability classification. Any soil and water-conservation project includes two distinct sets of operations, viz. (1) the mapping of land for classification according to its capability, and (2) planning and executing measures to check erosion, improve land productivity and reclaim wasteland. The farm plans for effective soil and water conservation are based largely on the capability of the land. The land-capability classification map is normally prepared by interpreting a standard soil-survey map.

Land-capability classification is a systematic arrangement of differeent kinds of land according to those properties that determine the ability of the land to produce crops on a virtually permanent basis.

The factors determining land-capability. These are the major soil characteristics of the land, e.g. the texture of the top soil, its effective depth, permeability of the top soil and subsoil, and associated land features, e.g. the slope of the land, the extent of erosion, the degree of wetness and susceptibility to overflowing and flooding.

The grouping of soils into capability classes is done primarily on the basis of their capability to produce common cultivated crops and pasture plants without deterioration over a long period.

Land-capability classes. The land-capability classes are based on the intensity of hazards and the limitations of use. The land-capability classes range from the best and most easily farmed land to that which has no value for cultivation, grazing or forestry, but which may be suited to wild-life, recreation or for watershed protection. They all fall into 2 broad groups : one suitable for cultivation and other land uses, and the other not suitable for cultivation, but suitable for other land uses.

LAND SUITABLE FOR CULTIVATION AND OTHER USES

There are four class of land which are suitable for cultivation and other purposes. Their details & limitations are as

CLASS I (GREEN COLOUR). Soils in class I have very few or no limitations that restrict their use.

This type of land is nearly level and the erosion hazard is low. The soils are deep, well-drained, easily worked, hold water well and are either fairly well supplied with plant nutrients or are highly responsive to the application of fertilizers. The soils are not subject to damage because of overflow. The local climate must be favourable for growing many of the common field crops. In irrigated areas, the soils may be placed in class I, if the limitation of the arid climate has been removed by relatively permanent irrigation works.

These soils need ordinary management practices to maintain productivity. Such practices may include the use of one or more of the following : fertilizers, lime, cover and green-manure crops, conservation of crop residues and crop rotations.

Soils in this class are suited to a wide range of plants, may be used for cultivated crops, pastures, forests and wild life, food and cover.

CLASS II (YELLOW COLOUR). Soils in class II have some limitations which reduce the choice of plants or require simple conversion practices.

The limitations of soils in class II may result from the effects of one or more of the following factors : (*i*) a gentle slope, (*ii*) a slight susceptibility to erosion, (*iii*) less than ideal soil depth, (*iv*) occasional damaging overflow, (*v*) wetness which can be corrected by drainage, but existing permanently as a moderate limitation, (*vi*) slight to moderate salinity or sodium, easily corrected but likely to re-occur, and (*vii*) a slight climatic limitation on soil use and management.

They needy terracing, strip-cropping, contour cultivation, water-disposal area, covered with vegetation crop rotation, cover and green-manure crops, stubble mulching, the use of fertilizers, manure and lime. These soils may be used for growing cultivated crops, raising pastures, forests and for wild-life food and cover.

CLASS III (RED COLOUR). Soils in class III have moderate limitations which reduce the choice of plants and require special conservation practices.

Limitations of soils in class III may result from the effects of one or more of the following factors : (*i*) moderately sloping land, (*ii*) moderately susceptible to water or wind erosion, (*iii*) frequent overflow accompanied with some crop damage, (*iv*) very slow permeability of the sub-soil, (*v*) wetness or continuing water-logging after drainage, (*vi*) shallow soil depth up to the bed-rock, hard-pan or clay-pan which limits the rooting-zone and the water storage, (*vii*) low moisture-holding capacity, (*viii*) moderate salinity or sodium, and (*ix*) moderate climatic limitation.

The soils can be used for raising cultivated crops, pastures, forests and wild-life food and cover.

CLASS IV (BLUE COLOUR). Soils in class IV have severe limitations that restrict the choice of plants and require careful management.

The restrictions in the use of these soils are greater than those in class III and the choice of plants is more limited. When these soils are cultivated, very careful management is required and the conservation practices and more difficult to apply and maintain.

The use of these soils for cultivated crops is limited as a result of the effect of one or more permanent features, such as (*i*) steep slopes, (*ii*) severe susceptibility to water and wind erosion, (*iii*) severe effect of past erosion, (*iv*) frequent over-flow accompanied with severe crop damage, (*vii*) excessive wetness with a continuing hazard of water-logging after drainage, (*viii*) severe salinity or sodium, and (*ix*) moderately adverse climate.

These soils can be used for crops, pastueres forests and wild life food and cover.

SOIL :

The word "*SOIL*" is derived from the latin word "*SOLUM*" meaning ground. Here, the word soil is used in a strictly scientific sense. To the scientist soil is not merely the product of simple rock decay - the result of both chemical decomposition and mechanical disintegration of rocks, whether igneous or sedimentary, but a reworked natural material having special and definite characteristics of its own.

Soil is defined as a thin layer of earth's crust which serves as a natural medium for the growth of plants. Soil is formed by weathering process of parental material rocks. Soil differ from parental material in the morphological, physical, chemical and biological properties. Generally it is said that, it is a mixture of mineral constituents and of organic matter, which later consists, more or less of the decomposed residues of plant materials and to a lesser extent of animal remains and extreta.

There are various types of soils like red, black, yellow - alluvial etc. Depth of soil is variable and it differ in texture like coarse, fine, soils serve as reservoirs of nutrients and soil's flora. The most characteristic constituents of soil are the colloidal organic matter and colloidal clay. Soil organic matter and particles conserve soil moisture for crops, plantation, soil living organisms. Soil provide even mechanical anchorage and favourable tilth. The components of the soil are mineral material, organic material,water and air the proportion of which vary.

Since, the main interest of soils to MAN is for growing his crops, it is necessary to learn also about the physical and chemical properties of the soils. Because for growing with success particular crop is dependent. In soil studies, the following physical properties are important

(*i*) Specific gravity (*ii*) Pore space (*iii*) Plasticity
(*iv*) Cohesion (*v*) Colour (*vi*) Texture
(*vii*) Soil temperature (*viii*) Permeability.

Soil Forming Materials

Rocks are the chief sources for the parent materials over which soils are developed. There are three main kind of rocks viz (*a*) igneous rocks (*b*) sedimentary rocks (*c*) metamorphic rocks. Soil differs in chemical composition. The main chemicals are

SiO_2, Al_2O_3, Fe_2O_3, MgO, FeO, CaO, Na_2O, K_2O, CO_3'', P_2O_5, MnO, TiO_3 etc. and water moisture.

Weathering, soil formation and development proceed simultaneously. The weathering may be chemical or physical. Physical weathering comprised of — water, temperature, wind, plants/animals, while chemical weathering comprised of solution, hydration, hydrolysis, carbonation, oxidation-reduction.

Soil types and Soil groups

Depending upon weather a particular soil had originated from the direct decomposition of the crystalline rocks or it had been formed from deposits which have already passed through a cycle or cycles of weathering, transport or consolidation, we have respectively the *Primary and Secondary soils.*

The important soil groups of the world are the following :

1. Soils of the Podsolic group : They are found in cold to temperate regions under humid conditions. The Pod sols and their congeners are completely leached soils in which Calcium Carbonate and Calcium sulphate are present only in very subordinate amounts. Soils of this include the Tunda soils, Podsols, Brown forest soils, Prairie soils and Mountain soils.

2. Tschernosems and their related groups : Such soils develop best in semi and climate. Hence leaching is rather incomplete and generally there is always the development of horizons of carbonate deposition. The Tschernosems are also known as Black Earths. Owing to the invariable presence of humified organic matter in these soils. It includes the typical Tschernosems, black and cotton soils, chestnut Earths and Desert soils.

3. **Groundwater soils including Peat :** Such soils are found throughout the middle, northern and western parts of Europe and their greatest development is in the Pripet Marsh region of Poland. In these areas, the ground water table is either at or near to the surface.

4. **Brown and grey soils of the semi-desert :** As one moves from regions covered with Chestnut soils, into more arid areas, brown and grey semi-desert soils are noticed. The latter have a lower organic matter content and calcareous and gypseous salts occur in such soils within a few inches of the surface.

5. **Saline, Alkaline and Soloth soils :** Such soils are formed in areas of inland drainage without an outlet to the sea such as the inland seas or salt lakes, there is always a characteristic accumulation of mainly the sodium salts with some amount of potassium salts in them. Alkaline soils are characterised by the presence of the carbonate of sodium.

6. **Soils of the Humid Tropics and the sub Tropics :** The tropical soils are made up almost entirely of the products of chemical weathering, together with a greater or less admixture of unweathering minerals such as quartz, magnetite etc. It is red in colour but modified by organic matter content. Red loams, red earths and the laterites belong to this group.

7. **Soil associated with Calcareous parent rocks :** Two types of this soils are

(*a*) Soils of a brownish or brownish grey colour, it belongs to Rendzimes.

(*b*) Soils predominantly red or reddish brown in colour, it belongs to Terra Rosa soils.

Classification of soils in India

Soils widely vary in their characteristics and properties. Understanding the properties of the soils is important in respect of the optimum use they can be put to and the best management requirements for their efficient and productive use. Classification helps to reduce the study of the number of individuals to a few well defined units. Different systems of soil classification have been used from time to time, but brief descriptions of major soil groups are given below—

1. **Alluvial soils :** This include the deltaic alluvium calcareous alluvial soils, Coastal alluvium and Coastal sands. It is the most important soil group of India contributing the largest share to its agricultural wealth. This soil are derived from the deposition laid by the numerous tributaries of the Indus, Ganges and the Brahmputra systems. These streams draining the Himalayas. Geologically, alluvium is divided in to Khadar generally light coloured sandy and Bhanger generally dark and full of "Kankar".

2. **Black soil :** These soils vary in depth from shallow to deep. The typical soil derived from Deccan trap is the "Regur" or black cotton soil. It is common in Maharashtra, Western parts of Madhya Pradesh, A.P., Gujarat, Tamilnadu. It is comparable with the "Chernozems of Russia and Spairie soil of U.S.

3. **Red soils :** The ancient crystalline and metamorphic rocks on meteoric weathering have given rise to the red soils. The red colour is due to the wide diffusion of iron rather then to a high proportion of it. The soils grade from poor thin gravelly and light coloured varieties of the plains and valleys. They are poor in nitrogen, phosphorus, lime, potash, iron oxide. The soils comprise vast areas of Tamil Nadu, Karnataka, Goa, Daman Diu, M.P. Orissa, Chhotanagpur, Bihar, West Bengal, U.P. etc.

4. **Laterites and Lateritic soils :** It is a formation peculiar to India and some other countries. It is a mixture of the hydrated oxides of aluminium and iron with small amounts of manganese oxide and tatania, etc. These soils are poor in lime, magnesia nitrogen, potassium oxide but humus and P_2O_5 is high.

5. **Desert Soils :** The most predominant component of this soil is quartz but feldspar and hornblende grains also occur with a fair proportion of calcareous grains. Some of these soils contain high percentage of soluble salts, possess high pH, a very high percentage of calcium carbonate and are poor in organic matter.

6. **Problem soils :** These are the soils which owing to land and soil characteristics, cannot be economically used for the cultivation of crops without adopting proper reclamation measures. High eroded soils, revive lands, soils on steeply sloping lands etc. constitute one set of problem soils.

7. **Acid soils :** Soils having pH below 7 are considered to be the acidic, but those which have pH less than 5.5 and which respond to liming may be considered to qualify to be designated as acid soils. It occurs widely in Himalayan region, peninsula, Gangetic delta.

SOIL EROSION

Soil erosion means the removal of material from the surface of the soil by the agency of running water, wind or even by gravity. Since the superficial layers of the soil are the richest in plant food and thus the feeding ground of plant roots, the process of soil erosion involves a definite loss of valuable plant nutrients and if it becomes sufficiently intense, may lead to the complete destruction of the soil as the seat of plant growth. Where soil erosion is intense, the natural soil profile is destroyed or truncated and indeed, may never attain full development.

In India soil erosion is a serious problem in certain parts as in Chhota Nagpur. Gully erosion is welmarked in Gwalior, Mandsaur, Shivpuri of Madhya Pradesh. The following types of erosion are observed in India.

Normal or geologic erosion. This is a normal feature of any landscape. Geologic erosion takes place steadily but so slowly that ages are required for it to make any marked alteration in the major features of the earth's surface. There is always an equilibrium between the removal and formation of soil, so that unless the equilibrium is disturbed by some outside agency, the mature soil preserves, more or less, a constant depth and character indefinitely.

Accelerated soil erosion. The removal of the surface soil from areas denuded of their natural protective cover as a result of human and animal interference takes place at a much faster rate than that at which it is built up by the soil-forming processes. This accelerated detachment rapidly ravages the land and it is with this type of soil erosion that we are so seriously concerned. Nature requires, on an average, about 1,000 years to build up 2.5 cm of top soil, but wrong farming methods may take only a few years to erode it from lands of average slope.

Wind erosion. Wind erosion takes place normally in arid and semi-arid areas devoid of vegetation, where the wind velocity is high. The soil particles on the land surface are lifted and blown off as dust-storms. When the velocity of the dust-bearing winds is retarded, coarser soil particles are deposited in the form of dunes and thus fertile lands are rendered unfit for cultivation. In other places, fertile soil is blown away by winds and the subsoil is exposed, as a result the productive capacity of the soil is considerably reduced.

Water erosion. Soil erosion caused by water can be distinguished in three forms, viz. (1) sheet erosion, (2) rill erosion, and (3) gully erosion.

Sheet erosion. Sheet erosion removes a thin covering of soil from large areas, often from entire fields, more or less, uniformly during every rain which produces a run-off. This type of erosion is very insidious, since it keeps the cultivator almost ignorant of its ill-effects. It is generally neglected, although the soil deteriorates slowly and imperceptibly. Its existence, however, can be detected by the muddy colour of the run-off from the fields.

Rill erosion. When sheet erosion is allowed to continue unchecked, the silt-laden run-off forms a well-defined, but minute finger-shaped grooves over the entire field. Such thin channelling is known as rill erosion.

Gully erosion. When rill erosion is neglected, the tiny grooves develop into wider and deeper channels, which may assume a huge size. This is called 'gully' erosion. Gullies are the most spectacular evidence of the destruction of soil. The gullies tend to deepen and widen with every heavy rainfall. They cut up large fields into small fragments and in course of time, make them unfit for cultivation.

Landslides or slip erosion. A landslide is defined as an outward and downward movement of the slope-forming material, composed of natural rocks, soil, artificial fills, etc. The fundamental causes of landslides are topography of the region and geological structure, the kinds of rocks and their physical characteristics. The immediate cause of a slide may be an earthquake, or a heavy rainfall, which unduly saturates the ground or a part of a road. However, these are accidents rather than fundamental causes.

Stream bank erosion. Torrents are defined as hill streams characterized by wide-spreading beds on emergence from the hills with ill-defined banks, flashy flows and swift currents. Usually, they are dry water-courses, except during the rainyseason when with every heavy downpour in their catchment, they get very much swollen with flood and subside almost to its normal tiny size immediately after the storm is over.

These sudden and violent flows are responsible for moving immense quantities of detritus, comprising boulders, shingle, sand and silt, depending upon the geology of the terrain. This debris gets deposited in the torrent bed in the form of scattered islands owing to the sudden widening of the torrent channel after it emerges from the hills, or owing to the flattening of the gradient in the lower reaches, or because of obstructions caused by wild vegetation and uprooted trees. The bed level of the torrent is raised by these deposites. These deposits, in turn, reduce the transporting capacity of the torrent, resulting in overflows and the meandering of the course and in the erosion of the banks.

MECHANISM OF EROSION

Water erosion. Soil erosion caused by rainfall is the result of the application of energy from two distinct sources, namely (*i*) the falling rain drops, and (*ii*) the surface flow. The energy of a falling rain drop is applied slantingly or vertically from above, whereas that of surface flow is applied more or less horizontally along the surface of the ground. The chief role of the falling rain drop is to detach soil particles, whereas that of the surface flow (outside the rills and gullies) is to transport the soil. The falling rain drop also makes a major contribution to the movement of the soil on unprotected sloping lands during the periods of heavy impact storms, by splashing large quantities downslope and by imparting transporting capacity to the surface water by keeping it turbid. More than 100 tonnes of soil per hectare can sometimes be lost yearly in this fashion from a bare and highly detachable soil on slopping land.

Wind erosion. Wind is responsible for three types of soil movement in the process of wind erosion. They are known as : (*i*) saltation, (*ii*) suspension, and (*iii*) surface creep.

(*i*) *Saltation.* The major portion of the soil carried by the wind is moved in a series of short bounces called "saltations". The soil carried in a saltation consits of fine particles ranging from 0.1 to 0.5 mm in diameter. Saltation is caused by the direct pressure of wind on soil particles and their collision with other particles. After being pushed along the ground surface by the wind, the particles leap atmost vertically in the first stage of saltation. Some grains rise only a short distance; others leap 30 cm or higher, depending directly on the velocity of the rise from the ground.

(*ii*) *Suspension.* Very fine soil particles, less than 0.1 mm in diameter, are carried into suspension, being kicked up into the air by the action of particles in saltation. The movement of fine dust in suspension is completely governed by the characteristics movement of the wind. Suspended material is carried long distances from its original location and is thus a complete loss to the eroded area, whereas the soil moved in saltation and surface creep usually remains within the eroded areas, especially when erosive winds are from different directions.

(*iii*) *Surface creep.* Soil particles, larger than about 0.5 mm in diameter but smaller than 1.0 mm, are too heavy to be moved in saltation but are pushed or spread along the surface by the impact of particles in saltation to form a surface creep.

About 90 per cent of the total soil movement in wind erosion is below the height of 30 cm, and about 50 per cent of it is within 5 cm of the ground level. The control of wind erosion is mainly based on the reduction or eliminaton of movement in saltation.

The remedial measures suggested to arrest soil erosion are

(*a*) Deep tillage, in order to facilitate the absorption of the excessive water in to the soil.

(*b*) Direct control processes such as "Contour-bunding".

(*c*) Under-draining of the lands – this consists of laying down drains in gullies to enable the natural drainage of water.

(*d*) Re-forestation.

Attention has already been drawn to the important role played by organic matter eq. humus in plant growth.

Land Degradation

The total land under agricultural use is around 58.4% i.e. grossed cropped area is 167.41 million hectares. The land not fit for cultivation i.e. barren land is around 9.9%. The area under forest is 21.6%, but it needs to be raised.

In addition to water, land resources are the precious resources. Food security depend on conservation and proper utilization of all resources. Due to use and over exploitation land resources are degraded. It is due to the more & more pressure with increasing population. Land degradation is a real alarm. Because soil formation is a very slow process. In millions of years we have a layer crust of fertile soil. In general, formation of 1.0 cm soil crust from parent material take 300 - 400 years. Fertile soil have high percentage of organic matter vis-a-vis microorganisms. Each gram of fertile soil have 30 billion micro-organisms.

Some 1.9 billion ha of agricultural land have been degraded to some an extant and 8 million ha are converted to non agricultural use such as homes, highways, shopping centres, factories, reservoirs etc. In India about 175 million land is effected by degradation problem. Land degradation is caused by impoverishment or erosion of soil, water runs off, vegetation by shrinking of lakes etc.

Significance of the problem : Agricultural production has witnessed dramatic rise in the last three decades or so in the countries world over. In India, green revolution brought about technological break through, which led to the use of short duration high yielding varieties helping intense use of land in a year, increasing area brought under irrigation and prolific use of Chemicals such as fertilizers and pesticides. India, being vastly agriculture oriented, historically has had policies in various phases for the development of agriculture with the expectation that development of agriculture would lead to overall development of the nation and help eradication of poverty. It has been of late recognised that the increasing efforts to raise agricultural growth has cost us clearly in the form of land & water degradation.

Large scale ecological losses were reported in cropland grassland and forest land, such as soil erosion, soil alkalinity and salinity, micronutrient deficiency, water logging and fast depletion and contaimination of ground water. These factors limit future gains from the land & water resources.

Department of land resources, ministry of rural development, Govt. of India has identified different types of degraded land (Table 2.9).

Table 2.9. Categorywise percentage of degraded land as on 2000
(Percentage of total degraded land)

1.	Gullied/ravinous land	3.22
2.	Upland with or without scrub	30.40
3.	Water logged & marshy land	2.58
4.	Land affected by alkalinity/salinity	3.22
5.	Shifting cultivation area	5.50
6.	Degraded notified forest	22.02
7.	Grazing land	4.07
8.	Degraded under plantation crops	0.90
9.	Sands - Inland/Coastal	7.84
10.	Mining/Industrial	0.20
11.	Barren Rocky/Stony	0.12
12.	Steep slopping area	1.20
13.	Snow covered/glacial	8.73

It is clear that soil erosion is the major problem of land degradation.

LANDSLIDES

A landslide is a sudden collapse of a large mass of hillside. There are many different types of landslides where not only earth, but rock, mud, and debris flow down the side of a slope.

Since the beginning of the monsoon season in June India has been hit by heavy rains and landslides affecting in particular, Arunachal Pradesh, Assam, and Bihar states. According to the latest information on the impacts of the landslide, more than 12 million people are affected and more than 270 people were killed in these three states.

Landslides mostly occur

1. Where landslides have occurred before.
2. On steep slopes.
3. On benches.
4. Where drainage is causing a problem.
5. Where certain geologic conditions exists.

Types of landslides

1. Shallow, disrupted landslide—Example of this type is the Santa Susana Mountains and the mountains north of the Santa Clara River Valley. Here more than 75% of the slope area was denuded by landslides triggered by strong shaking.

2. Deep, Coherent Landslides—These triggered by the earthquake were far less numerous than disrupted slides, they contributed significantly to the total volume of landslide material because they tended to be much larger. Some of these landslides are

(*a*) San Martinez Graude Landslide

(*b*) Rancho Camulos Landslide

Factors causing landslides

There are following main causes of landslides—

1. Landslides are the sudden downhill movements on earth or other solid materials and are usually caused by rain thaws or forces either increasing the top material layers or making the slope

too steep. They can be triggered by earthquakes, saturation with heavy rain or crashing waves.

2. Excessive rainfall or snowmelt, however is also known to saturate and lubricate soil on steep angles. Rapid temperature changes can also cause land to slide by alternatively shrinking and expanding soil formations, or forming ice heaves between layers of rock.

3. Forest fires are indirectly responsible for landslides because they take away slope vegetation making erosion easier.

4. Man can also cause slides by mining the earth, underground excavation, pumping and draining groundwater levels or overdeveloping hillsides. Man induced landslides are generally done for the development purposes i.e. industrial, forming roads, agricultural use, homes, etc. They use heavy explosives for that. In this case no serious casualties or damage occur because proper warned earlier to shift in safer places.

Effects : No heavy damage occur in man induced landslides but thousands of people affected and killed due to landslides. Many houses can be damaged and the loss of public properties is also noticed. Roads and rail communication may remain cut off from rest of the regions. Thunder storms cause debris flows on hill slopes leading to deposits of mud. Heavy rains at the same time may worsen the situation.

DESERTIFICATION

Desertification is a process by which productive potential of arid or semiarid land falls. The decrease in productivity is varies from 10%—50%. Thus desertification leads to the conversion of irrigated crop land to desert (where productivity is minimum). It is characterized by devegetation loss of vegetal cover, depletion of ground water, salinization and soil erosion.

Water have a binding capacity as well as it provide vegetal cover. Draught in three consecutive years in Central India in recent years has accelerated the process of wind erosion and desertification. In Rajasthan and Kutch-Saurashtra area, extension in the area of desert is assuming serious proportion. Climatic change and anthropogenic activities are also responsible for desertification.

During last so many years large area has destroyed (agriculture land) by Sahara desert. In India, also, so many places which affected by desertification. An extreme example of sand movement in the Coastal area of Saurashtra has hampered activities at the ports.

Deforestation is also one of the cause of desertification. Because, after forests grasslands are used by human. So human activities are also responsible for desertification. Govt. and worldwide ecologists are seriously thinking about this problem. United Nations Environmental Programme (UNEP) organised a conference in 1977 at Nairobi to discuss about the change of desert land in to productive agriculture land. The increasing cattle population heavily graze in grasslands or forests and denude the land area, which is not suitable for seed germinations. The dry barren land becomes loose & more proper to soil erosion. Thus overgrazing is also one of the cause for desertification.

Mining and quarrying activities are also responsible for conversion of productive land in to desertification. It is studied that, the last 50 years about 900 million ha of land have undergone desertification over the world. Due to mining disappearance of vegetation takes place and defacting of land occurs. Therefore, it is advised to use the eco-friendly mining technology.

Salinization is also one of the cause for conversion of agricultural land to desert. Govt. planning to take such projects, through which conservation of land and desertification can be checked. For salinization farmers are advised to use gypsom, organic fertilizers, manures with proper irrigation. Then sodium will go deeper in the soil. Awareness programmes should be conducted for affected people. It will supplement in the check of desertification.

Role of an individual in conservation of Natural resources :

Planning of a suitable strategy for the conservation of our natural resources and most judicious execution of planned strategies is called as Conservation Management. Environmental planning, evaluation, monitoring, and impact assessment are methods of conservation management. The Indian philosophy of conservation is to keep "Harmony with Nature". Therefore, we have to learn to live with nature. For this every individual has to play his role to conserve the nature & natural resources.

It is well known that people destroying, over utilising natural resources for their own interest. Resources are limited, if they will not be properly used, they will exhaust. Therefore, before doing anything, awareness should be aroused by various methods. People should understand the importance of resources i.e. land water, air, forest, minerals,energy etc. that these are precious and should be used with great care. Every individual must be made to realise that he has to make a personal contribution to the programme to conserve the nature. For this we can take help from media, education and Govt. institutions by organising small programmes, documentary, seminars, conferences etc.

It should not be optional but for all i.e. old, young, rich, poor, industrialist, common man, consumer, businessman, resident of a posh colony, slum dweller. Every body should take part in this work. Voluntary organisations are doing some work in this regard but it is not sufficient. Some important roles of individuals in maintaining peace, harmony and equity in nature are as

1. People should at once stop the over utilization of natural resources instead they must be properly used.
2. Instead of deforestation, representation should keep in mind. We should take help from the Govt. for plantation programmes. Everybody should take part in plantation and care the plants.
3. We should protect wildlife. Though hunting is not allowed even then the persons are doing so. For this educated young should teach the lesson of wild life act.
4. Mixed croping, crop rotation, and proper use of fertilizer insecticide, pesticides should be taught to farmers. Encourage the use of manures, biofertilizers organic fertilizers.
5. We should make habit for waste disposal, compose and to restore biodiversity.
6. Try to educate local people for the protection and judicious use of natural resources.
7. We should use light, fans and other domestic appliances when it is needed.
8. Maintain a balance between resources and human needs.
9. Maintain the essential ecological processes and the life support systems.
10. Install rain water harvesting system in houses, colonies.
11. We should recycle the waste and waste water for agriculture purposes.
12. The fossil fuel should be used only when no other alternative source is available.
13. We must develop energy saving methods to avoid wastage of energy. We should remember *energy saved is energy produced.*
14. Prevent soil erosion.
15. Use drip irrigation and sprinkling irrigation to improve irrigation efficiency and reduce evaporation.
16. Utilize renewable energy sources as much as possible. Encourage use of solar cooker, pump etc.
17. Discourage the frequent use of car, bike, encourage the walk and bicycle.

We know "collecting drop-drop-drops form a big ocean", similarly if each of us will aware about the judicious use of natural resources, all of us will conserve the nature.

Equitable use of resources for sustainable life Style : The equal distribution of natural resources should be for all irrespective of rich or poor. There must be balance between the need and consumption particularly for drinking water, food, fuel etc. The developed countries are utilizing more resources as compared to developing countries. This imbalance is responsible for rich become richer and poor gone poorer. This is due to sharp increase in population in developing countries. But it does not mean that people of developed countries are rich and having good life style, and less developed countries people are poor. Less developed countries also have rich and poor both but facing the problem of population and available natural resources.

Developed countries like USA, Canada, Japan, Australia etc. have 22% of world's population utilising 86% of natural resources. Thus it is needed to divert the resources to poor countries to narrow down the gap between the two. To achieve sustainable life style, there should be equal distribution of global resources and income to meet everyone's need. But in the long process of economic development only the powerful and strong people exploited most of the environmental resources even at the cost of migration of poor people already using those resources. Now stress is put on our natural resources and on the poor and inefficient people.

Owing to these conditions our Govt. inclined to change the existing model of economic development which should be based on the principles of peace, harmony and equity i.e. sustainable development.

QUESTIONS

Short answer type questions:–

1. What do you understand by "Natural Resources" ? Give its types.
2. State and explain the renewable and non renewable resources.
3. What are Fossil fuels ? Give examples.
4. Write note on the "problems of natural resources".
5. Discuss the environmental effects of deforestation.
6. Describe the benefits of dams.
7. What are effects of mining ?
8. Why is water regarded as a precious resources ?
9. What is Chipko movement ?
10. Differentiate between Surface and Ground Water.
11. What are the consequences of over exploitation of water ?
12. Write note on world food problem in India.
13. Explain overgrazing.
14. What do you mean by Bio-energy ?
15. How you will differentiate Tidal and Ocean energy ?
16. "Nuclear Power is the important source of energy", comment upon.
17. Write about the desertification in brief.
18. Write about non-conventional energy sources.
19. Write an essay on mineral wealth of India.
20. What are the alternative services of energy ?

Long answer type questions :–

1. Write an essay on "Natural Resources".
2. Explain deforestation. Describe its causes and control measures.
3. What are the India's water resources ? Differentiate in to surface and ground water.
4. "It is said, that next world war will be on water" comment upon the statement giving facts.
5. What are mineral resources in India ? Give its types. How mining effects the environment ?

6. How the modern agriculture effects the environment ? Explain.
7. What do you mean by (*i*) water logging and (*ii*) salinity ?
8. What are energy sources ? Are they fulfilling the growing need of energy ?
9. Write an essay on "Coal and Petroleum are the main sources of energy".
10. What is soil ? How it is classified ?
11. Explain soil erosion. How it happened ? Give its types.
12. What do you mean by land degradation and land slides ?
13. Describe the role of an individual in conservation of natural resources.
14. Write about the availability of minerals in India. Explain the effect of mineral mining on environment.
15. To explain the importance of water resources, describe the problems from over exploitation, drought and floods.
16. Write on essay on forest resources.
17. Define the role of an Individual in conserving natural resources.
18. Describe the impact of modern agricultur practices on environment.

UNIT 3

Ecosystems

INTRODUCTION

Every living organism is surrounded by materials and forces which constitutes its environment and from which it must derive its needs. Thus, for its survival, a plant, an animal or a microbe can not live completely sealed in an impervious skin or shell, but requires from its environment a supply of energy, a supply of materials and a removal of waste products. For these basic requirements each living organism has to depend and also to interact with different non-living or abiotic and living or biotic components of the environment. The scientific study of the interactions of organisms with their physical environment and with each other is called ECOLOGY (Helena Curtis 1975). It mainly concerns with the directive influences of abiotic and biotic environmental factors over the growth, distribution, behaviour and survival of organisms (Herreid II 1977). The credit goes to German Zeologist E. Haeckel in 1966, who used the term as *Oekologie* to the inter-relationships of living organisms and their environment.

The word ecology comes from two Greek words "*oikos*" meaning 'household' or 'home' or 'place to live or habitation' and "*logos*" meaning "discourse" or 'study'. Thus ecology deals with the organism and its place to live i.e. environmental Biology.

CONCEPT OF ECOSYSTEM

Now, we can say *Ecology* deals with interrelationships between the biotic and abiotic compounds of an *Ecosystem.* Ecology is both Science and Art. The term ecosystem was first coined by *A.G. Tausley 1935.* It is derived by two words '*eco*' means environment and '*system*' implies a complex of co-ordinated units.

Ecosystem is defined as a community of organisms interacting with one another and the environment in which they live i.e. **study of home.** A home can be a drop of water for an amoeba or so, a home for a lion may be many miles of land over which it searches for its food. A pond, forest, lake a river, an ocean, a dam, a cropland, a garden, an orchard, a city, aquarium, wood lots and fields may be an example of ecosystem. Further an ecosystem may be natural, man-made or artificial like an aquarium. There are two types of systems open and cybernetic. The ecosystem is an open system for it receives energy from an outside source, the sun, fixes and utilizes it, and ultimately dissipates heat to space i.e. flow into and out of it.

But the ecosystem also an abstraction and has no particular size. In a sense, asking how big an ecosystem is would be, like asking how big a city, an engine or a triangle is. All of these things are abstractions. No two ecosystems are exactly alike. **An ecosystem concept is that the living organisms of a community not only interact among themselves but also have functional relationship with their nonliving environment. This structural and functional system of communities and their environment is called an ecosystem.**

FUNCTIONING AND TYPES OF ECOSYSTEM

Functioning of the ecosystem is self-regulating and self-sustaining. This depend upon flow of energy, cycling of materials and perturbations both intrinsic and extrinsic. It is recognised as a dynamic concept with structural hetergeneity based on at least four functional phases. A rapid release phase consisting of tightly bound resources is replaced by a reorganisation phase followed by a exploitative phase, which gradually transformed into conservation phase or climax phase. The control of ecosystem function by nutrient flux and the condition of the physical environment is called Bottom-up-control but function via tropic interactions is called Top-down control.

Depending upon the species, diversity and the manner in which they are organised, Ecosystems are of following types —

1. Permanent and Natural ecosystem. These operate under natural conditions without any interference (even by human beings). These can be further classified in to—

(*i*) Terrestrial ecosystem

(*ii*) Aquatic ecosystem

Terrestrial ecosystems operate on land hence Forest, Desert and grassland and Agro-ecosystems included in this type. While Aquatic ecosystem operates in water. It can be devided in two

(*a*) Fresh water ecosystem

(*b*) Marine ecosystem

Freshwater ecosystems are usually named after the size and nature of the fresh water body such as pond, lake & river.

Marine ecosystem is largest ecosystem on earth, which consists of several sub-divisions each having its physico-chemical and Biological characteristics. For example, in the deepest ocean producers are absent but in many other organisms survive which dependent for food on the dead organic matter coming from the upper layers of the ocean.

2. Temporary and Natural ecosystems. These are short lived but operate under natural conditions.

3. Artificial or Anthropogenic ecosystems. These are man-made like fishery tanks dams, croplands and space ecosystems also. Fish aquarium is also come under this head.

These typologies are determined not only by the species composition but also by the physiognomic characteristics and soil and climatic conditions.

STRUCTURE OF AN ECOSYSTEM

It is a description of the species of organisms that are present (including information on their life histories, populations and distribution in space). The structure of ecosystem provides information about the range of climatic conditions that prevail in the area, composition and organization of Biological Communities and Abiotic compounds constitute the structure of an ecosystem. According to *Odum,* from the trophic (Food) point of view, an ecosystem has the following components:

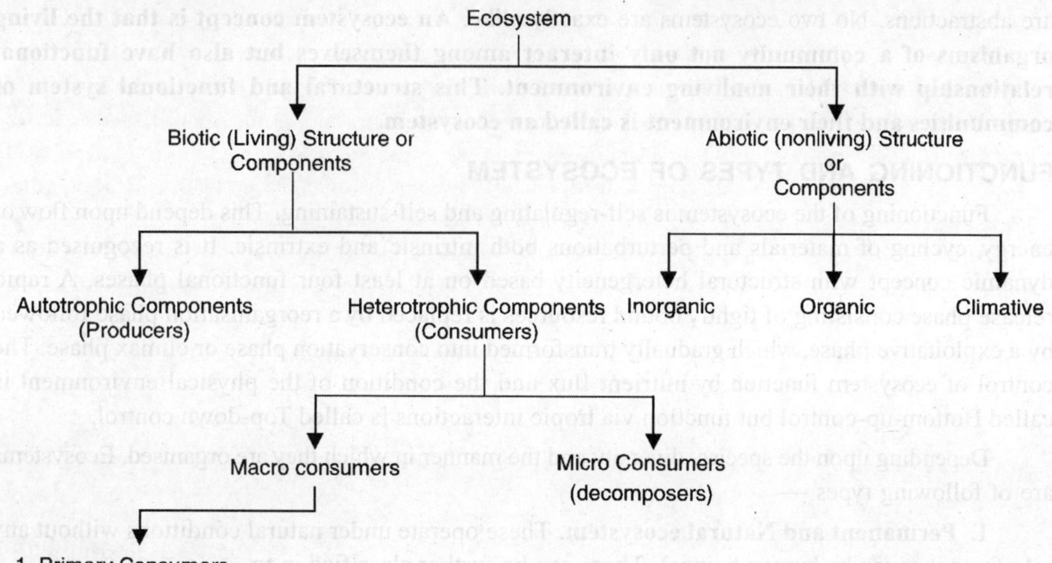

1. Biotic Structure. Producers, consumers and decomposers are components of biotic ecosystem. Living things are made of control and other chemicals with a lot of water added. Living organisms exchange expel convert, assemble, disassembles, organise and otherwise manipulate the constituents of earth, air and water. Biotic structure includes plants, animals and micro organisms present in an ecosystem. We have identified producers, decomposers and consumers are the basic components of biotic ecosystem. These can be distinguished on the bases of their source of energy and material.

(*a*) **Autrotrophic components** (Autotrophic = self nourishing)

In which the fixation of light, energy, the use of simple inorganic substances and manufacture of complex material predominates. These are also called **producers**.

(*b*) **Heterotrophic Components** (Heterotrophic = other nourishing)

These utilizes, rearranges and decomposes the complex materials synthesized by the autotrophs. The most intense heterotrophic activity takes place where the organic matter accumulates in the soils and sediments. These are also called **consumers**.

PRODUCERS

All green plants are producers. They are also called "**converters**" or "**transformers**". They are living members of the ecosystem that utilize sunlight as their energy source and single inorganic materials from soil, air and water to transform them by photosynthesis in to more complex energy rich chemicals as their own food. Producers are largely photosynthetic plants and their kind varies with the kind of ecosystem. In dense forest the trees are the most important producers. In lakes and ponds, the producers are rooted or large floating and microscopic plants (phytoplankton) usually the algae. They are also known as "**photo-autotrophs**" (*Photo = light, auto = self, troph = food.*)

Recently, scientists have found ecosystems based on chemical energy at great ocean depth (more than a km), where there is no light. The producers in these systems are bacteria that are able to gain energy from the oxidation of H_2S that seeps from volcanic rents in the ocean. Since these organisms get their energy from chemical reactions rather than sunlight, they are called "**Chemotrophs**".

CONSUMERS — As we have seen earlier, consumers are heterotrophs, the living organisms which ingest other organism. They derive their food directly or indirectly from the producers. The food is then digested i.e. broken down to simple substances which are metabolized in the consumers body and released the waste product to the environment. Consumers are of following types—

(*i*) *PRIMARY CONSUMERS.*— These are also called 'HERBIVORES' which feed directly on the producers. They vary with the kind of ecosystem. For example a deer and giraffe is a primary consumer in forest ecosystem, while cow or a goat is in a grassland or crop ecosystem. Protozoans and certain crustaceans which feed floating algae are also primary consumers.

(*ii*) *SECONDARY CONSUMERS.*— They are also called "*CARNIVORES"* (meat eaters). For example insects gamefish in a pond eat primary consumers.

(*iii*) *TERTIARY CONSUMERS.*— In most of ecosystem some organism that eat other carnivores like — they are tertiary consumers.

(*iv*) *OMNIVORE.*— A person or animal eating plants and animals is called omnivores.

(*v*) *TOP CARNIVORES.*— Some ecosystem have animals like lion and vulture, which are not killed or rarely killed and eaten by other animals are called top carnivores.

(*vi*) *DETRITIVORES.*— These are the bottom living which subsist on the rain of organic detritus from autotrophic layers e.g. beetles, termites, ants crabs etc.

3. *DECOMPOSERS.*— They are also the living components, mainly bacteria and fungi which breakdown complex compounds of dead protoplasm of producers and consumers to simple organic compounds and ultimately in to inorganic nutrients. In all the ecosystems, this biotic structure prevails. Molds and mushrooms of the forest are the largest of the decomposers that are visible. The role of decomposers in ecosystem is very important. They are responsible for the completion of ecosystem mineral cycles. They are also called *microconsumers* or *saprobes* or saprophytes or saprotrophs. (Sapros = rotten, trophs = feeder) other examples are bacteria and fungi.

2. *ABIOTIC STRUCTURES OR COMPONENTS.*— The physical and chemical components of an ecosystem constitute its abiotic structure. It includes two things—

(*i*) *MATERIALS OR CHEMICAL FACTOR*—The materials are like water, minerals, atmospheric gases and other inorganic salts. They also include some organic matter such as amino acids, decay products, lipids, carbohydrates, proteins etc. The quantity of abiotic materials like the minerals present at any given time in an ecosystem is termed as the '*standing state*' or '*standing crop.*'

(*ii*) *ENERGY OR PHYSICAL FACTOR.* This is in the form of light, heat and stored energy in chemical bonds. Annual rainfall, wind latitude and altitude etc. are also some physical factors, which have a strong influence on ecosystem. For proper functioning of an ecosystem there must be a continuous 'flow of energy' and 'cycling of minerals' among the organisms of the ecosystem.

ENERGY FLOW IN THE ECOSYSTEM

Energy is needed for every biological activity. Solar energy is transformed in to chemical energy by a process of photosynthesis. This energy is stored in plant tissue and then transformed in to mechanical and heat form during metabolic activities. In the biological world the energy flows from sun to plants and then to all heterotrophic organisms like nitroorganisms, animals and man i.e. from producers to consumers. 1% of the total sunlight falling on the green plants is utilized in photosynthesis. This is sufficient to maintain all life on this earth. There is no 100% flow of energy from producers to consumers. Some is always lost to environment. Because of this, energy can not be recycled in an ecosystem '*it can only flow one way*.'

The flow of energy follows the two laws of thermodynamics.

Ist law of thermodynamics. The law states that energy can neither be created nor be destroyed but it can be transformed from one form to another. Similarly, as we have read earlier, solar energy utilized by green plants (Producers) in photosynthesis converted into biochemical energy of plants and later into that of consumers.

IInd law of thermodynamics. The law states that energy transformation involves degradation or dissipation of energy from a concentrated to a dispersed form. We have seen dissipation of energy occurs at every trophic level. There is loss of 90% energy, only 10% is transferred from one trophic level to the other.

SUN AS THE SOURCE OF ENERGY. Sun is the source of energy which extends radiations from high frequency to low frequency. Approximately 99% of total energy is in the region between UV and IR. The visible spectrum spreads over 0.38 μ to 0.77 μ involving about 50% of solar radiations. Some autotrophs however utilize energy released from oxidation processes for the synthesis of organic food.

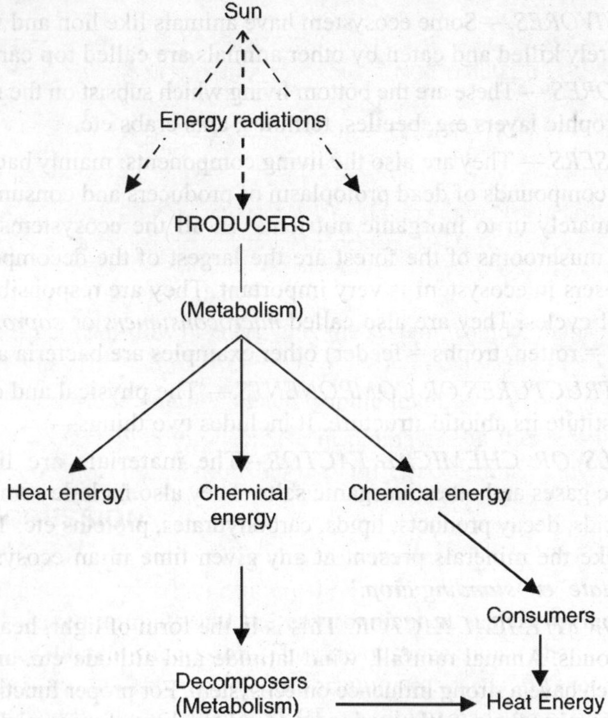

Fig 3.1 Sun as the source of energy

ENERGY FLOW MODELS

As we have seen, that there is unidirectional flow of energy from sun to the producers and then various types of consumers. Therefore, behaviour of energy in ecosystem can be termed **Energy flow**. About 34% of the sunlight reaching the atmosphere is reflected back in to its atmosphere. 10% is held by ozone layer, water vapours and other atmospheric gases. Rest 56% reaches the earth surface. Out of this 1-5% is used by green plants for photosynthesis.

$$6CO_2 + 6H_2O \xrightarrow[\text{Chlorophyll}]{\text{Sunlight}} C_6H_{12}O_6 + 6O_2$$

Rest is absorbed as heat by ground vegetation or water. The flow of energy in an ecosystem can be explained with the help of various energy flow models—

1. ODUM'S ENERGY FLOW MODEL

E.P. Odum 1963 explained flow of energy involving three trophic levels with the help of his universal energy flow model. (Fig.3.2) As the flow of energy takes place, there is gradual loss of energy at every level thereby resulting in less energy available at next trophic level.

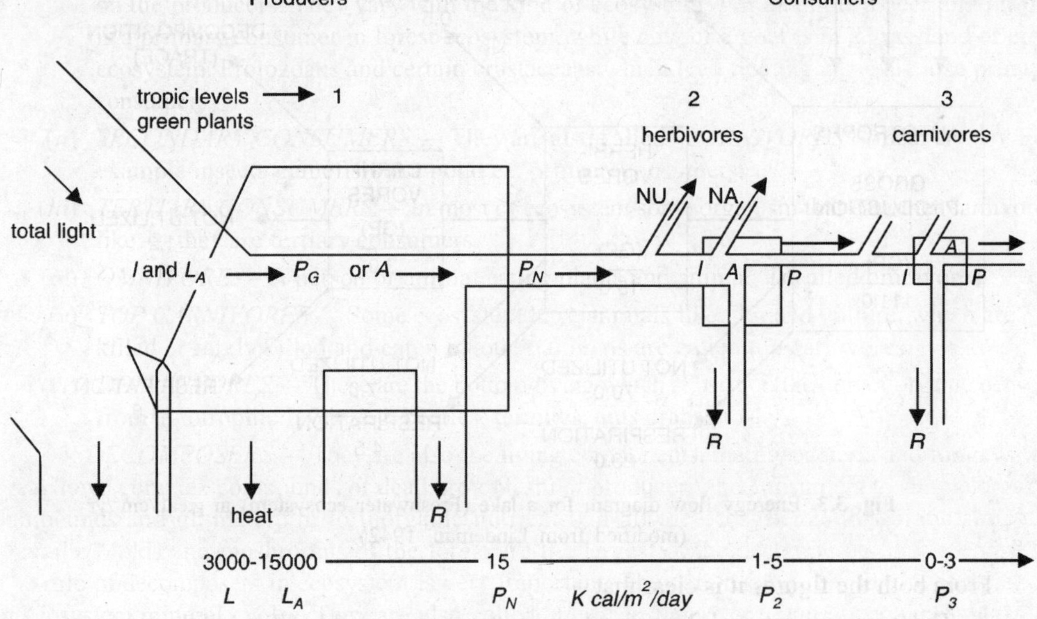

Fig. 3.2. Depicting three tropic levels (Adopted from, Odum, 1963)

Out of total 3000 KCal of light falling (L) on producers level, only 50% i.e. 1500 KCal is absorbed by autotrophs with an efficiency of energy capture is only 0.1 - 1%. As reported 21% of this energy is consumed in metabolic reactions of autotrophs for their growth, development, maintenance and reproduction. Thus gradual decline in energy at the second trophic level i.e. herbivores and then at third trophic level i.e. carnivores in grazing food chain is observed.

2. LINDEMAN'S ENERGY FLOW MODEL

Lindeman in 1942 gave the unidirectional energy flow model of fresh water ecosystem. Model shows that out of total 118,872 g.cal/cm^2/year incident solar radiations, producers can utilize only 1% (111.0 g cal/cm^2/year) in their photosynthesis (Fig.3.3)

About 21% of this gross production (GP) is utilized in metabolic functions of producers, 3% is utilized in decomposition and 63% remains unutilized in decomposition and 63% remains unutilized. Thus only 13-14% i.e. 15 g.cal/cm^2/year of GP is available to herbivores. At this level about 30% of it (i.e.15 g.cal/cm^2/year) is utilized in metabolic functions i.e. respiration, growth and reproduction etc. This is more than the autotrophs consumed i.e. 21%. Again 3% of it is utilized in decomposition while 47% is remains un-utilized. Thus only 20% energy of the autotrophs is available (i.e. 3 g.cal/cm^2/year) to carnivorous. It is also reported that about 70% energy is available for carnivores, which is not utilized and only 28-6% of net production passes to carnivorous. Carnivores utilized 60% energy at this level in metabolic activities and rest is remains as un-utilized.

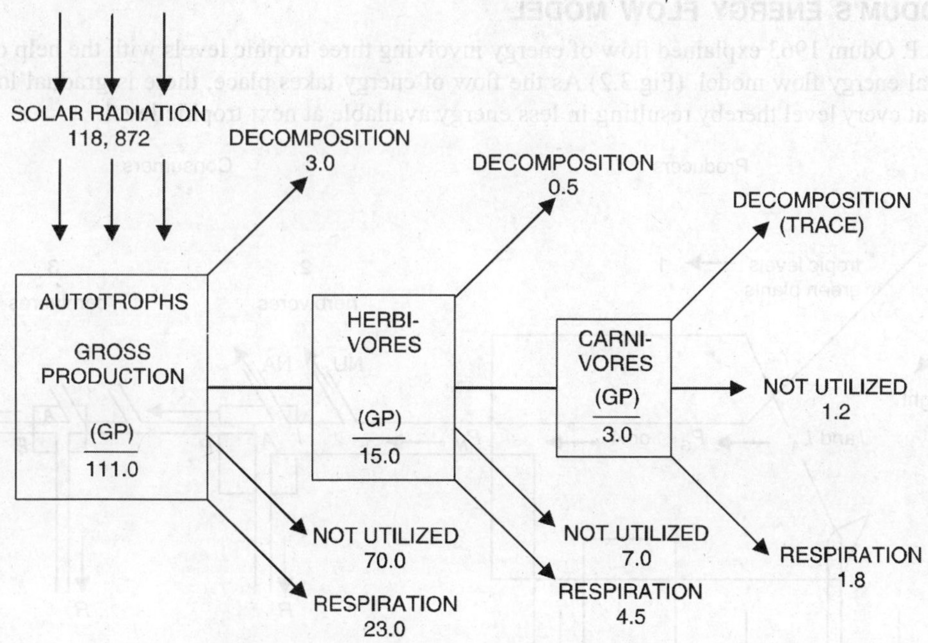

Fig. 3.3. Eneregy flow diagram for a lake (Freshwater ecosystem) in g.cal./cm^2/yr (modified from Lindeman, 1942)

From both the figures it is clear that —

1. There is unidirectional flow of energy i.e. the system would collapse if the primary source, the sun were cut off.
2. Progressive decrease in energy at each trophic level.
3. There is also a corresponding decrease in biomass. But there is no correlation between biomass and energy. This relationship may differ according to situations.

ECOLOGICAL SUCCESSION

Biotic communities are not static, they change with time. This change can be understood on several levels. Changes take place continuously in the community structure, organization, physiognomy, the associated animals and the environment at a place in the course of time, this phenomenon is called **ecological succession.** The rate of successional changes is rapid initially and gradually it slows until a point of dynamic equilibrium is reached, and the community is more or less stable. A complete succession is called a **SERE.** A sere is made up of a number of seral stages.

Seral communities are large in number depending on minor differences in edaphic, topographic or other local conditions. These communities are open to a large scale invasion of other species. Seral communities react to the prevailing environment in such a way that the environment is modified sufficiently to become less favourable to the existing sere. Seral communities are therefore, short lived and quick in development. As against these the climax communities are only a few and similar in identical climatic conditions. These are more or less closed to invasion of outside floristic elements. Climax communities attain a dynamic equilibrium with their environment. They cannot change their environment to such an extent as to make it less suited to the existing communities. Climax communities are formed after a long period of time and once formed they maintain their structure and composition. It is the final or the last seral stage.

Succession is the "**birth**" of an ecosystem, and subsequent "**aging**" process of its abiotic and biotic features. **ODUM (1971)** has rightly included the following three parameters in his definition of ecological succession.

(1) It is an orderly process of community development that involves changes in species structure and community processes with time, it is reasonably directional and therefore predictable.

(2) It results from modifications of the physical environment by the community, i.e. succession is community controlled even though the physical environment determines the pattern, the rate of change and often sets limits as to how for development can go.

(3) It culminates in a stabilized ecosystem in which maximum biomass (or high information content) and symbiotic function between organisms are maintained per unit of available energy flow.

In any of the basic environments such as terrestrial, fresh water or marine, the succession may be of following two types—

(i) PRIMARY SUCCESSION. It is the process of species colonization and replacement in which the environment is initially virtually free of life, i.e. the process starts with base rock or sand deme or river delta or glacial debris and it ends when climax is reached. The sere involved in primary succession is called **PRESERE.** Primary succession occurs when a community begins to develop on a site previously unoccupied by living organisms.

(ii) SECONDARY SUCCESSION. The term secondary succession refers to community development on locations or sites previously occupied by well developed communities. It occures where a community has been disrupted and the surface is completely or largely devoid of vegetation. It may be due to earthquake, fire or even clearing of forests by man. In each case organism modify the environment in a way that allow one species to replace another. The sere involved in secondary succession is called **SUBSERE.**

Depending on the moisture contents, the primary and secondary successions may be of the following types—

(A) HYDRACH or HYDROSERE. The succession when starts in the aquatic environment such as ponds, lake, streams, swamps, bogs etc.

(B) MESARCH OR MESOSERE. It is an intermediate type with adequate moisture. The succession when begin in such an area is called mesarch.

(C) XERACH OR XEROSERE. The succession when starts in Xeric or dry habitat having minimum amounts of moisture, such as rocks, dry deserts etc is called xerach. A temporary community in an ecological succession on dry and sterile habitate is called Xerosere. It may be of three types—

(i) LITHOSERE. i.e. succession initiating on rocks.

(ii) PSAMMOSERE. i.e. succession initiating on sand.

(iii) HALOSERE. i.e. succession initiating on saline water or soil.

Some times succession is also classified into two on the basis of community metabolism.

(a) AUTOTROPHIC SUCCESSION. It is characterised by early and continued dominance of autotrophic organisms like green plants. It begins in a predominantly inorganic environment and the energy flow is maintained indefinitely.

(b) HETEROTROPHIC SUCCESSION. It is characterised by early dominance of heterotrops such as bacteria, actinomycetes, fungi and animals. It begins in a organic environment and there is a progressive decline in energy content.

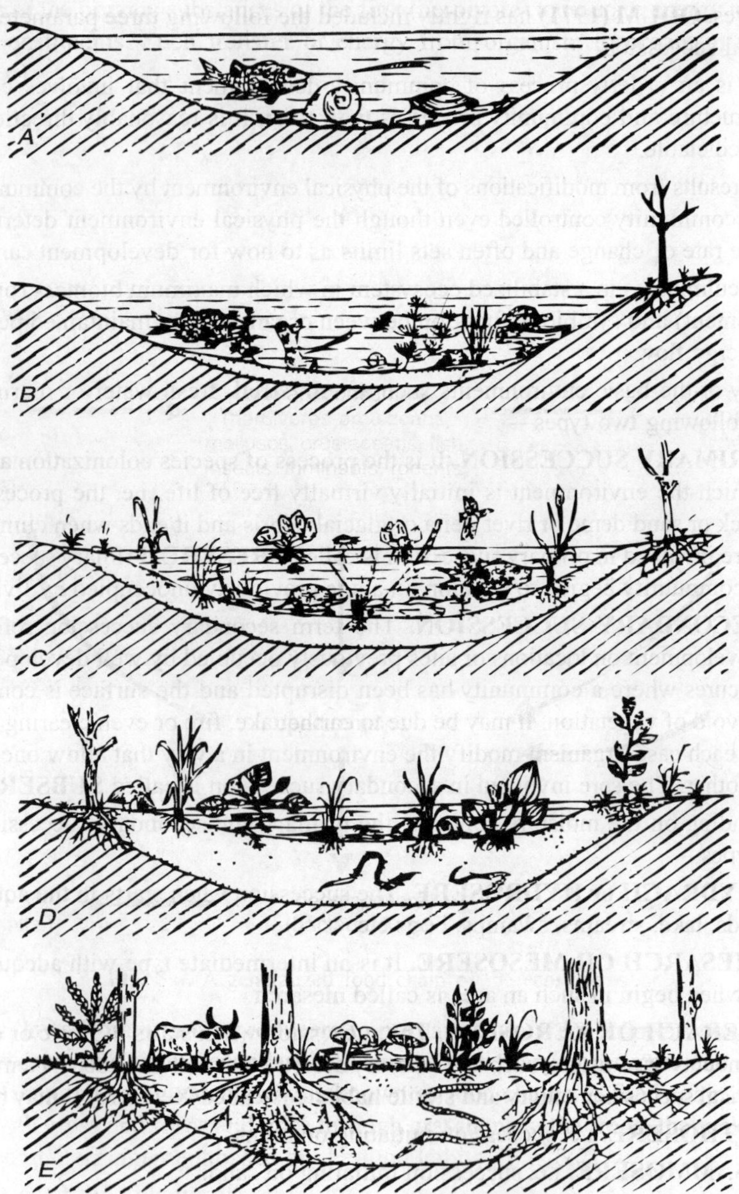

Fig. 3.4 Hydroseric succession showing through stages A to E formation of different communities at the same place. Gradual silting of pond bed reduces the depth. Finally a terrestrial community has formed.

In some cases the replacement of one type of community by another is due to modification of the environment by communities themselves. Such a successional process is called **Autogenic succession.** But in some cases, if replacement is largely due to forces other than the effects of communities on the environment, this is called **Allogenic succession.** It occur in a highly disturbed or eroded area or in ponds nutrients and pollutants enter from outside and modify the environment and in turn the communities.

GENERAL PROCESS OF SUCCESSION

The complete process of a primary autotrophic ecological succession involves the following sequential steps which follow one another.—

1. NUDATION. The process of succession begins with the formation of a base area or nudation by several reasons such as volcanic eruption, flood, landslide, erosion deposition, fire, disease etc. Some base areas are also created by man *e.g.* walls, burning, digging etc.

2. INVASION. The invasion is the arrival of the reproductive bodies or propagules of various organisms and their settlement in the new or base area. Plants are the first invaders (pioneers) in any area because the animals depend on them for food. It include the three steps—

(i) **Dispersal or migration.** It is the process in which propagule leaves the parent plant and arrives the bare area. The seeds, spores or other propagule of the species reach the bare area through the agency of air, water or animals.

(ii) **ECESIS.** This is the successful establishment of migrated plant species in to new area. It includes germination of seeds, growth of seedlings and starting of reproduction.

(iii) **AGGREGATION.** This is the final stage of invasion where immigrant species increase their number by reproduction and aggregate in a large population in the area.

3. COMPETITION. As the number of individuals grows, there is competition both inter-specific (between different species) and intra specific (within the same species) for space, water and nutrition. They influence each other in a number of ways known as **COACTION.**

4. REACTION. When living organism grow, use water and nutrients from the substratum and in turn they have a strong influence on the environment which is modified to a large extent and is known as reaction. When they become unsuitable for the existing species, favour some new one, which replace them. Thus, reaction leads to several seral communities.

5. STABILIZATION OR CLIMAX. Eventually a stage is reached when a final terminal community becomes more or less stabilised for a longer period of time and it can maintain itself in the equilibrium or steady state with climate of that area. This last seral stage is mature, self maintaining, self reproducing through development stages and relatively permanent. This final stable community of the sere is the **CLIMAX COMMUNITY** and the vegetation supporting it is the **CLIMAX VEGETATION.**

FOOD CHAINS

A sheep may eat some grass and in turn it may be eaten by a person. The algae of a lake will be eaten by many Zooplanton, such as crustaceans and insect larvae. They are eaten by small fish and in turn they are eaten by bars, which may be caught by a bear. From all these examples it is clear that plants form the link between biotic and abiotic components of the ecosystem. They draw water and minerals from the soil and combine them with sunlight and carbondioxide from the air to make carbohydrates, fats, proteins, vitamins and usable minerals through photosynthesis.

Small harbivorous organisms such as Caterpillars field mice etc. Consume this vegetable material and convert it to animal material, which serve as food to meat eating animals. They are eaten by larger carnivores. This sequence of eaten and being eaten, with the resultant transfer of energy is known as **FOOD CHAIN.** Thus in food chains organisms of an ecosystem are linked together. Each step is known as **trophic level** and the study of the energy flow through these steps is called **trophic ecology.** Food chains are not isolated from each other.

Primary producers trap radiant energy of sun and transfer that to chemical or potential energy of organic compounds such as carbohydrates proteins and fats. When herbivore eats a plant and these compounds are oxidised. As we have read earlier the energy liberated is just equal to the amount of energy used in synthesizing the substances. When this animal is eaten by another one, along with transfer of energy from a herbivore to carnivore a further decrease in energy occurs as the carnivore oxidise the organic substances of the first (herbivore) to liberate energy to synthesize its own cellular constituents. Such transfer of energy from organism to organism sustains the ecosystem (Fig. 3.6)

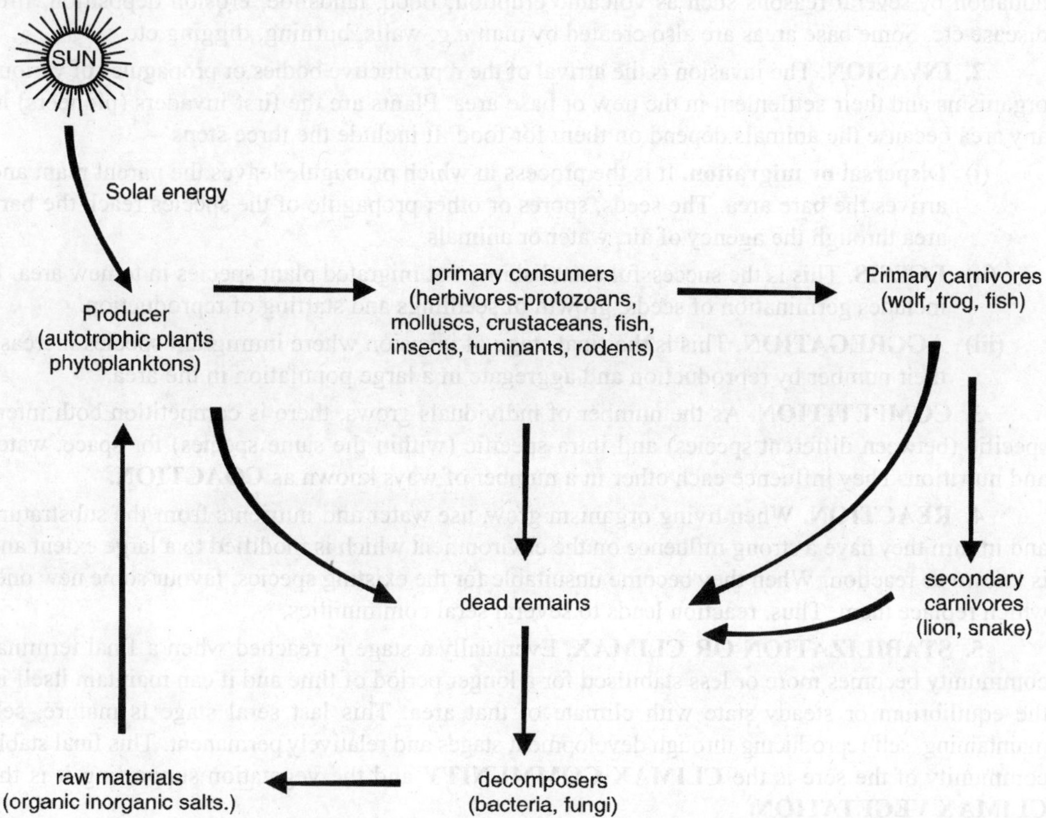

Fig. 3.6. A generalized food chain of the ecosystems.

Thus energy flows from primary producers to primary consumers, from primary consumers to secondary consumers and from secondary consumers to tertiary consumers and so on. This simple chain of eating and being eaten away is known as food chains. A food chain in grassland ecosystem starts with grass and forbs and goes through grasshopers, the frogs, the snake, the hawk in an orderly sequential arrangement based on the food habits. (Fig. 3.7)

Other examples of food chains are :

1. Grass → Rabbit → Fox → Wolf → Lion.
   (Grass land ecosystem)
2. Phytoplanktons → Waterfleas → small fish → Tuna
   (Pond ecosystem)
3. Lichens → reindeer → Man
   (Arctic Tundra)

In nature, we can distinguish two types of food chains.

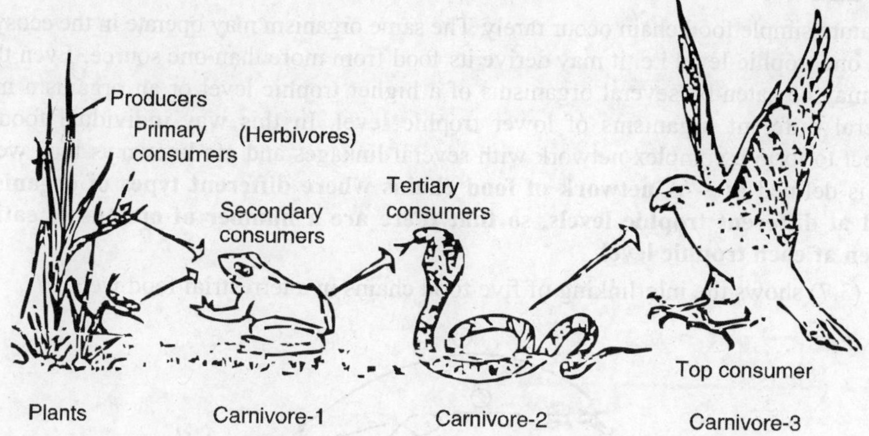

Fig. 3.5. Food and a food-chain

1. GRAZING FOOD CHAIN

This type of food chain starts from green plants and ends to carnivores by passing through herbivores. All examples cited above show this type of food chain. The primary carnivores or secondary consumers eat herbivores or primary consumers of the ecosystem. And likewise, secondary carnivores or tertiary consumers eat primary carnivores. The total energy assimilated by primary carnivores or gross tertiary production and its disposition in to respiration, decay and further consumption by other carnivores is entirely analogous with that of herbivores. Thus much of the energy flow in these chains can be described as follows —

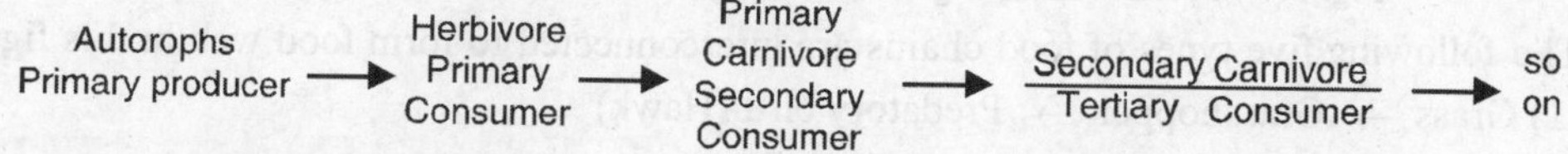

2. DETRITUS FOOD CHAIN

The term **detritus** is given to organic wastes, exudates and dead matter derived from grazing food chain. The energy contained in this detritus is not lost to the ecosystem as a whole, rather it serves as the source of energy for a group of organisms (**Detritivores**), they differ from grazing food chain called the detritus food chain. These food chains are less dependent on solar energy, but chiefly depend on the influx of organic matter produced in another system. Such food chains operates in the decomposing accumulated litter in a temperate forest.

In some ecosystems, considerably more energy flows through the detritus food chains than through the grazing food chains. The organisms of the detritus food chains are, algae, bacteria, slime molds, fungi, actinomycetes protozoa, insects, mites, crustaceans, molluses worms, nematodes etc. Some species are highly specific in their food requirements and some can eat almost anything. All these are detritus consumers. They ingest large amounts of the vascular plant detritus. These animals are in turn eaten by some minnows and small game fish i.e. small carnivorus which in turn serve as food for larger game fish and fish eating birds i.e. top carnivores.

Heald (1969) and **Odum (1971)** have studied the detritus food chain of mangrove leaves (Rhizophora mangle) of Southern Florida. Conclusively, we can understand detritus food chain as follows —

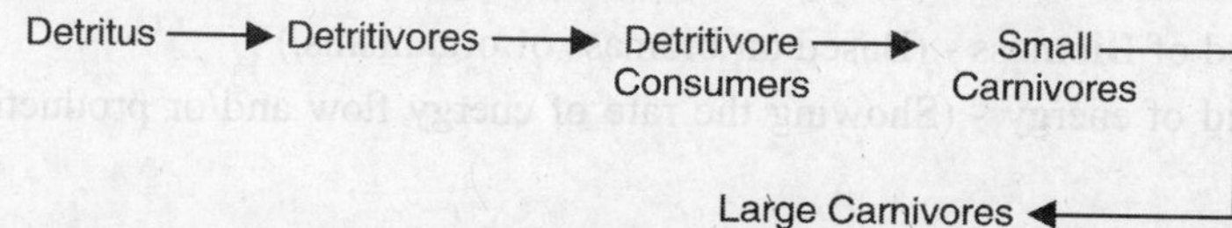

FOOD WEBS

In nature simple food chain occur rarely. The same organism may operate in the ecosystem at more than one trophic level i.e. it may derive its food from more than one source. Even the same organism may be eaten by several organisms of a higher trophic level or an organism may feed upon several different organisms of lower trophic level. In this way individual food chains interconnect to form a complex network with several linkages and are known as food web. Thus food web is defined as—**"A network of food chains where different types of organisms are connected at different trophic levels, so that there are a number of options of eating and being eaten at each trophic level.**

Fig. (3.7) shows the interlinking of five food chains in a terrestrial food web.

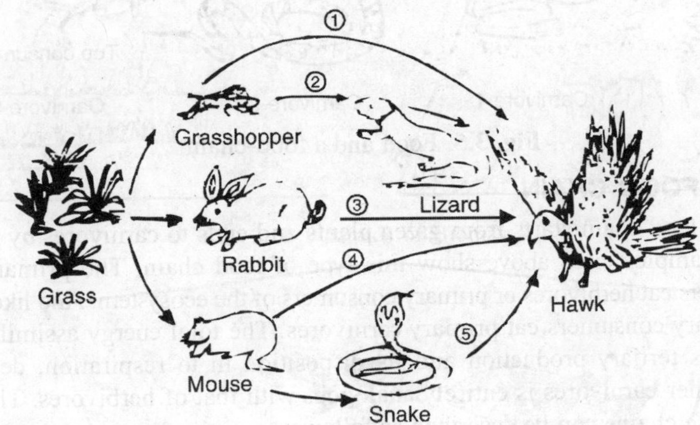

Fig. 3.7 A food web in a grassland ecosystem with five possible food chains.

The following five types of food chains are interconnected to form food web in this figure.

(1) Grass → Grasshopper → Predatory bird (Hawk)

(2) Grass → Grasshopper → Lizzard → Hawk.

(3) Grass → Rabbit → Hawk (or vulture or man)

(4) Grass → Mouse/Rat → Hawk

(5) Grass → Mouse/Rat → Snake → Hawk.

This shows, food chains in natural conditions never operate as isolated sequences but are interconnected with each other forming some sort of interlocking pattern (which is referred to as a food web).

ECOLOGICAL PYRAMIDS

Charles Elton in 1927, noted that the animals at the base of the food chain are relatively abundant, while those at the end are relatively few in number i.e. there is progressively decrease in between the two extremes. Secondly, there is some sort of relationship between the numbers, biomass and energy content of the primary producers, consumers of the first and second orders and so on to top, Carnivores in any ecosystem. These relationships may be represented in diagrammatic (Graphic) ways and are referred to as **ecological pyramids or Eltonian Pyramids.**

Ecological pyramids are of three general types —

1. Pyramid of numbers - (Based on number of organisms at each level.)
2. Pyramid of Biomass - (Based on biomass of organisms.)
3. Pyramid of energy - (Showing the rate of energy flow and/or productivity at successive trophic levels.

The pyramids of numbers and biomass may be upright or inverted depending upon the nature of the food chain in the particular ecosystem whereas pyramids of energy are always upright.

1. Pyramid of numbers. This deals with the relationship between the number of producers, herbivores and carnivores at successive trophic levels. At the base of such figure (pyramid) is always the number of primary producers and the subsequent structures on this base are represented by the number of consumers at successive levels. In figure 3.8 a grassland ecosystem, the producers which are mainly grasses are always many in number. This number then shows a decrease towards apex, as the primary consumers or herbivores like rabbits are less in number than the grasses. The secondary consumers are lesser in number than primary consumers. Finally the top consumers (tertiary) like hawks or other animals are least in number. Thus the pyramid becomes upright. In a pond ecosystem, the pyramid is also upright.

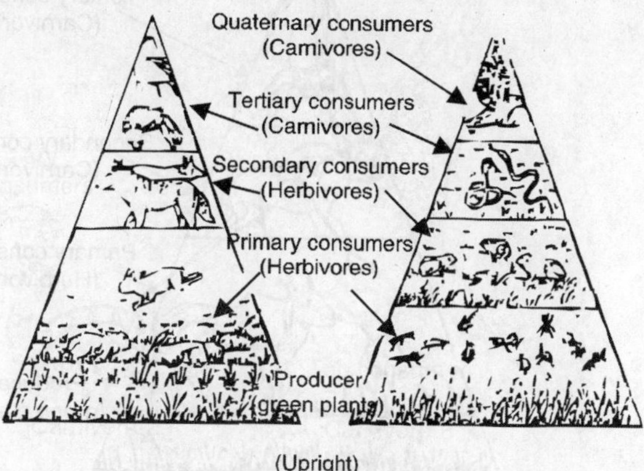

Fig. 3.8 Pyramids of Numbers

Here the producers which are mainly the phytoplanktons as algae, bacteria etc. are maximum in number, the herbivores which are smaller fish are lesser in number than producers. The secondary consumers are lesser in number than herbivores. Finally the top consumers (tertiary) are least in number.

In a forest ecosystem, however the pyramid is inverted, Fig. 3.9 illustrates an instance where the number of primary producers (a tree) is less than that of herbivore birds feeding upon the tree fruits. The number of parasites like bugs and lice living and feeding upon the birds body is still higher. Thus depending upon the size and biomass the pyramid of numbers may not be always pyramidal, it may even be completely inverted in shape.

Odum (1971) has studied the grassland and forest ecosystems by collecting the data from USA and England respectively.

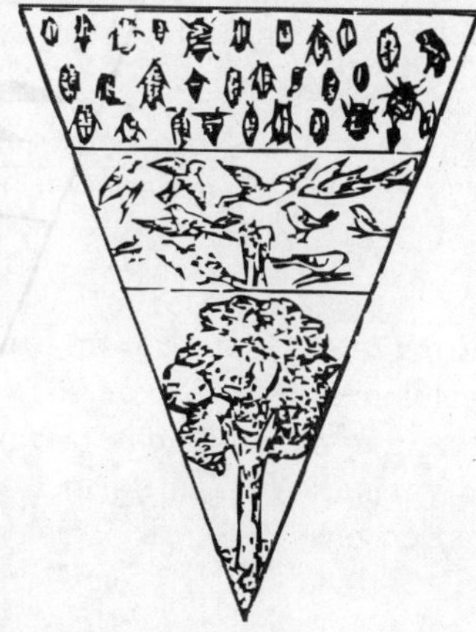

Fig.3.9 Pyramid of numbers

2. Pyramid of Biomass. Pyramids of biomass are comparatively more fundamental, as they instead of geometric factor, show the quantitative relationships. In order to explain the inverted nature of a pyramid of numbers, the idea of pyramid of biomass is given where the weight of primary producers forms the base. In figure 3.10 the ecosystem is shown where the pyramid of biomass is upright. The biomass of one tree is very high. The biomass of a number of birds feeding upon the tree is far less than that of the tree. Similarly, the biomass of even a very large number of parasite in and on the body of the birds is far less. Thus the pyramid of biomass, therefore, becomes upright. But there can be instances where the pyramid of biomass also get inverted as shown in figure 3.11. The biomass of phytoplanktons is quite negligible as compared to the small herbivores i.e. fish that feed on them. The biomass of large carnivores (Fish) feeding on small fishes is still higher. Harvery (1950) studied that in English Channel the biomass of primary producers is only 4 g/m^2 whereas that of the consumers is 21 g/m^2. This is the case in most aquatic bodies.

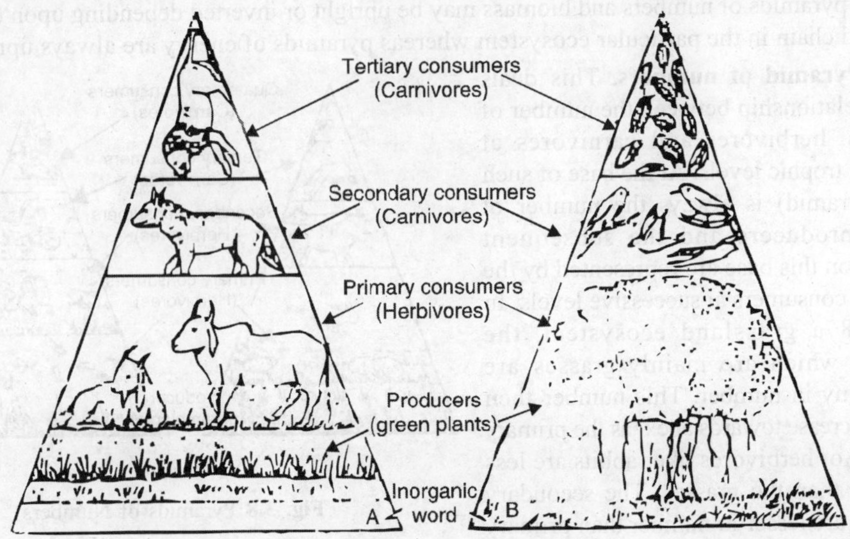

Fig. 3.10 Pyramids of Biomass

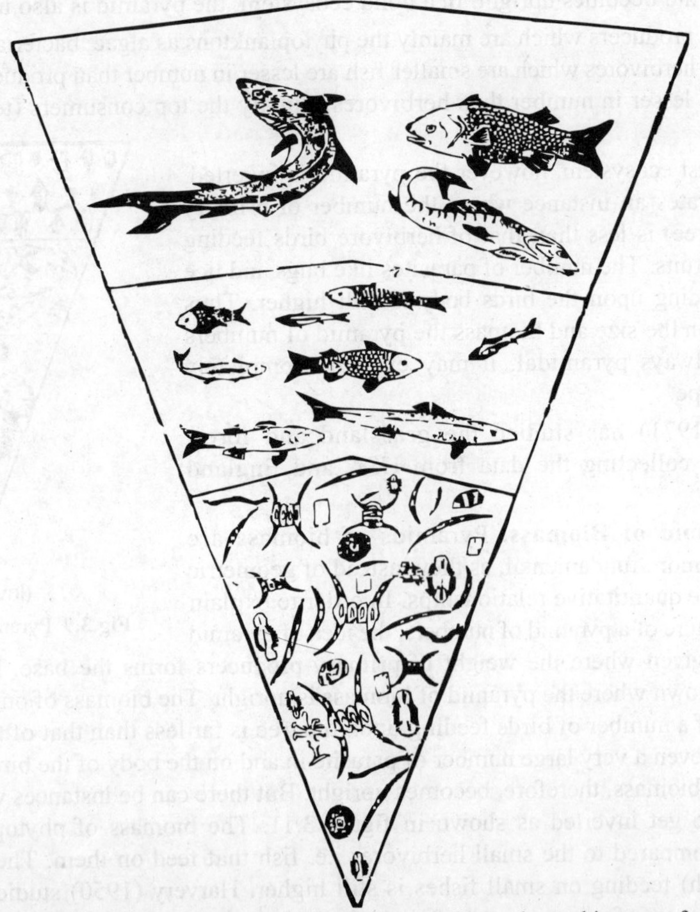

Fig. 3.11 Inverted pyramid of biomass in an aquatic ecosystem due to lower biomass of phytoplanktons than of consumers in unit volume of water at any one time.

3. Pyramid of energy

Of the three types of ecological pyramids, the energy pyramid give the best picture of overall nature of the ecosystem. As against the pyramids of numbers and biomass the shape of the pyramid of energy is always upright, because in this the time factor is always taken in to account. The pyramid of energy represent the total quantity of energy utilized by different trophic level organisms of an ecosystem per unit area over a set period of time. The base upon which the pyramid of energy is constructed is the quantity of organisms produced per unit time or the rate at which food material passes through the food chain. Energy pyramids are always slopping (upright) because less energy is transferred from each level than was paid into it. In figure 3.12 organisms of the terrestrial and an aquatic ecosystems are shown. The quantity of the energy trapped by green plants in an area over a period is highest compared to that of organisms of other trophic levels and therefore the base of the pyramid

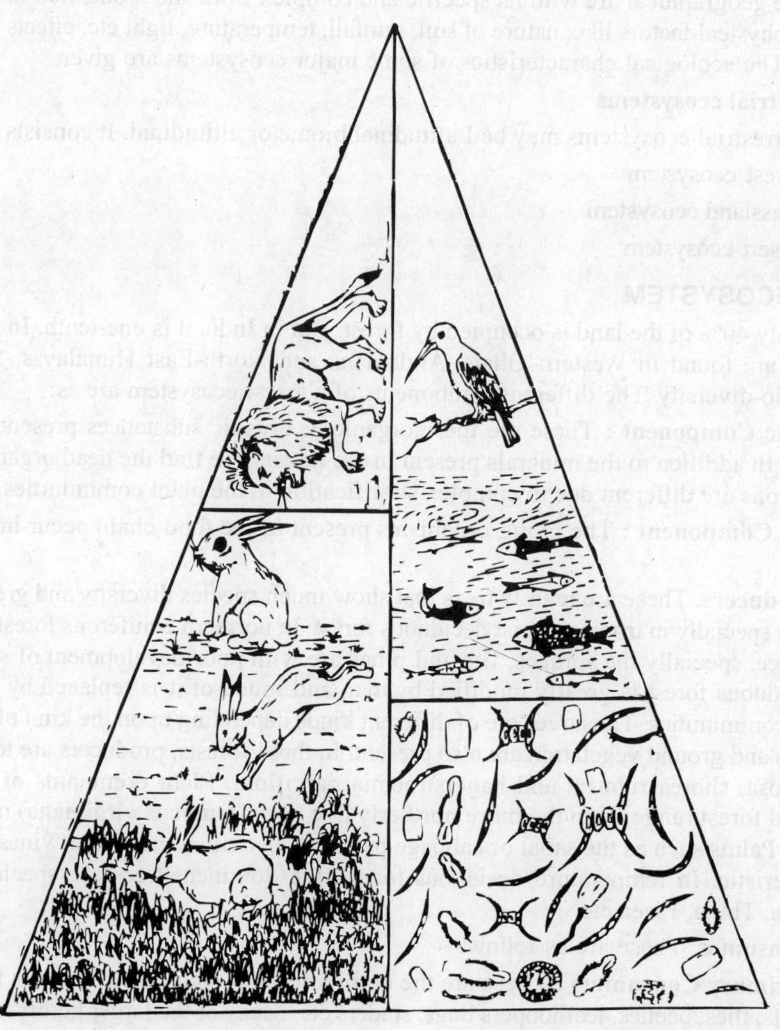

Fig. 3.12 Pyramids of energy in grassland and aquatic ecosystems. The cumulative energy contents utilized by primary producers is always higher as compared to energy utilization by successive trophic levels, over a period of time in a given area.

is broad. The population of phytoplanktons in aquatic ecosystem also complete their life cycle and sets of new generation in every few hours or days. The cumulative energy content of these generations of phytoplanktons trap in course of a year is certainly much more than that of only a few generations of herbivore fishes in the corresponding time and space. The energy content of top carnivores (utilized in one year) is the least. Therefore, the pyramid of energy is upright. The ratio of the amount of energy absorbed and the amount of energy which would be retained in biomass is known as ecological efficiency.

SOME MAJOR ECOSYSTEMS

There are three types of ecosystems in nature—

1. Terrestrial ecosystem
2. Freshwater ecosystem
3. Marine ecosystem

A large geographical are with its specific and complex flora and associated fauna is called a **biome.** The physical factors like, nature of soil, rainfall, temperature, light etc. effects the vegetation of a biome. The ecological characteristics of some major ecosystems are given.

Terrestrial ecosystems

The terrestrial ecosystems may be Latitudinal biome or altitudinal. It consists of

(*i*) Forest ecosystem

(*ii*) Grassland ecosystem

(*iii*) Desert ecosystem

FOREST ECOSYSTEM

Roughly 40% of the land is occupied by forest. But in India it is one-tenth. In India, tropical rain forests are found in Western Ghats, Andamans and North-East Himalayas. So these have maximum bio-diversity. The different components of a forest ecosystem are as:

Abiotic Component : These are the inorganic & organic substances present in the soil & atmosphere. In addition to the minerals present in the forests, we find the dead organic debris. The light conditions are different due to complex stratification in the plant communities.

Biotic Component : The living organisms present in the food chain occur in the following order—

1. Producers. These are mainly trees that show much species diversity and greater degree of stratification specially in tropical moist deciduous forest. In northern coniferous forest needle leaved evergreen tree, specially the spruces, firs and pines are with poor development of shrub and herb layers. Deciduous forest is greatly modified by man and much of it is replaced by cultivated and forest edge communities. Thus trees are of different kinds depending upon the kind of forest. Beside trees, shrubs and ground vegetation are also present. In these forests, producers are tectona grandis, butea frondosa, shorea rubusta and hagerstroemia parviflora. Plant dominants of the evergreen broad leaved forest range from the more northerly live oaks (Ouercus virginiana) magnolias bays and hollies. Palms such as the sabal or cabbage palm are also often prominent vines and epiphytes are characteristic. In temperature deciduous forests the dominent trees are species of Ouercus, Acer, Betula, Thuja, Picea etc.

2. Consumes. These are as follows—

(a) Primary Consumers : These are the herbivores that include the animals feeding on tree leaves as ants, flies, beetles, leafhoppers bugs, spiders etc. Many of the larger herbivorous vertibrates like moose, snowshoe have grouse are found on broad leaved developmental communities. Similarly some animals like elephants, nilgai, deer, moles, flying foxes, fruitbats, mongooses etc. are grazing on shoots and/or fruits.

(b) Secondary Consumers : These are the carnivores like snakes, birds, lizards, fox etc. feeding on herbivores.

(c) Tertiary Consumers : These are the top carnivores like lion, tiger etc. that eat carnivores of secondary consumers level.

3. Decomposers. These are wide variety of micro-organisms like actinomycetes (streptomyces), bacteria (Bacillus, clostridium, Pseudomonas etc.), Fungi (species of Aspergillus, Coprinus, Polyporus, Fusarium, Trichoderma etc.) Rate of decomposition in tropical and subtropical forests is more rapid than that in the temperate ones.

2. Grassland Ecosystem

This type of terrestrial ecosystem occupy roughly 19% of the earth surface. Grasslands dominated by grass species but some times also allow the growth of a few trees and shrubs. Rainfall is average but erratic. These are three types of grasslands depending upon climatic regions—

(i) Tropical grassland. Tropical Biomas (grasslands with scattered trees or clumps of trees) are found in warm regions with 40-60 inches of rainfall but with a prolonged dry season when fires are an important part of the environment. The largest area of this type is in Africa (Fig.3.13) but sizable tropical biomes or grassland also occur in South America and Australia. In Africa these are known as **Savannas.** Grasses belonging to such genera as Penicum, Pennisetum, Andropogon and Imperata. A view of this African Savanna country, including grass scattered trees and mammalian herbivores is in fig. 3.14.

(ii) Temperate grassland. In US and Canada, these grasslands are known as **prairies,** in South America as **pumpas**, in Africa as **Velds** and in Central Europe and Asia as **Steppes.** These occur where rainfall is too low (between 10 - 30 inches) to support the forest life form but is higher than that which results in desert life forms. However grasslands also occur in regions of forest climate where edaphic factors favor grass in competition with woody plants. Temperate grasslands generally occur in the interior of continents. In North America tall grass prairies have now been replaced by grain agriculture.

(iii) Arctic Tundra. There are two tundra biomes covering large areas of arctic, one in the Palearctic and other in the Nearctic region. In both continents the boundary between tundra and forest lies further north in the west where climate is moderated by warm westerly winds. The ground remains frozen except for the upper few inches during the open season. The permanently frozen deeper soil layer is called **permafrost.** The tundra is a wet arctic grassland consists of lichens, grasses, sedges and dwarf woody plants.

The various components of the grasslands are

Abiotic Components. The elements like C, H, O, N, P, S, etc. are supplied by CO_2, water, nitrates, phosphates, sulphates etc. present in soil & atmosphere. In addition, some other elements are also present in traces.

Biotic Components

(i) Primary Consumers. The herbivores feeding on grasses are grazing animals as cows, buffaloes, deers, sheep, rabbit, mouse etc. Besides them some insects like leptocorisa, dystercus, oxyrhachis, etc. termites and millipeds etc also feed on the leaves of grasses.

(ii) Secondary Consumers. The animals like fox, jackals, snakes, lizards, birds etc. (Carnivores) feed on herbivores.

Sometimes the hawks feed on secondary consumers.

Decomposers. The microbes active in the decay of dead organic matter are different species of fungi, some bacteria and actinonycetes. They bring about the minerals back to the soil, thus making them available to the producers.

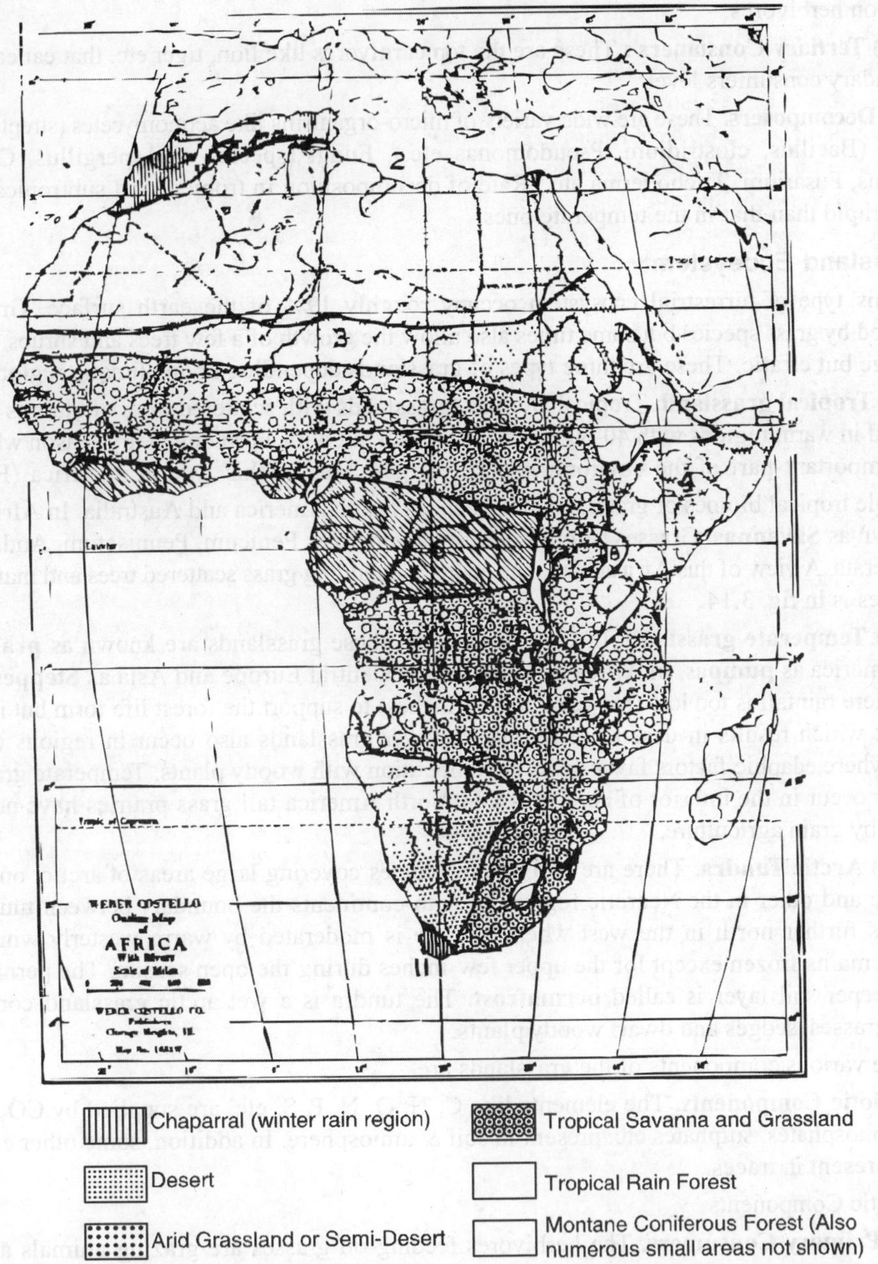

Fig. 3.13 The biotic regions of Africa. The six major biomes are indicated by the legend. Note that only a reltively small part of the continent is "jungle" (i.e., tropical rain forest). The very extensive tropical savanna and grassland is the fabulous "big game" country. Some of the larger faunal areas are numbered as follows: 1, Barbary; 2, Sahara; 3, Sudanese Arid; 4, Somali Arid; 5, Northern Savanna; 6, Congo Lowland; 7, Abyssinian Highland; 8, Kenya Highland; 9, Southern Savanna; 10, Southwest Arid; 11, Cape Town "winter rain" region. All of the faunal areas are part of the Ethiopian or African biogeographic region (see Figure 3.14), except the Barbary, which is Palearetic. Map based on sketch-map of Moreau, 1952.)

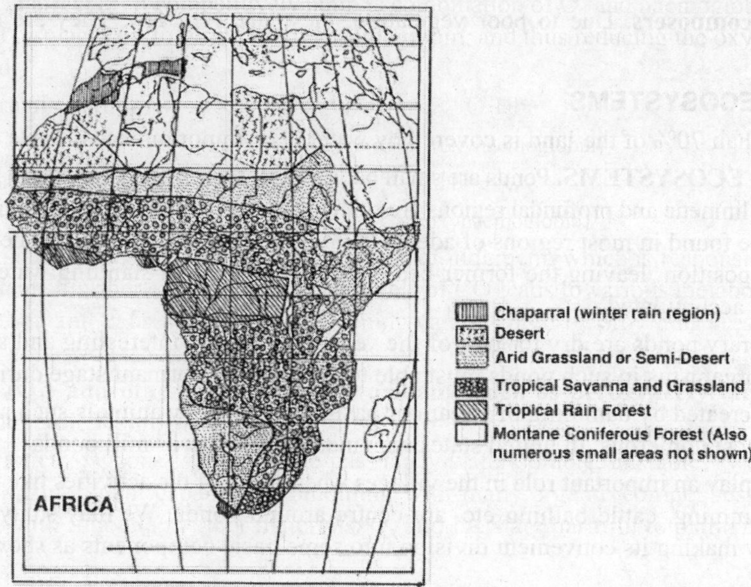

Fig. 3.14 A view of the tropical savanna of Africa. Grass, scattered trees, which have picturesque shapes, dry season fires, and numerous species of large mammalian herbivores are unique features of this photograph by Donald I. Ker, Ker & Downey Safaris Ltd., Nairobi, East Africa.)

DESERT ECOSYSTEM

Desert generally occur in regions having less than 10 inches of rainfall. Scarcity of rainfall may be due to—

(1) High subtropical pressure as in the Sahara and Australian desert.

(2) Geographical position in rain shadow

(3) High altitude.

About 1/3rd of our world's land area is covered by deserts. There are three life forms of plants that are adapted to deserts :

(*i*) The annuals, which avoid drought by growing only when there is adequate moisture.

(*ii*) The succulents, such as cacti, which store water

(*iii*) The desert shrubs

Based on the climatic conditions, deserts may be classified as

(*a*) Sahara, Namib in Africa, Thar, Rajasthan (India) are called **tropical desert,** which are driest.

(*b*) Mojave in Southern California is called **temperature desert** where days are very hot and cool in winters.

(*c*) Gobi desert in China is called **cold desert** where cold winters and warm summers.

The biotic components are

(1) Producers. As we have seen, these are shrubs bushes, some grasses and few trees. Some times, Cacti are also present. Some lower plants like lichens and Xerophytic mosses may also be present.

(2) Consumers. The most common animals are reptiles and insects. In addition to them, some nocturnal rodents and birds are also found. Camels "the ship of desert" feed on tender shoots of the plants.

(3) Decomposers. Due to poor vegetation, these are very few. They are some fungi and bacteria.

AQUATIC ECOSYSTEMS

More than 70% of the land is covered by water. The important ecosystems are

POND ECOSYSTEMS. Ponds are small bodies of water in which the littoral zone is relatively large and the limnetic and profundal regions are small or absent. Stratification is of minor importance. Ponds may be found in most regions of adequate rainfall. They are continually being formed, as a stream shift position, leaving the former bed isolated as a body of standing water where organic materials are accumulated.

Temporary ponds are dry for part of the year are specially interesting and support a unique community organisms in such ponds must able to survive in a dormant stage during dry period.

Ponds created by damming of stream or basin by man or by animals such as the bearer, are among the most numerous. In united states man made ponds were "mill ponds".

Ponds play an important role in the villages where most of the activities like washing clothes, bathing, swimming, cattle bathing etc. are centre around ponds. We may study the pond as an ecosystem by making its convenient division into some basic components as shown in fig. 3.15.

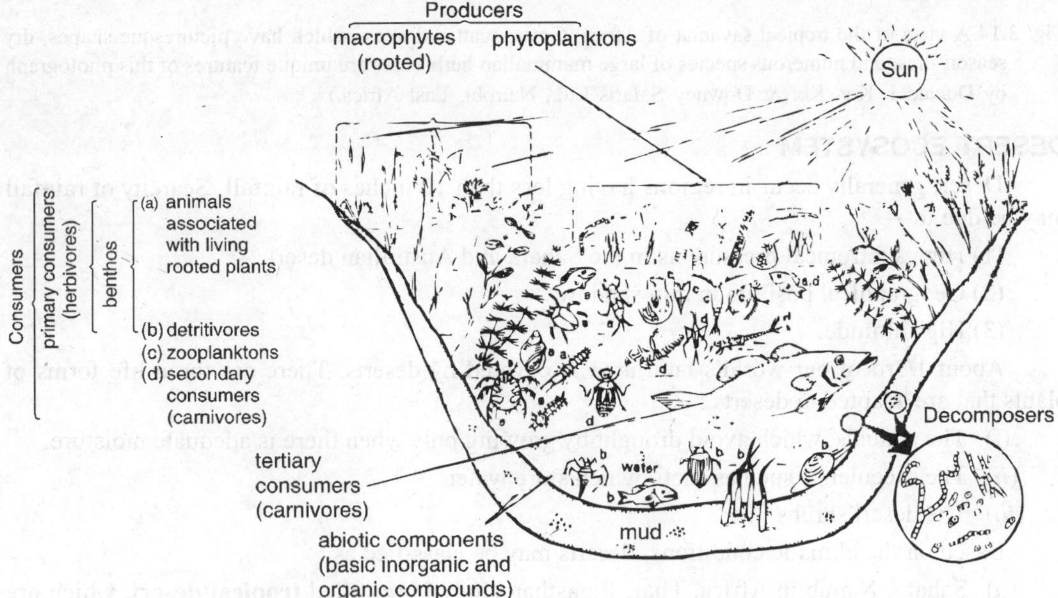

Fig.3.15 Diagram of pond ecosystem, showing its basic structural units—the abiotic (inorganic and organic compounds) and biotic (producers, consumers and decomposers) components.

These components are as

Abiotic Component. Apart from heat, light the basic inorganic and organic compounds, elements are water, CO_2, oxygen, calcium, nitrogen, phosphorus, amino acids etc. The amount of the minerals present at any time in the physical environment of the pond. "Standing state" may be estimated by appropriate methods. Light intensity and turbidity index of water at different depths can also be measured by lux-photometer and Sechhi disc respectively. pH of water and mud is determined by Electrical pH meter. DO, CO_2, phosphate, nitrogen, etc. can also be measured by appropriated methods. Carbohydrates, proteins, lipid etc. are also estimated for biomass determination.

Biotic Components They are as follows—

1. Producers. These are autotrophic, green plants and bacteria. They fix radiant energy and with the help of minerals from water & mud form complex organic substances like Carbohydrates, proteins & lipids. Producers are of the following types—

(a) Macrophytes. These are mainly rooted larger plants which include partly or completely submerged floating and emergent hydrophytes. The common species of the plants are Trapa, Typha, Sagittaria, Nymphaea, Chara, Hydrilla, Utricularia, Marsilea, Azolla, Sylvinia, Spirodella, Lemna etc.

(b) Phytoplankton. These are minute, floating or suspended lower plants like Ulothrix, Spirogyra, Cladophora, Oedogonium, Cosmarium, Eudorina Pandorina, Volvox, Chlamydomonas etc. and some flagellates. Biomass is estimated as weight of standing crop per unit area or volume. Generally, biomass and energy content of the vegetation decreases from the margin of the pond towards its centre. Energy content is generally expressed in terms of cal/gm dry wt.

2. Consumers. Most of the consumers are herbivores except insects and some large fish. But generally are heterotrophs. In pond consumers are distinguished as—

(*i*) Primary Consumers. These are herbivores, also known as "primary macro consumers" feeding directly on living plants. They may be large or in small size. They are further differentiated as—

(*a*) Benthos. These are the animals associated with living plants labelled as 'a' in fig and those bottom forms which feed upon the plants remains at the bottom labelled as 'b' in fig. Benthic population include fish, insect larvae, mites, molluses, crustaceans etc. Besides there some animals like cows, buffaloes and birds also visit the pond.

(*b*) Zooplanktons. These are chiefly the rotifers, (Brachionus, Lecane etc.), protozonas (Euglena, Coleps etc.) and Crustaceans (Cyclops, Stenocypris etc.). They feed on phytoplanktons lebelled as 'c' in fig.

(*ii*) Secondary Consumers. These are Carnivores like insects and fish which feed on primary consumers (herbivores) like Zooplanktons lebelled as 'd' in fig.

(*iii*) Tertiary Consumers. These are some large fish feed on smaller fish as shown in fig. In pond fish may occupy more than one trophic levels as shown in figure.

3. Decomposers. These are microconsumers, which absorb only a fraction of the decomposed matter. They decompose organic matter of both producers as well as microconsumers in simple forms. Thus they play an important role in return of mineral elements again to pond. The bacteria, actinomycetes and fungi (species Aspergillus, Cladosporium, Pythium, Penicillium, Circinella etc.) are most common decomposers in water and mud of the pond.

MARINE (OCEAN) ECOSYSTEM

The marine environment of seas and oceans is large occupying 70% of the earth surface. The volume of the surface area of marine environment lighted by sun is small in comparison to the total volume of water involved. All the seas are interconnected by currents, dominated by waves, influenced by tides and characterised by saline water. Each ocean indeed represents a very large and stable ecosystem. Oceans play an important role in regulating many biogeochemical and hydrological cycles, thereby regulating the earths climate. They have some major life zones i.e. coastal, Euphotic, Bathyal and Abyssal zones.

The **biotic components** of an ocean are as follows—

1. PRODUCERS

These are autotrophs, which are mainly the phytoplanktons. They trape radiant energy from sun through their pigments. A number of macroscopic seaweeds (Brown and red algae) are also come in this category. They are in distinct zones at different depths of water.

2. CONSUMERS

These are heterotrophic macroconsumers being dependent for their nutrition on the primary producers. These are

(*i*) The herbivores like Crustacions, molluscs, fishes etc. which feed directly on producers are called primary consumers.

(*ii*) The carnivores fishes like shad, herring etc. feeding on herbivores are called secondary consumers.

(*iii*) The top carnivores fishes like cod, haddock, halibut etc. that feed on secondary consumers are called tertiary consumers.

3. Decomposers.—The microbes active in the decay of dead organic matter are chiefly bacteria and some fungi.

ESTUARIES (ESTUARINE ECOLOGY)

Estuarine is derived from the word *aestus means tide.* Pritchard in 1967 defined as a semi-enclosed coastal body of water, which has a free connection with the open sea. It is thus strongly affected by tidal action and within it sea water is mixed with fresh water from land drainage. River mouths, coastal bays, tidal marshes and bodies of water behind barrier beaches are examples. Estuaries could be considered as transition zones or ecotones between the fresh water and marine habitats.

Not all rivers open in to estuaries, some simply discharge their run off in to ocean. Estuaries differ in size, shape and volume of water flow, all influenced of the region in which they occur. *Deltas* are by accumulation of sediments. When silt and mud accumulations becomes high enough to be exposed at low tide, then *tidal flats* are developed.

To illustrate estuaries, the different classifications will be represented based on

(1) Geomorphology (2) Water Circulation and stratification (3) Systems energetics.

According to Pritchard 1967, four subdivisions of estuaries are from zeomorphological point of view—

(*i*) Drowned river valleys

(*ii*) Fjord type estuaries

(*iii*) Bar - built estuaries

(*iv*) Estuaries formed by tectonic processes.

River - delta estuaries found at the mouths of large livers such as Mississippi or the Nile. It is different from formers. On hydrographic basis esturaries can be placed in three broad categories.

(*a*) Highly stratified or salt wedge estuary.

(*b*) The partially mixed or moderately stratified estuary

(*c*) The completely mixed or vertically homogenous estuary

The Hypersaline estuary is a special type.

Physico Chemical Aspects of Estuaries :

Current and salinity both are important here. Estuarine currents result from the interaction of a one direction stream flow which varies with the session and rain fall with oscillation ocean tides and with wind. The salinity varies vertically and horizontally and fluctuates amazingly between 0.5 to 0.35%. The water level in the estuary fluctuate regularly unlike that of river. No Echinodermates, Cephalopoda and other molluscs could survive in the estuaries. The temperature in estuaries fluctuate considerably biannually and sessionably.

The sessional and tidal cycles cause changes in nutrient concentration in the estuary. Any how, all estuaries have high productivity. The concentration of nutrients and fix carbon is very high level of production within the detritus food chain.

Biotic Communities of Estuaries

Carrikar in 1967 has classified the regions of estuaries in to upper, middle & lower reaches with increasing range of salinities and the mouth with salinity nearly equal to the sea. He has also classified the animals inhabiting the estuarine region into - oligohaline (.5 to 5%), mesohaline (5-18%). Krishnamoorthy has reported the extent of penetration of the polychacts in Madras from the Bay of Bengal. In Hoogly - Matla estuarine, Gopalkrishnan 1971 has reported an abundance of phytoplanktonic forms, several species of diatous synedra, Naricula etc. and blue green algae like microcystis, oscillatoria, adyar estuary of Madras coast is found rich in invertebrates and vertebrate fauna. In short, the esturine ecosystem is a complex and interesting one. It is also very vulnerable environment, because it has served as condiuts for shipping and as sites for cities.

LAKE ECOSYSTEM

Lakes are inland depressions containing standing water. They vary in size and depth (few feet to 5000 feet). Some lakes wave outlet streams. In lake there are three to five well recognized horizontal strata namely.

(*i*) **Littoral zone** — Shallow water near the shore forms this zone.

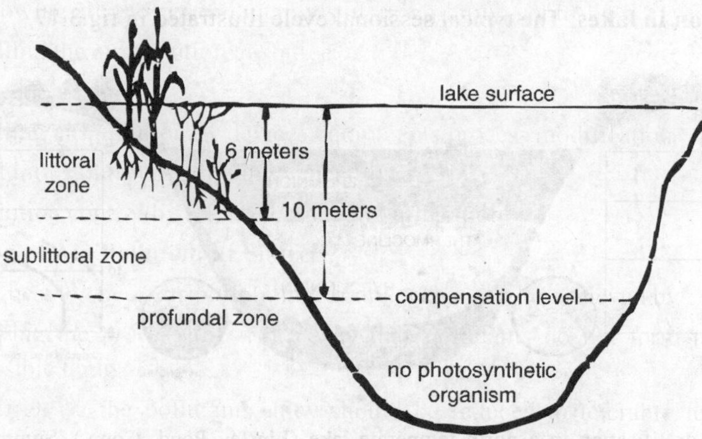

Fig. 3.16 Different zone of a deep freshwater lake.

It contains upper warm and oxygen rich circulating water layer, which is called *epilimnion.* It include rooted vegetation.

(*ii*) **Sublittoral zone.** It extends from rooted vegetation to the non circulating cold water with poor oxygen zone i.e. hypolimnion.

(*iii*) **Limnetic zone.** It is the open water zone away from the shore. It is up to the depth of effective light penetration where rate of photosynthesis is equal to the rate of respiration.

(*iv*) **Profundal zone.** It is the deep water area beneath limnetic zone and beyond the depth of effective light penetration.

(*v*) **Abyssal zone.** It is found only in deep lakes since it begins at about 2000 meter from the surface.

Kinds of lakes. Based on the physical factors, productivity etc. different classifications of lakes are given. Based on temperature, Hutchinson (1957) classified into *dimictic, monomictic* and *polymictic.* Based on Humic acid contents, the lakes are classified in to clear water lakes and Brown water lakes.

Physico-chemical properties of lakes

Lakes have the tendency to become thermally stratified during summar and winter to undergo definite seasonal periodicity in depth. Light too penetrates only to a certain depth, depending on turbidity.

Biotic Communities of lakes

Organisms depending on substratum are called *pedonic forms* and that are free from it called limnetic *forms*. The lakes have several type of organisms.

(*i*) **Neuston.** These including floating plants such as duckweeds and many type of animals. Animals are called **epineuston** while others including insects called **hyponeuston.**

(*ii*) **Plankton.** These are small plants and animals whose powers of self locomotion is very limited. Certain zooplanktons are very active some planktons are called as **nektoplanktons.**

(*iii*) **Nekton.** These animals are swimmers.

(*iv*) **Bethos.** These includes the organisms living at the bottom of the water mass. These living above the sediment water interface are termed **benthic epifauna** and those living in sediments itself are termed as **infauna.**

Stratification in lakes. The typical sessional cycle illustrated in fig.3.17.

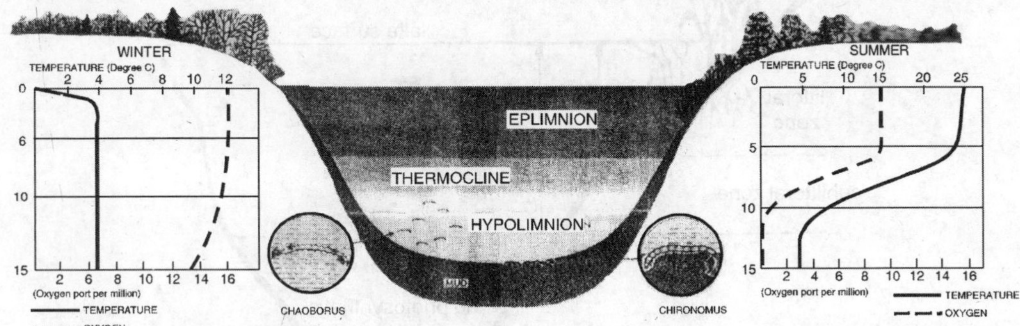

Fig. 3.17 Thermal stratification in a north temperate lake (Linsley Pond. Conn.). Summer conditions are shown on the right, winter conditions on the left. Note that in summer a warm oxygen-rich circulating layer of water, the epilimnion, is separated from the cold oxygen-poor hypolimnion waters by a broad zone, called the thermocline, which is characterized by a rapid change in temperature and oxygen with increasing depth. Two typical hypolimnion organisms are shown (see also Figure 3.18). (After Deevey, 1951).

During the summer the top water become warmer than the bottom waters, as a result only the warm top layer circulates and it does not mix with the more viscous colder water, called thermocline. The upper water layer is epilimnion. Colder noncirculating water is the hypolimnion. Subtropical lakes having surface temperatures that never fall below 4°C. In terms of water circulation patterns most of the lakes of the world can be conveniently assigned to one of the following categories (Hutchinson 1957).

(a) Dimictic (mictic = mixed) Two sessional periods of free circulation.

(b) Cold monomictic. Water never above 4°C (polar regions), seasonal overturn in summer.

(c) Warm monomictic. Water never below 4°C. One period of circulation in winter.

(d) Polymictic. More or less continually circulating with only short, if any, stagnation period.

(e) Oligomictic. Rarely mixed.

(f) Micromictic. Permanently stratified.

STREAMS

Biotic community in streams is quite different from that of ponds. Most streams in the vicinity of urban areas are polluted. Streams are fresh water aquatic systems where water current is a measure controlling factor, oxygen and nutrients are in water. Differences between streams and ponds revolve around a triad of conditions.

(i) Current. It is a major controlling and limiting factor in streams. Velocity of the current varies greatly in different parts of the same stream and from one time to another. In large streams

Fig. 3.18 Two streams illustrating the close interdependence of streams and the terrestrial watershed. A small woodland stream whose biota is almost entirely dependent on the import of leaf and other organic detritus from the forest B. A salmon stream in Alaska at a time of a salmon "run" when biomass and mineral nutrients are being "exported" by air-breathing predators (bears and gulls) that are feeding on the fish & soil Conservation Service Photo B. U.S. Dept. Interior, Fish and Wildlife Service Photo.)

the current may be so reduced that virtually standing water conditions result. Current is the most important primary factor which (1) makes for a big difference between stream and pond life and (2) governs differences in various parts of a given stream. The velocity of the current is determined by steepness of the surface gradient, the roughness of depth and width of the stream bed.

(ii) Land-water Interchange. The land-water surface junction is relatively great in proportion to the size of the stream habitat. This means that streams are more intimately associated with the surrounding land. (Fig. 3.18) than are most standing bodies of the water. Most streams depend on land areas and on connected ponds, back waters and lakes for a large portion of their basic energy supply. Streams have producers of their own such as fixed filamentous green algae encrusted diatoms and aquatic mosses. Some time plankton and detritus coming in to stream from quietor water. Streams form an open ecosystem that is interdigitated with terrestrial and leutic systems.

(iii) Oxygen. Stream organisms face more extreme conditions in regard to current and temperature. Because of the small depth large surface exposed to the air and constant motion, streams generally contain an abundant supply of the oxygen even when there is no green plants. For this reason stream animals generally have a narrow tolerance and are specially sensitive to reduced oxygen. Therefore, stream communities are specially susceptible to and quickly modified by any type of organic pollution, which reduces the oxygen supply, fig. 3.19

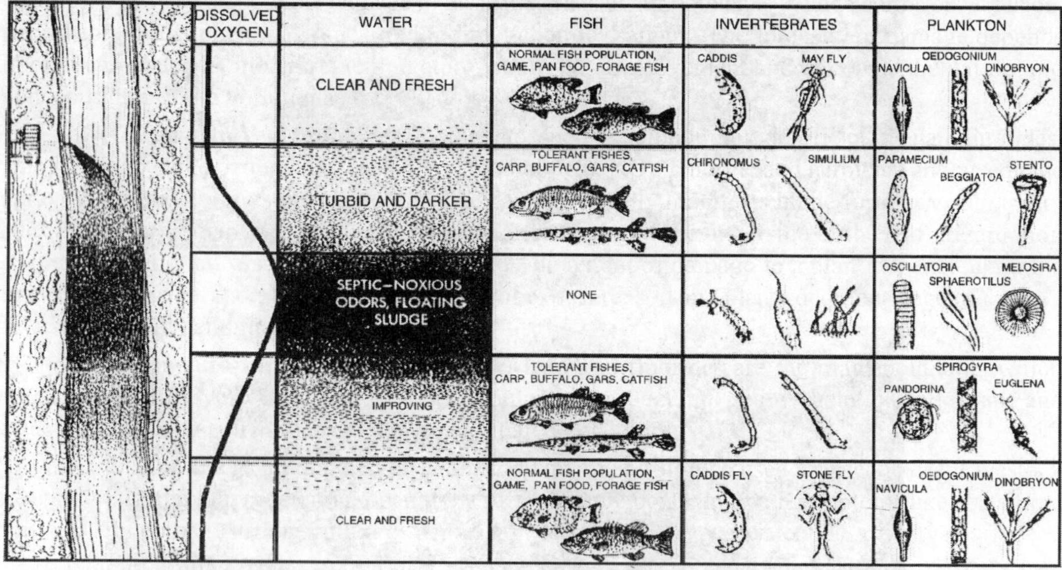

Fig. 3.19 Pollution of a stream with untreated sewage and the subsequent recovery as reflected in changes in the biotic community. As the oxygen dissolved in the water decreases (curve to the left), fishes disappear and only organisms able to obtain oxygen from the surface (as in Cutex mosquito larvae) or those which are tolerant of low oxygen concentration are found in zone of maximum organic decomposition. When bacteria have reduced all of the discharged material the stream returns to normal. (After Eliassen, Scientific American, Vol. 186, No. 3, March, 1952.)

ZONATION IN STREAMS

In streams zonation is longitudinal. In streams we find zones increasingly older stages from source to mouth. Charges are more pronounced in the upper part, because of gradient, volumes of flow and chemical composition charges rapidly. The change in composition of communities is likely to be more pronounced in the first mile than in the last fifty miles. The longitudinal distribution of fish in a stream may be selected as a specific example. Thompson & Hunt found that the number of individuals decreased down stream but the size of fish increased so that biomass density remained same.

STREAM COMMUNITIES

Streams generally exhibit two major habitats i.e. rapid & pools. Some within these categories the type of the bottom whether sand, pebbels, clay, bedrock or rubble_rock is very important in determining the nature of the community and population density. Current is the major factor for rapids but hard bottom may offer favourable surfaces for organisms. Benthic invertebrates have higher density in rapids but claws, burrowing odonata and ephemeroptera are more abundant in pools. Stream fish find refuges in pools and feed in or at the base of rapids. Planktons are at the mercy of the current organism in rapids and pools (in lesser extent) show adaptations for maintaining position in swift water - Some of these are Cladophora, Fontinalis, Diptera larvae Simulium, Blepharocera, Snails, flatworms, insect larvae etc.

QUESTIONS

Short answer type questions:–

1. Explain the concept of ecosystem.
2. Describe the structure of an ecosystem.
3. What do you mean by pyramid of energy?
4. Explain pond ecosystem.
5. Discuss Lake ecosystem in brief.
6. Write a note on ocean Ecosystem.
7. Write about food web.
8. Give a definition of ecosystem.

Long answer type questions

1. What do you mean by producers, consumers and decomposers ?
2. Taking example, describe the energy flow in ecosystems.
3. What do you mean by Food Chains & Foodwebs ?
4. Discuss the different types of ecological pyramids.
5. How many types of ecosystems are ? Discuss the grassland type of ecosystems.
6. Write an essay on “aquatic ecosystems”.
7. Write short notes on –
   (*i*) Ecological succession
   (*ii*) Pyramids
   (*iii*) Estuaries
   (*iv*) Streams
8. Write an essay on food chain.
9. Write an essay on structure of ecosystem.
10. Types of ecological succession.
11. Write detailed note on various proplic level of an ecosystem.

UNIT 4

Biodiversity and Its Conservation

INTRODUCTION

The term "Biodiversity" is short form of "Biological Diversity" and was coined by Walter G. Rosen in 1986 (Wilson 1994). According to Article-2 of Convention on Biodiversity (CBO). Biodiversity may be defined as, "Biological diversity means the variability among living organisms from all sources including, interalia, terrestrial, marine and other ecosystems and the ecological complexes of which they are part, this includes diversity within species between species and of ecosystem.

Biodiversity is neither the numbers of organisms present in any natural ecosystem nor a resource, but a property of living systems. According to Harvey B. Lillywhite (2002) it refers to "the variety and variability among living organisms and the ecological complexes in which they occur.

No one knows exactly how many species occur an our planet. Scientists believe that the total member of species on earth is in between 10 million to 80 million (Stork - 1988, Wilson 1988). We have been able to enlist only 1.4 million species so far. Nature has taken more than 600 million years to develop this exceedingly complex spectrum of life on this planet. The existence of human race depends on health and well being of other life forms in the biosphere.

Biological diversity is the total variety of life on our planet. Total number of races, varieties or species i.e. the sum total of various types of microbes, plants, animals present in a system is referred as *Biological diversity or simply as Biodiversity.*

GENETIC, SPECIES AND ECOSYSTEM DIVERSITY

Biodiversity is usually analysed at three levels i.e. species, genetic and ecosystem, each of which has its own significance.

1. Diversity of Biotic Communities and Ecosystems : Depending largely upon the availability of abiotic resources and conditions of the environment an ecosystem develops its own characteristic community of living organisms. A small pond, for example, constitutes an ecosystem and possesses a set of flora and fauna different from a river which is another type of ecosystem. Different types of forests, grass-lands, lakes, ponds, rivers, wet-lands etc. represent diverse ecosystems each with a characteristic biotic community.

2. Diversity of Species Composition within a Community : The biotic component in an ecosystem may be composed of a few species only or a large number of species of plants, animals and microbes, which react and inter-act with each other and with the abiotic factors of the environment. The richness of species in an ecosystem is usually referred to as **Species diversity.**

3. Diversity of Genetic Organization within a Species : Within a species there are often found a number of varieties or races or strains which slightly differ from each other in one, two or a number of characters such as shape, size, quality of their product, resistance to insects, pests and diseases, ability to withstand adverse conditions of environment etc. These differences are due to slight variations in their genetic organisation. This diversity in the genetic make up of a species is referred to as **Genetic diversity.** A species with a large number of races, strains or varieties is considered to be rich and diverse in its genetic organization.

A species is usually the unit of classification in most of the taxonomic works and can be defined as a group of organisms genetically so similar to each other that they can interbreed and produce fertile off-springs. For example, Horses and Zebra are different species. However, they are genetically similar and can interbreed but the off-springs produced are infertile. A species is usually recognisably different from each other in appearance but sometimes the differences are so minute that it is difficult to distinguish a species from another.

Genes which are biochemical packages passed on by parents to their off-springs are similar in organisms belonging to a given species. However, subtle differences occur which are expressed as differences in size, colour, or sometimes invisible characters such as susceptibility to some disease, resistance to drought or low temperatures etc. These differences are used to classify individuals of a species into different varieties or races.

BIOGEOGRAPHICAL CLASSIFICATION OF INDIA

India is one of the 12 mega biodiversity countries in the world. The country is divided in to 10 biogeographic regions. The wide variety in physical features and climatic conditions have resulted in a diversity of ecological habitats like forests, grasslands, wetlands, coastal and marine ecosystems and deserts which harbour and sustain immense biodiversity.

Biogeographically India is situated at the trijunction of three realms namely Afro-Tropical, Indo-Malayan and Paleo-Arctic realms and therefore has characteristic elements from each of them. This assemblage of three distinct realms makes the country rich & unique in biological diversity. With only 2.4% of the land area, India accounts for 7-8% of the recorded species of the world. The studies of the distribution of biota (flora + fauna) are collectively called biogeography. There are two major approaches to the study of biogeography (Kendeigh - 1974), (I) descriptive or static biogeography (II) Interpretative or dynamic biogeography. India is characterised by variety of climate types and flora of different types are in its different parts. India has sufficient number of biomes which represent a sum total of the biological community interacting within single life Zone where climate is similar. The following 13 biogeographical regions have been identified in India :

1. Himalaya
2. The Desert
3. Deccan Peninsula
4. Malabar
5. Andaman Islands
6. Nicobar Islands
7. Gangetic Planes
8. Laccadive Islands
9. Maldive/Chagoas Island

10. Western Ghats
11. Burman/Bangalian forest
12. Marine Coast
13. Coromondal Mahanandian

Floristic (Botanical) Regions of India

The country has been divided in to the following nine floristic regions with respect to floral diversity :

(i) Western Himalayas : It extends from Kumaon to Kashmir and has annual rainfall up to 200 cm. Correspond to three climatic belts, there are three zones of vegetation.

(a) Submontane zone. It is constituted of tropical and sub tropical parts and extends up to 1500 meters altitude. It comprises mostly of Siwalik ranges. Snowfall does not occur. The plants like Shorea robustica, Dalbergia sissoo, Cedrela toona, Eugenia jambolano, Acacia Catechu, Butea monosperma (Dhak) Zizyphus etc. are found in this region.

(b) Temperate Zone. Above submontane zone extend temprate zone forests up to 3500 meter altitude. They are dominated by plant species like Acer, Ulmus, Rhododendron, Betula, Salix, Populus, Cornus, Bumus, Pinus, Taxus, Picea etc.

(c) Alpine Zone. It extends from 3500 – 4500 metres altitudes and is characterized with alpine forest vegetation. Most common tree species are Betula, Juniperus, Rhododendrous etc. and herbs like Primula, Potentilla, Polygonum etc.

(ii) Eastern Himalayas. It includes regions of Sikkim and NEFA and is characterised by more rainfall, less snow and higher temperature. This is also divided into the following three zones altitudinally.

(a) Tropical zone. Upto 1800 metres altitudes, this zone has tropical semi-evergreen or moist deciduous forests. These forests comprise the plants like *Shorea robusta, Acacia catechu, Delbergia sissoo, Terminalia, Albizzia, Cedrela, Dendrocalamus* (bamboo) etc.

(b) Temperate zone. This zone extends between 1800 metres to 3800 metres altitudes and has typical montane temperate forests which are dominated by oaks like *Michelia, Quercus, Pyrus, Symplocos, Eugenia,* etc., at lower levels and by conifers as *Juniperus, Cryptomeria, Abies, Pinus, Larix.*

(c) Alpine zone. Beyond the temperate zone, extends alpine zone upto 5000 meters altitudes. It has alpine vegetation including *Juniperus* and *Rhododendron* with its other typical flora.

(iii) Indus plains. This zone includes the arid and semiarid regions of Punjab, Rajasthan, Kutch, part of Gujarat and Delhi. The rainfall is less than 70 cm. The vegetation is tropical thorn forest in semi-arid region and is typical desert in the arid region. The plants of this zone are primarily xerophytic. The common plant species of this zone are *Acacia nelotica, Prosopis sp., Salvadora, Tecomella, Capparis, Tamarix, Zizyphus, Calotropis, Panicum, Saccharum, Cenchrus, Euphorbia.* etc.

(iv) Gangetic plains. This region extends over Uttar Pradesh, Bihar, Bengal and part of Orissa and is characterised by moderate amount of rainfall and most fertile (*i.e.*, alluvial) soils. Vegetation of this zone is chiefly of tropical moist and deciduous and dry deciduous forest type. The common plants of this zone are *Dalbergia sissoo, Acacia nelotica, Saccharum munja, Butea monosperma, Madhuca indica* (mahua), *Terminalia arjuna* (arjuna), *Buchanania lanzan* (chiraunji), *Diospyros melanoxylon* (tendu), *Cordimyxa* (lisora), *Acacia catechu* (khair), *Azadirachta indica* (neem), *Ficus bengalensis* (bergad), *Ficus religiosa* (pipal), *Mangifera indica,* etc.

(v) Central India. It comprises Madhya Pradesh, parts of Orissa and Gujarat. The rainfall is 150–200 cm and its vegetation is thorny, mixed deciduous and teak type. The chief plants of this region are *Tectona grandis, Madhuca, Diospyros, Butea, Dalbergia, Terminalia, Carissa, Zizyphus, Acafia, Mangifera,* etc.

(vi) Malabar (west coast). This region include western coast of India from Gujarat to Cape Comorin and has heavy rainfall. The forests are tropical evergreen in extreme west, semi-evergreen towards interior subtropical or montane temperate evergreen forests in Nilgiris and mangroves near Bombay and Kerala coast.

(vii) Deccan Plateau. This region extends all over peninsular India (*i.e.*, Andhra Pradesh, Tamil Nadu and Karnataka) and has rainfall upto 100 cm. Its central hilly plateau has tropical dry deciduous forests of *Boswellia serrata, Tectona grandis* and *Hardwickia pinnata,* while, the low eastern dry Coromandal Coast has tropical dry evergreen forests of *Santalum album* (chandan), *Cedrelā toona* and plants like *Acacia, Prosopis, Euphorbia, Capparis, Phyllanthus,* etc.

(viii) Assam. This region is characterised by heavy rainfall (200 to 1000 cm). The vegetation is either dense evergreen forest or sub-tropical. The evergreen forests include trees like *Dipterocarpus macrocarpu, Mesua ferrca, Shorea robusta, Ficus elastica,* etc., bamboos as *Bambusa pallida, Dendrocalamus hamiltonii,* etc., grasses like *Imperata cylindrica, Saccharum* sp., *Themedasp*., insectivorous plants as *Nepenthes* sp., and also epiphytes (ferns and orchids).

(ix) Andmans. This region possesses a varied type of vegetation: mangroves and beech forest at its coasts and evergreen forests of tall trees in the interior. Important plant species of this island are *Rhizophora, Mimusops, Calophyllum, Lagerstroemia,* etc.

India has large numbers of wetlands, mangroves and coral reefs to its credit.

VALUES OF BIODIVERSITY

Biodiversity is a valuable natural resource for the survival of man kind. Man has domesticated a number of economically important plants and animal species. Old traditional varieties and the mid relatives of domesticated plants and animals constitute a vital genetic resource for us. Many plants and animals including wild life are of very important for human being. They can be used directly or indirectly to have consumptives, productive, social, ethical, aesthetic & ophons values i.e. in terms of money.

Consumptive value. Most of the developing countries obtain fuel wood from forests. Still more than 1500 million people cook their food by burning wood. About 1000 million cubic meter wood is used for fuel across the globe. This imposes heavy pressure on forests. Hunting of wild life, use of grass with some commercially important plants as fodder are of only comptive.

Various tribal societies fully depend on forests (biodiversity) for their habitation and livelihood. They used tubers, roots, fruits, seeds and meat of wild animals as their food.

Productive Value : Bamboos, grasses, canes, essential oils, tanning material, dyes, gums, resin, drugs, spices, poisons, insecticides, soap substitutes, rudraksha, lac, honey wax, tusser, Mahua seeds, Mahua flower and other seeds are forest products, they have their high commercial values. In addition to these, various herbs and animal body parts are sold in commercial market, both at national and international levels. Some benefits like, water quality, recreation, education, scientific research, regulation of climate etc. are indirect values to biodiversity that provide economic advantages to the people without consumption of the resource.

Some plants have been found to have immense medicinal properties, few of them with their curative properties are given below.

Name of medicinal plant	*Medicinal derivative*	*Curative property*
Cinchona	Quinine	Treatment of malaria
Isabgoal	husk and seeds of Isabgoal	Laxative, useful in chronic diarrhoea and dysentery
Opium poppy	Morphine, codeine and narcotine	Mental problems and cough
Brahmi	Juice of leaves and and stems	Repairs loss of memory
Ashwagandha (*Rauolfia serpentina*)	Serpentine	Urine problems
Basil (Tulsi)	Leaf-extract	Cough and cold, fever
Chalmogra	Seeds	Leprosy
Jambul	Bark-extract	Asthma and Bronchitis
Kalmegh	Root-extract	Liver tonic
Doob	Leaf-extract	Antiseptic

The lack of marketing facilities, lack of technical and financial support, involvement of middlemen in the business and large range of variation in the selling price of medicinal plants are keeping away farmers from the main trade.

Many pharmaceuticals have traditionally been derived from plants and animal sources. World-wide medicines from plants are worth over 40 billion dollars a year (Govt. of India 1991). Eighty percent of the people in tropical areas depend upon traditional medicines. Penicillin and tetracyclin and amongst the 3000 antibiotics extracted from micro-organisms. Guggal is an oleo-gum resin, long used in ayurvedic medicines for its anti-inflammatory, anti-rheumatic and hypo-cholesterolemic activity.

SOCIAL VALUES : Social value is one of the instrumental values where some thing has as a means to another's end. Materialistic uses of biodiversity are the core of instrumental values. The biodiversity has distinct social value attached with different societies. Goods and services provided by ecosystems to our society include

(1) Provision of food, fuel and fibber.

(2) Provision of shutter and building materials.

(3) Purification of air and water.

(4) Detoxification and decomposition of wastes.

(5) Generation and renewal of soil fertility, including nutrient cycling.

(6) Control of pests and diseases.

(7) Stabilization and moderation of earth's climate.

(8) Maintenance of genetic resources as key inputs to crop varieties.

(9) Live stock breeds, medicines and other products etc.

The charismatic species that have captured the public's heart and won their support for conservation. Biological resources are the pillars upon which we bind civilization. The loss of biodiversity threatens our existance i.e. social life. Thus protecting biodiversity is in our self interest.

These are the social values of biodiversity because biological resources provide the basis for life on earth including men. Fig. 4.1.

Human/Biodiversity Interrelationships

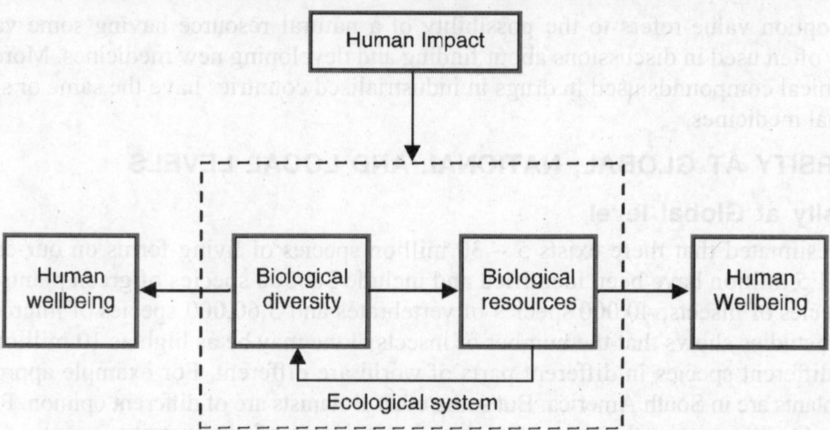

Ethical values : Ethical or religious values is also one of the indirect values of biodiversity. The ethical and religious value of biodiversity is rooted in the understanding that huminity is part of nature and that we are just one species among others. All species have an inherent right to exist. Future generations also have an inherent right to know them and to have the choice of using them or not.

Religions also have an significant impact on our attitude towards natural resources. The Buddhist perception of nature, for example, is based on different practices and approaches than that of Christian belief, though both are consistent with conserving biodiversity. Ethical value is one of the non-use values, which derive from human ethical considerations relating to matters such as the extinction of species and ecosystems.

Aesthetic value : The aesthetic value of biodiversity has been expressed in many ways through art, poetry, songs, literature, music and dance. Forests are closely linked with our relegion and culture. Human race has a great evolutionary attachment with forests as our ancestors lived in forests. Forests are nature's laboratories, where Scholars study natural sciences. Many types of trees are worshipped in tribal and Hindu societies i.e. Peepal, Bargad, Tulsi etc. Some animals like cow is worshipped by Hindus in all over India. In series of this many birds, colourful butterflies, mammals have great aesthetic value for human beings. Eco-tourism generate large amount of revenue annually that gives the aesthetic value of biodiversity. In this tourism people far and wide spend a lot of money and time to visit wilderness areas, where they enjoy the aesthetic value of diversity.

OPTION VALUES : Biological resources existed in this biosphere are very important for human beings. The option value of biodiversity suggests that any species may prove to be a miracle species. It is the precious gifts of nature presented to us. Option value is the indirect value of a species to provide an economic benefit to human society at some point in near future.

Option value is the value or a person's willingness to pay (WTP) to preserve the option of having an irreplaceable resource available for future use. The concept of option value had infinitive appeal, but was less defined. Attempts have been made to integrate option value into the main body of consumer theory. Most literature on option value is primarily concerned with technical issues such as the sign/direction. But few studies like Greenley et al 1981, Chopra 1993, Pearce and Moran 1994 on the topic show that option value for biodiversity is high. In this value, the insurance benefit that is provided to society through the protection that a resilient ecological system provided.

Biodiversity is natural capital, and therefore supplies a stream of value to current & future generations. Secondly lowers the risk of adverse outcomes. This too involves a future perspective in that the risks being considered confront both current & future generations. This is the option value of biodiversity protection.

The option value refers to the possibility of a natural resource having some value in the future. It is often used in discussions about finding and developing new medicines. More than 70% of the chemical compounds used in drugs in industrialised countries have the same or similar uses in traditional medicines.

BIODIVERSITY AT GLOBAL, NATIONAL AND LOCAL LEVELS

Biodiversity at Global level

It is estimated that there exists 5 – 30 million species of living forms on our earth and of there only 1.5 million have been identified and include 300000 species of green plants and fungi, 800000 species of insects, 40,000 species of vertebrates and 3,60,000 species of microorganisms. But present studies shows that the number of insects alone may be as high as 10 million. The data related to different species in different parts of world are different. For example approx. 200000 species of plants are in South America. But some other Botanists are of different opinion. Precipitation and temperature are among the most important deterninants of biodiversity.

Terrestrial biodiversity of the earth is best described as biomes, which are the largest ecological units present in different geographic areas. It is also estimated that about 125000 flowering plant species in tropical forests, but only about 1 – 3% are of these are known. The table 4.1 shows the estimated number of species worldwide. Millions of species of plants, birds, amphibians, insects as well as mammals are in the tropical rainforests. It is said that, they are the earth's largest store house of biodiveristy. About 70% of global biodiversity lies in these rainforests. Many plants of these rainforests are of medicinal use. Tropical deforestation is reducing the biodiversity by half a percent every year.

Table 4.1 : Estimated number of species worldwide

Taxnomic group	*No. of spp.*
Bacteria	3600
Blue green algae	1700
Fungi	46983
Bryophytes	17000
Gymnosperms	750
Angiosperms	250000
Insects	750000
Sponges	5000
Custaceans	9000
Molluscs	38000
Star fishes	50000
Fishes	6100
Amphibians	19056
Reptiles	6300
Birds	9036
Mammals	4008

The tropical forests are regarded as the riches in biodiversity. According to the opinion of the scientists more than half of the species on the earth live in moist tropical forests, which is only 7% of the total land surface. Insects (80%) and primates (90%) make up most of the species.

The species diversity in tropics is high as :

In tropics as the conditions for evolution were optimum and for extinction fewer.

In tropics species diversity was conserved over geological time. Due to low rates of extinction prevailing there; and

Biological diversity is the result of interaction between climate, organisms, topography, parent soil materials, time and heredity.

However, these explanations need experimental observations and confirmation.

IUCN and several other world Authorities have identified 12 Megadiverse countries and made comparative studies on flora-fauna, their endemism and protection efforts etc. The countries identified are

1. Brazil
2. Colombia
3. Venezuala
4. Peru
5. Ecquador
6. Indonesia
7. Democratic Republic of Congo (Zaire)
8. India
9. China
10. Malaysia
11. Australia
12. Mexico.

BIODIVERSITY AT NATIONAL LEVEL

India is located in south Asia, between latitude 6° and 38° N and longitudes 69° and 97° E. The Indian landmass extending over a total geographical area of about 3029 million hectares, is bounded by Himalayas in the north, the bay of Bengal in the east, the Arabian sea in the west, and Indian Ocean in the South. The wide variety in physical features and climatic situation have resulted in a diversity of ecological habitats. This richness in biodiversity is due to immense variety of climatic and altitudinal conditions coupled with varied ecological habitats. The Indian region having a vast geographical area is quite rich in biodiversity with a sizable percentage of endemic flora and fauna. These vary from the humid tropical Western Ghats to the hot desert of Rajasthan, from the cold desert of Ladakh and the icy mountain of Himalayas to the warm costs of peninsular India.

Table 4.2 : Number of recorded biota in India

	Taxon	*No. of Species*
FLORA	Bacteria	850
	Algae	2500
	Fungi	23000
	Lichens	1600
	Bryophyta	2700
	Pteridophyta	1022
	Gymnosperms	64
	Angiosperms	17000
	TOTAL	**48736**

FAUNA		
	Protozoans	2577
	Porifera	519
	Cnidaria	237
	Ctenophora	10
	Platyhelminthes	1622
	Nematoda	2350
	Rotifera	310
	Kinoryncha	10
	Gastrotricha	88
	Acanthocephala	110
	Sipuncula	38
	Mollusca	5042
	Echiura	33
	Annelida	1093
	Onychophora	1
	Arthropoda	57525
	Phoronida	3
	Bryozoa	170
	Entoprocta	10
	Brachiopoda	3
	Chaetognatha	30
	Echinodermata	765
	Hemichordata	12
	Protochordata	116
	Fishes	2546
	Amphibians	204
	Reptiles	428
	Birds	1228
	Mammals	372
	TOTAL	**126188**

(based on available data)

In India, about 1,15,000 species of plants and animals have been identified and described. For example, the following crops arose in the country and spread throughout the world : rice, sugarcane, asiatic vignas, jute, mango, citrus, banana, several species of millets, several cucurbits, some ornamental orchids, several medicinal and aromatics. Infact, the country has been recognized as one of the world's top 12 megadiversity nations. This region is also a secondary centre of diversity for grain amaranthus, maize, red pepper, soybean, potatoes and rubber plant.

Table 4.3 : Biodiversity in animal species

Group	*Number of species*		*Percentage of*	*World*
	World	*India*	*Endemism*	*Percentage*
Mammals	4,231	372	8	8.79
Birds	12,450	1,200	4	9.63
Reptiles	6,300	435	33	6.90

Amphibians	4,184	181	62	4.32
Fishes	23,000	2,000		8.69
Insects	8,00,000	60,000		7.50
Molluscs	1,00,000	5,000		0.50

In flora, the country can boast of 45,000 species which accounts for 15 per cent of the known world plants. Of the 15,000 species of flowering plants, 35 per cent are endemic and located in 26 endemic centres. Among the monocotyledons, out of 588 genera occurring in the country, 22 are strictly endemic.

The North Eastern region boast of being unique treasure house of orchids in the country. The important Indian Orchids are *Paphiopedilum fairieyanum, Cymbiadium aloiflium, Aerides crispum,* etc.

India is very rich in faunal wealth and has nearly 75,000 animal species, about 80 per cent of which are insects. The distribution of major animal groups are shown in the table 4.3.

In animals, the rate of endemism in reptiles is 33% and in amphibians 62%. Further there is wide diversity in domestic animals, such as buffaloes, goats, sheeps, pigs, poultry, horses, camels and yaks. Domesticated animals too have come from the same cradles of civilization as the major crops.

There are no clear estimates about the marine biota though the coastline is 7,000 km long with a shelf zone of 4,52,460 sq km and extended economic zone of 20,13,410 sq km. There is an abundance of seaweeds, fish, crustaceans, molluscs, corals, reptiles and mammals.

Information regarding other flora and fauna are patchy. Hundreds of new species may be present in the country awaiting discovery. The Western Ghats in Peninsular India, which extend in the southern states, are a treasure house of species diversity and has about 5,000 species. It is estimated that almost one-third of the animal varieties found in India have taken refuse in Western Ghats of Kerala alone.

BIODIVERSITY AT LOCAL LEVEL

The biodiversity at local level can be well understand by demarcating the points, places, zones rich in biodiversity. This can be understood as compositional i.e. rich in plants & animals of same habitats and genetic make up.

We can also study the local biodiversity on following lines

1. Richness of species at a given place.
2. Physical characteristics of habitat and vegetation in particular area.
3. Change in species composition across different habitats.
4. Local diversity based on climate, geographical, ecological and other processes responsible for creation.
5. Rate of change across gradients and conditions.

It is said that environmental variables are responsible for diversity but temperature play an important role in affecting the biodiversity of an area. Thus local areas are well affected in heterogenious and homogenious habitats.

INDIA AS A MEGA DIVERSITY NATION

Megadiversity concept covers the broad frame of biodiversity concept, but emphasizes more on species richness, threatened species and endemic species. The hot-spots concept emphasizes more on the exceptional concentration of endemic species besides the imminent threats of habitat destruction (**Normal Meyer**).

Megadiversity, a term used by International organization (refer World Bank Technical paper no. 343). Megadiversity nations are Maxico, Colombia, Ecuador, Peru, Brazil, Zaire, Medagaskar, China, Malaysia, Indonesia, Australia and India (12 countries). Thus India is one of the 12 megadiversity nations. These countries have been considered to have the largest fraction of the world's species and with a high concentration of endemic species.

Two areas in India have been identified as megadiversity hot spot areas, which are western ghat forests and Eastern Himalayan forests, although India as a whole has been marked as a megadiversity nation. Miller Meier says, "India is remarkable in both species richness and endemism although it ranks 10th position."

The World Bank in their publication (Paper 343) made their observations based on terrestrial vertebrates, butterflies and higher plants. In case of fresh water fauna India holds a position among the first ten countries of the world.

The Eastern Himalayas from a humid region having high monsoon rainfall, milder temperature and less snowfall. The mighty mountains with their snow-peaks and extremely rich forests exert a tremendous influence on the flora and fauna of the region. Arunachal Pradesh is a land of mighty rocks and luxuriant forests, gentle streams and raging torrents. It presents a breath-taking spectacle of nature in her glory, beauty of gorges and galaxy of ethnic people make the area as one of the best in the world. The mountain ranges in Sikkim, Bhutan, Arunachal Pradesh, Nagaland, Manipur, Mizoram, Tripura, Meghalaya and the Darjeeling hills are symbol of celestial splendour where a good number of peaks rise well over 7000 m, the highest being the Kanchinjongha 8335m. which is very close to Mt. Everest, the World's highest peak.

The marshes and wetlands of this region have more than 200 species of aquatic and semi-aquatic plants, besides innumerable species of phytoplankton, zooplankton and benthic flora and fauna that cater to the need of man, animals and birds.

This zone has many more such rare qualities.

About 46000 plant species and 81000 animal species have been described from the 70% of total geographical area. Huge diversity found in India is mainly due to favourable environmental conditions. India has 10 biogeographical regions.

India's biosphere are well placed, cultural diversity is well examplified in its different religions, languages, traditions, festivals etc. Ayurveda, Unani, Homeopathys & herbal preparations (for pharmoceutical & cosmetic purposes) all are based on plants. They are parts of traditional biodiversity in India. Biodiversity is an important strength of India. Out of world's one lac species of insects, 60 thousands are found in India. The two hot-spots of India have been identified habitat for 5332 endemic species of mammals, reptiles, amphibians, birds and higher plants. Many crops like rice, sugarcane, mango, jute, citrus, banana, bazra, jwar etc. arose in India and spread throughout the world. A large proportion of the Indian biodiversity is still unexplored. 7500 km long coastline of our country rich in coral reefs, mangroves, estuaries, molluscs, crustaceans, polychaetes, marine algae etc. Nearly 5000 species of flowering plants had their origin in India.

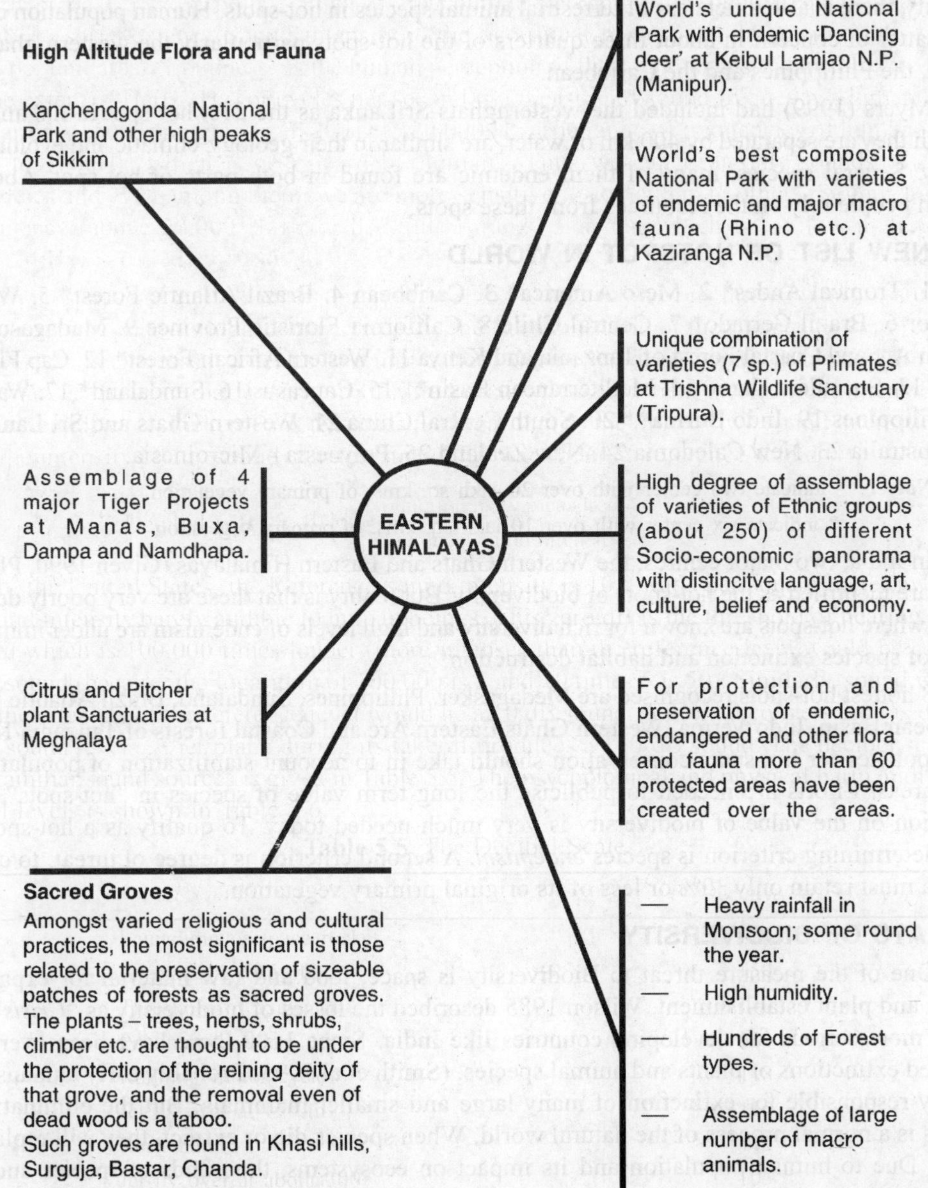

HOT-SPOTS OF BIODIVERSITY

The explosive population growth is one of the prime causes of biodiversity loss (Soule – 1991). The most threatened of all biologically rich areas are called Hot-Spots. Scientists of Conservation have mapped 25 global biodiversity hot-spots. Hot-spots are identified by two main criteria, first plant endemism and then degree of threat. Within them, are housed at least half of the world's terrestrial species. Population growth and migration in these species rich regions have made conservation efforts more urgent. In 1995 more than one billion people were living in hot-spots. Population growth in these areas have been proceeding at two and a half times the rate of the world's population. Ecosystems within the hot-spots cover less than 2% of earth's land surface, over 131,000 species are found within them as endemics. Because of the prevalence of high plant diversity, there is also a richness of terrestrial animal species in hot-spots. Human population density is a matter of concern in about three quarters of the hot-spots particularly the western ghats, Sri Lanka, the Philippines and the Caribbean.

Myers (1999) had included the westernghats SriLanka as the 21st hot spot in his analysis. Though they are separated by 400 km of water, are similar in their geology, climatic and evolutionary history. Several species, many of them endemic are found in both parts of hot-spot. About 18 million people rely on the resources from these spots.

THE NEW LIST OF HOTSPOT IN WORLD

1. Tropical Andes* 2. Meso America* 3. Caribbean 4. Brazil Atlantic Forest* 5. Western Ecuador 6. Brazil Cerrado* 7. Central Chile 8. California Floristic Province 9. Madagascar 10. Eastern Arc and Coastal Forest of Tanzania and Kenya 11. Western African Forest* 12. Cap Floristic Forest 13. Succulent Karoo 14. Mediterranean Basin** 15. Caucasus 16. Sundaland* 17. Wallacea 18. Philippines 19. Indo Burma** 20. South Central China 21. Western Ghats and Sri Lanka 22. SW Australia 23. New Caledonia 24. New Zealand 25. Polynesia - Microinesia.

Note : ** indicate two center with over 20 lakh sq. kms. of primary vegetation.
* indicate six center with over 10 lakh sq. kms. of primary vegetation.

In India, two major centres, the Western Ghats and Eastern Himalayas (Given 1990, Platnick 1991) are identified as the hot-spots of biodiversity. But reality is that these are very poorly defined. In fact where hot-spots are known for rich diversity and high levels of endemism are under immediate threat of species extinction and habitat destruction.

8 hottest hot-spots recognised are Medagasker, Philippines, Sundaland, Brazil Atlantic Coast, Caribbean Basin, Indo Verma, Western Ghats Eastern Arc and Coastal forests of Tanzania/Kenya. Govt. policies for hot-spots conservation should take in to account stabilization of population in these areas. Efforts are needed to publicise the long term value of species in "hot-spots". Thus education on the value of biodiversity is very much needed today. To qualify as a hot-spot, the main determining criterion is species *endemism*. A second criterion is degree of threat, to qualify an area must retain only 30% or less of its original primary vegetation.

THREATS OF BIODIVERSITY

One of the measure threat to Biodiversity is space, food and raw material for expanding human and plant establishment. Wilson 1985 described the losses of biodiversity as "*Crisis*", and this is more serious for developing countries like India. Since 1600 there have been over 1000 recorded extinctions of plants and animal species. (Smith et al 1993) Probably early humans were directly responsible for extinction of many large and smaller mammals. But the elimination of species is a normal process of the natural world. When species die or extinct, they will replace by others. Due to human population and its impact on ecosystems, thousands of species and sub-species become exinct every year. According to E.O.Wilson, we are losing 10,000 organisms a year i.e. 27 per day. If this will continue, we may destroy millions of plants, animals and microbes in next few decades. It is studied that 99% of all species of fossil that ever existed are now extinct.

Before man's appearance on this planet the rate of extinction was one species per thousand years. However, the pressure of human activity has drastically changed the picture. Between 1650 AD and 1950 AD about 30 species of higher animals were lost. Studies show that about 50,000 invertebrates species are losing every year. Almost one recorded as threatened. Indian wild life act 1972 schedule - I provides a list of about 150 endangered species. Disappearance of Dinosaurs along with about 50% of exististing species at the end of Cretaceous period is the best example of extinction. In India 33% of reptiles and 42% of bird species are endemic. It is said that, the current extinction rates are possibly 4 or 5 times more than the rates in the fossil record.

The following are the measure causes and issues related to threats to biodiversity.

HABITAT LOSS : Habit loss due to human activities and other disturbances are welknown factor. Varying human disturbances are changing ecosystems and are thus threatening the biodiversity. Due to habitat degradation wild populations become more volnerable to predators and diseases. This is especially true for wild life, which suffer due to habitat loss and fragmentation. Habitat loss is in instalments so that the habitat is divided into small and scattered patches i.e. *habitat fragmentation*. The natural forests and grasslands, which were the natural homes of thousands species including wild life species, are going cleared day by day for conversion into agriculture lands, pastures, settlements or for developmental projects. Thus these species are perished due to loss of their habitat.

Due to pollution and the presence of toxic and hazardous pollutants, our fresh water resources have suffered and many species of aquatic birds, fish and mammals have been threatened. Electric power plants, which causes thermal pollution in biosphere affected all aquatic communities and their natural food chains. Marine biodiversity is also under serious threat due to human intervention. If the present rate of deforestation continue, there will be loss of about 12% birds species and about 15% plant species in south and Central America (Simberloft 1984). Huge amount of habitat are lost each year as the world's forests are cut down. Rain forests, tropical dry forests, wet lands, mangroves and grasslands are threatended habitats and leading to desertification. Problems of acid rains and global climate change are also welknown for habitat loss.

POACHING OF WILDLIFE

Poaching is another threat to wildlife. As an ancient period, hunters, collectors, and smugglers (traders) are the measure threat to a number of species including endangered speices. They collected furs, hides, horns, tusks, and some live specimens, herbal products and smuggled to others for millions of dollars. The alarming point in this case is that for one animal they killed more than one. It is an illegal trade and internationally banned.

The cost of these animal parts are surprising. The cost of Bengal tiger coat is more than one lac dollars. South American ocelot cost more than 50,000 dollars, a single orchid cost more than 5000 dollars, horns of rhinoceros cost their weight in gold. These are some examples by which we can understand the situation of trading wildlife products, which is highly profit making for poachers.

Over collection and over exploitation are the main causes of disappearance of plants of scientific and medicinal value. The reduction of genetic diversity among the cultivated species drastically limit possibilities of creating new cultivar in the future, which could be disastrous for human race. It is advisable that do not purchase the parts and products made from wild animals specially endangered species.

MAN-WILDLIFE CONFLICTS

Struggle for existance. This is applicable for both, man and wild animal. Due to habit loss animals come out of the forest and destroy the crops later on they become danger to human being. Villagers and affected people kill them. There are so many cases of conflict between man and wild life. In these cases forest department could not pacify, resulted to lack of non co-operation for wild life conservation from affected people.

Animals are prone to infection when they are under stress. Animals held in captivity are also more prone diseases. The elephants and other wild animals suffer pain and turn violent when they come to destroy the electric fancied crop field. It is noted that ill, weak and injured animals have tendency to attack man. Man and wild life conflicts also occur during human encroachment into forest area.

There are member of cases, when man eating tiger was reported several men because they like human flesh rather than animal flesh.

ENDANGERED AND ENDEMIC SPECIES OF INDIA

ENDANGERED SPECIES OF INDIA

According to Red Data Book and Botanical Survey of India about 427 endangered plant species are identified. This is about 20% of India's total floristic wealth of higher plants. Examples of endangered plants species occuring in the North - Eastern region including the eastern Himalayas are as—

Eastern Himalayas	Acer laevigatum, A. molle, Anglica nubigena, Bunium nothum, Carum villosum, Pimpinella wallichii (Northeastem region also), Panax pseudogineseng, Calamus inermis, Phoenix rupicola (Northeastern region also), Lactuca cooperi, Berberis affinis, Engelhardtia wallichyiana, Coptis teeta, Aquilaria agallocha (Northerastern region also). Boehmeria tirapensis.
Northeastern region	Heracleum burmanicum, Peucedanum sikkimensis, Pimpinella evoluta, P. flaccida, Trachelospermum auritum, llex embeloides, I. khasiana, I.venulosa, Amorphophallus bulbifer, A. sylvaticus, Dioscorea laurifolia, D. arbiculata, Hopea shingkeng Musa velutina, Hedychtum aurantiacum, H. gracilimum, H. dekianum, H. gratum. H. greenii, H. hookeri, H. marginatum, H. rubrum.

The array of biological resources including their genetic resources are renewable in nature or in similar *ex-situ* conservation with proper management can support human needs indefinitely. Thus, it is appropriate to treat these as the fundamental sources for sustainable development. Their perpetuation in the existing forms and to the extent feasible on existing basis is therefore, essential to suitable explore factors from such a big reservoir for tomorrow's need.

Wetlands are transitional zone that occupy intermediate position between dryland and open water. It is estimated that India has about 4.1 million of wetlands, including natural and man made. Wetlands harbour enormous diversity of floral and faunal species, many of which are endangered. Some of the endangered animal species are listed in table 4.4. In India the number of official endangered animal species has risen from 13 (In 1952) to 250 (now). Some Indian bustard crocodile, muskdeer, blue whale, black buck, chinkara, wolf, nilgai, ante lope, tiger, flamingo, pelican white crane etc. Coral reefs are also under great threat. A species said to be endangered when its number reduced to a critical level or whose habitats have been drastically reduced.

Table 4.4 Some Endangered Animal Species in Wet Lands

	Species	*Common name*	*Wetland*
1	*Cervus eldii eldii*	Manipur brow-antlered deer or Sangai	Kalibul lamjao National Prak
2.	*Dugong dugon*	Dugong	Gulf of Mannar Andaman and Nicobar Islands
3.	*Cervus duvaucelii*	Swamp-deer or barasingha	Wetlands in Terai Assam
4.	*Prionailurus viverrinus*	Fishing cat	Swamps of Terai Himalayas, Sunderbans

5.	*Platanisia gangetica*	Gangetic dolphin	Ganges, Chambal and Brahmaputra rivers
6.	*Rhinocerõs unicornis*	Indian one-horned rhinoceros	Kaziranga National Park, Manas
7.	*Bubalus bubalis*	Water buffalo	Kaziranga National Park
8.	*Panthera tigris*	Bengal tiger	Sunderban National Park
9.	*Anser indicus*	Bar-headed goos	Wetlands of Ladakh
10.	*Grus leucogeranus*	Siberian crane	Keoladeo Ghana National Park
11.	*Houbaropsis Bengalensis*	Bengal florican	Wetlands of Manas National Park
12.	*Francolinus gularis*	Swamp Partridge	Wetlands of Manas National Park
13.	*Ceryle lugubris*	Crested Kingfisher	
14.	*Leptoptilos dubius*	Greater adjutant stork	
15.	*Leptoptilos javanicus*	Lesser adjutant stork	
16.	*Ardea insignis*	White-bellied heron	Rivers of Assam and Arunachal Pradesh
17.	*Phoeniconaias minor*	Asian lesser flemingo	Rann of Kutch, Sundarbans
18.	*Carina scutulata*	While-wingled wood duck	Assam and Arunachal Pradesh
19.	*Megapodius nicobarensis*	Megapode	Nicobar Islands
20.	*Anas gibberifrons albogularis*	Andaman grey teal	Andaman Islands
21.	*Crocodylus palustris*	Marsh crocodile or Mugger	Hiran lake in Gir National Park
22.	*Gavialis gangeticus*	Gharial	National Chambal Wildlife Sanctuary
23.	*Crocodylus porosus*	Estuarine crocodile	Bhitarkanika Wildlife Sanctuarv
24.	*Lepidochelys olivacea*	Olive ridley turtle	
25.	*Dermochelys coriacea*	Leather back turtle	Andaman and Nicobar Islands

Some important plants which have medicinal, ornamental, forestry, economical, scientific etc. values are also in endangered conditions due to their over exploitation.

ENDEMIC SPECIES OF INDIA

Endemism of India biodiversity is significant. About 4,900 species of flowering plants are 33% of the recorded floras are endemic to the country. These are distributed over 141 genera belonging to 47 families. These are concentrated in the floristically rich areas of North East India, the Western Ghats, North West Himalayas and the Andaman and Nicobar Islands.

The Western Ghats and the Eastern Himalayas are reported to have 1,600 and 3,500 endemic species of flowering plants, respectively. These constitute two of 18 hot spots identified in the world. It is estimated that 62% of the known amphibian species are endemic to India of which a majority occur in Western Ghats. Nearly 50% of the lizards found in India are endemic with a large number being found in Western Ghats.

Endemic species are the plants, which are limited in their distribution i.e. they are restricted to a small area and not found elsewhere in the world. It may be due to

- Poor adaptability of a species in a wide range of ecological
- Presence of some geographical barrier,e.g. Sea, Mountains etc.
- Failure of dispersal of reproductive organs (propagules, seeds, runners etc.)
- The species might have been comparatively young and not have enough time to spread.

Some examples of endemics include *Metasequoia* living gymnosperm endemic in China, *Sequoia* (red wood tree) endemic in coastal valleys of California, USA, *Primulla* and *Potentilla,* at high altitudes of Himalayas, *Ginkgo biloba* endemic in Japan and China.

Some of the endemic species in India are listed in table 4.5 :

Table 4.5 : Endemic species in India

	Group	*No. of Species*
Plants	Pteridophyta	200
	Angiosperms	4950
Animals		
Protozoa	Parasitic	500
	Free living	90
	Lepidoptera	9
Mollusca	Land and Freshwater	967
Pisces	Freshwater	64
	Marine	14
	Amphibia	123
	Reptilia	182
	Aves	60
	Mammalia	44

Indian sub continent has about 62% endemic flora restricted mainly to Himalayas, Khasi Hills and Western Ghats.

CONSERVATION OF BIODIVERSITY

The geological history of biodiversity is about 3.5 to 4 billion years old. The first appearance of multicellular organism was perhaps a mile stone in the history of biodiversity which did not diversify until about 600 million years ago. It is true that biodiversity is a source, once extinct can not be regenerated again. Due to human activities and over exploitation of ecosystem, the most severe extinction has occured. This is our one of the first and most important requirement i.e. conservation. From biological point of view, its our responsibility to conserve plants, animals (including cultivated plants and domestic animals, and their wild relatives.

In view of the importance of biodiversity (economic, environmental, scientific and medical) the urgent need for conservation of biodiversity was felt in Rio conference (1992). The World Summit on sustainable development held in Johannesburg in August 2002 reiterated that the conservation of biodiversity was necessary for the survival of human race on this earth. In the Convention on Biodiversity (CBD), which has 42 Articles, Article 8 and 9 are about in-situ conservation and ex-situ conservation respectively. The objectives of CBD clearly stated the conservation of biological diversity, the sustainable use of its components, and the fair and equitable sharing of the benefits arising from the utilization of genetic resources.

The conservation has to be ex-situ as also in-situ. Both are complimentary to each other.

INDIA'S EFFORTS FOR BIODIVERSITY CONSERVATION

Dr. M. S. Swaminathan (1983) reviewed the scientific aspects of conservation. He suggested that the first step in conservation should be defining the categories of materials (plants/genes) for preservation and the major methods preserving them. He suggested that the following categories should be usually regarded as important.

1. Cultivated varieties in current use.
2. Obsolete cultivars.
3. Primitive cultivars or land races.
4. Wild species and weedy species closely related to cultivated varieties.
5. Wild species of potential value to man.
6. Special genetic stock developed by man.

The principle of any technology designed for germplasm conservation should be to preserve the maximum possible genetic diversity of a particular plant or genetic stock for future use. Diversity in plants is at the level of species, varieties and individuals. Special genetic stocks may include the material (mutant or breed lines with identified gene or gene combination) developed and used in on going breeding programme. It has been estimated that the survival of approximately 9,000 wild species of plants is some way threatened and that the majority of these are from tropical regions. This further highlights the need for a positive approach to conservation of endangered plants. When new cultivars replace the primitive or conventionally used agricultural crops, it becomes especially important that the crop be properly documented and conserved.

According to Swaminathan (1983) conservation methodologies can take the following three forms :

1. Entire biomass
2. *In situ* preservation
3. *Ex situ* preservation

Many ways are being suggested for preserving biodiversity. Some of them include following :

1. No undisturbed land use.
2. Catalogues of genetic resources and national biological inventories be prepared.
3. Measures should be taken to reduced emission of green house gases and ozone destroying compounds.
4. Effective measures for the conservation of biodiversity be developed and strengthened in all countries.

The details of these conservation methods made a separate sub-discipline of biodiversity studies called *conservation biology.* These techniques of conservation of biodiversity include :

1. *In situ* conservation
2. *Ex situ* conservation

Conservation of biodiversity can be achieved in a number of complementary ways. All these methods, are within the broader concepts of gene banks.

EX-SITU CONSERVATION

Many protected areas and habitats are too small and subject to frequent changes to sustain the viable populations of species they seek to preserve. As human enterprize expands and wild life habitats shrink numerous life forms will be lost. Large land vertebrates which come in conflict with man shall be ultimately confined to wild life preserves. Whatever we do species shall continue to disappear and biosphere shall continue to lose its constituents one after the other in near future.

With wild life failing to survice in protected areas and natural habitats, conservation under human care appears to be the only solution. No doubt it is impossible to preserve all of the species. Thus affected but prudence demands that we should intensify our efforts for ex-situ conservation as well since it shall save at least some of the fragments of once extensive wild life for the future.

WHAT IS EX-SITU CONSERVATION

"Ex-situ" conservation : It means the wild-life conservation in captivity under human care. In this, the endangered plants and animals are collected and bred under controlled conditions in gardens, zoos, sanctuaries etc. wild-life management in captivity have the following advantages :

(*i*) The organisms are assured of food, water, shelter and security and hence can have longer life span and longer span of breeding activity, thereby increasing the possibility of having more number of offsprings.

(*ii*) The chances of survival of endangered species increase because of human care under secure conditions.

(*iii*) This offers the possibility of using genetic techniques to improve the species concerned.

However, there are some disadvantages and limitations of wild-life management in captivity :

(*i*) Since maintenance and breeding of plants and animals under captivity is very expensive, it can be adopted only for a few selected species.

(*ii*) Wild-life captivity only under a set of favourable environmental conditions deprives the organisms the opportunity to adopt to ever-changing natural environment. Therefore, new life forms cannot evolve and thus the gene-pool gets stagnant.

Ex-situ conservation, using sample populations, is done through establishment of gene banks, which include genetic resource centres, zoo's, botanical gardens, culture collections etc.

Ex-situ conservation is the chief mode for preservation of genetic resources, which may include both cultivated and wild material. Generally seeds or in-vitro maintained plant cells, tissue and organs are preserved under appropriate conditions for long term' storage as gene banks. This requires considerable knowledge of the genetic structure of population sampling techniques, methods of regeneration and maintenance of varietal genepools, particularly in cross pollinated plants.

The Methods of Ex-situ Conservation

The practice of ex-situ conservation involves technique which are essentially meant to maintain, multiply or help the species to survive under natural conditions. These include—

(*i*) Long term captive breeding

(*ii*) Short term propagation and release

(*iii*) Animal translocations

(*iv*) Animal Reintroduction

Instead of the above methods there are three techniques also used for ex-situ conservation. They are —

(*i*) Seed banks

(*ii*) Gene banks

(*iii*) In-vitro

(*iv*) In-vivo

Certain conservation agencies, botanical gardens and many research institutions are also working in this field. Establishment of "Central Zoo Authority" by Govt. of India is another milestone in the way of Ex-situ conservation.

CONVENTIONAL METHOD OF EX-SITU CONSERVATION BY SEEDS

Germplasm conservation in the form of seed is most convenient since seeds occupy a relatively small space. Their transport to various introduction centres and gene banks is also economical but the draw backs in conservation by seeds are :

1. Loss of viability over passage of time and susceptibility to insect or pathogen attack.
2. Inability to maintain distinct clones except for inbreed and apomict species.
3. Non-applicability to vegetatively propagated crop (*Dioscorea, Ipomoea, Potato* etc.).

Ex-situ is also an old method plants and animals are breeded under human care. However these captive breedings were rarely done for wild life conservation. Now, captive breeding and maintenance of wild animals and plants is important method for some life forms. In Ex-situ conservation first we identify the species then method is applied for conservation. In India there are three important gene bank/seed bank facilities—

1. National Bureau of Plant Genetic Resources (NBPGR) Delhi.
2. National Bureau of Animal Genetic Resources (NBAGR) Karnal, Haryana.
3. National Facility for Plant Tissue Culture Repository (NFPTCR)

In addition to these, National Bureau of Fish Genetic Resource (NBFGR), National Crop Genetic Resource Centre, International Centre for Genetic Engineering and Dept. of Biotechnology are also established for this purpose.

IN-SITU CONSERVATION

In-situ conservation is the conservation of ecosystem where all the flora, fauna and wildlife survive in tandem with nature. Development of gene sanctuaries, biosphere reserves, national parks, protected areas and reserve forests, where wildlife could grow and multiply, constitutes the vital forms of in-situ conservation. Development of biosphere reserves (in-situ conservation) has more than one advantage. It allows natural agencies of creating variation to act unabatedly, thus the range of natural variation continues to reshuffling and replenishing the plantation is never exhausted. The reserves are instrumental in preserving much of wildlife in the vicinity of human establishments.

Approximately 42% of the total geographical area of the country has been earmarked for extensive in-situ conservation of habitats and ecosystem. A protected area network of 45 National Parks and 448 wildlife sanctuaries has been created. The results of this network have been significant in restoring viable population of large mammals such as tiger, lion, rhinoceros crocodiles, elephants etc. Ten biodiversity rich areas of the country have been designated as biosphere reserves.

WHAT IS IN-SITU CONSERVATION

In-situ means in the natural, original place or position, as in the location of the explant on the mother plant prior to excision. In-situ conservation which include conservation of plant and animals in their native ecosystems or even in man made ecosystem, where they naturally occur. This type of conservation applies only to wild fauna and flora and not to the domesticated animals and plants, because conservation is achieved by protection of populations in nature. This method of conservation mainly aims at preservation of land races with wild relatives in which genetic exists and/or in which the weedy/wild forms present hybrids with related cultivars. These are evolutionary systems that are difficult of plant breeders to stimulate and should not be knowingly destroyed.

The in-situ conservation of habitats has received high priority in the world conservation strategy programmes launched since 1980. Institutional, arrangement, especially in countries of the developing world, have been emphasized. This mode of conservation has some limitations however; there is risk of material being lost due to environmental hazards. Further the cost of maintaining a large portion of available genotypes in nurseries or fields may be extremely high.

In-situ conservation includes a system of protected areas of different categories e.g. National parks, Sanctuaries, National Monument, Cultural landscapes, Biosphere Reserves etc. One of the best methods to save wildlife species, which is on the road to extinction, is to put it in a special enclosure to reproduce. Sanctuaries and National Parks, whose legal definition varies from country to country, best illustrate this.

NATIONAL PARKS, WILDLIFE SANCTUARIES AND BIOSPHERE RESERVES

National Parks or a Sanctuary may be defined *as an area, declared by state, for the purpose of protecting, propagating or developing wild life therein, or its natural environment for their scientific, educational and recreational value. Human activities like hunting, firewood, collection, timber harvesting etc. are restricted in these areas.*

The creation of National Parks, Wildlife Sanctuaries and Biosphere Reserves is an attempt to manage wildlife by defining protected areas. Wildlife therein is regularly monitored and necessary management strategies for their perpetuation and preservation are formulated and implemented. These protected areas not only benefit wildlife, but indirectly humans too. Their protection means the protection of entire ecosystem, which is necessary to continue to enjoy the benefits that we may now receive from it.

A National Park is an area dedicated to conserve the scenery (or environment) and natural objects and the wildlife therein. In national parks, all private rights are non-existent and all forestry operations and other usages such as grazing of domestic animals are prohibited. However, the general public may enter it for the purpose of observation and study. Certain parts of the park are developed for tourism in such a way that enjoyment will not disturb or scare the animals.

Sanctuaries are also of range between 100 sq. kms to 500 sq. kms. The boundaries of sanctuaries are often not well defined. Controlled biotic interference is permitted in sanctuaries which allow tourist activities as well.

The national reserve or biosphere reserve are usually large protected areas with boundaries circumscribed by legislation and are usually more than 5000 sq. kms in area. Except for a limited degree of biotic interference is permitted in biosphere. Here due attention is accorded to research and conservation of gene pool.

Table 4.6 : Important National Parks and Wildlife Sanctuaries in India

States	*National Parks and Wildlife Sanctuaries*
Andhra Pradesh	Pakhal Wildlife Sanctuary, PocharamWildlife Sanctuary, Kawal Wildlife Sanctuary, Kolleru Pelicanary
Arunachal Pradesh	Namidapha Wildlife Sanctuary*
Assam	Kaziranga National Park*, Manas Wildlife Sanctuary*
Bihar	Hazaribagh National Park, Betla National Park
Goa	Mollen Wildlife Sanctuary
Gujarat	Gir National Park, Velavadar National Park, Wild Ass Sanctuary, Nal Sarovar Bird Sanctuary
Haryana	Sultanpur Lake Bird Santuary
Jammu & Kashmir	Dechigam Wildlife Sanctuary
Karnataka	Bandipur National Park*, Nagarhole National Park, Ranganthitto Bird Santuary*, Silent Valley National Park
Kerala	Periyar Wildlife Sanctuary*, Wynad Wildlife Sanctuary
Madhya Pradesh	Kanha National Park*, Shivpuri National Park, Bandhavgarh National Park*, Panna National Park

Maharashtra	Tadoda National Park, Yawal Wildlife Sanctuary
Manipur	Keibul Lamjao National Park
Meghalaya	Balpakram Sanctuary
Mizoram	Dampa Wildlife Sanctuary
Nagaland	Intangki Wildlife Sanctuary
Orissa	Simlipal National Park*, Chilka Lake Bird Sanctuary
Punjab	Abohar Wildlife Sanctuary
Rajasthan	Ranthambore National Park*, Sariska Wildlife Sanctuary*, Ghana Bird Sanctuary
Sikkim	Kanchenjuga National Park
Tamil Nadu	Mudumalai Widlife Sanctuary, Vedanthangal Water Bird Sanctuary
Uttar Pradesh	Corbett National Park*, Rajaji National Park, Dudhwa National Park*
West Bengal	Jaldapara Wildlife Sanctuary

Table 4.7 : Biosphere Reserves in India

Biosphere Reserves	*State(s)*
Nilgiris	Tamil Nadu, Kerala and Karnataka
Namdapha	Arunachal Pradesh
Nanda Devi	Uttar Pradesh
Uttarkhand (Valley of Flowers)	Uttaranchal
North Islands of Andamans	Andmans & Nicobar
Gulf of Mannar	Tamil Nadu
Kaziranga	Assam
Sunderbans	West Bengal
Thar Desert	Rajasthan
Mannas	Assam
Kanha	Madhya Pradesh
Nokrek (Tura range)	Meghalaya
Little Rann of Kutch	Gujarat
Great Nicobar Island	Andmans & Nicobar

QUESTIONS

Short answer type questions:–

1. Define the term "Biodiversity".
2. What do you mean by genetic, species and ecosystem diversity ?
3. Define aesthetic value.
4. Discuss the biodiversity at local level.
5. What do you mean by poaching of wild life ?
6. What are the measure threats to biodiversity ?
7. What is Red Data Book ?
8. What is meant by alpha, beta and gamma richness ?

9. Write about Botanical regious of India.
10. Write an note on, "India is known as mega diversity. Nation."
11. Explain the importance of biodiversity.
12. Write note on endangered species.

Long answer type questions :–

1. Descrioe in detail the biogeographical classification of India.
2. Explain values of biodiversity. Give an account of different types of values.
3. Discuss the biodiversity at various levels.
4. Write an essay on "Hot-spots of biodiversity".
5. What do you mean by endangered and endemic species of India ? Give examples also.
6. What is conservation of biodiversity ? Describe in-situ and ex-situ conservation of biodiversity.
7. Explain "biosphere reserves". Write about the wild life sanctuaries also.
8. Write short notes on–
   (*a*) India as mega diversity nation.
   (*b*) Man-wild life conflicts
   (*c*) National parks
9. Write about the Endangered species of India.
10. Describe Biodiversity at Global level.
11. Describe various biogrographical regious of India.
12. Write an accourt of ecosystem diversity.

UNIT 5

Environmental Pollution

AIR POLLUTION

INTRODUCTION

According to U.S Public Health Service, "Air pollution may be defined as the presence in the outdoor atmosphere of one or more contaminants or combination thereof in such quantities and of such duration as may be, or may tend to be injurious to human, plant or animal life, or property, or which unreasonably interfere with the comfortable enjoyment of life, or property, or the conduct of business."

The atmosphere is a dynamic system, which steadily absorbs various pollutants from natural as well as man-made sources, thus acting as a natural sink. Gases such as CO, CO_2, H_2S, SO_2, and NO_x as well as particulate matter, such as sand and dust, are continually released into the atmosphere through natural activities such as forest fires, volcanic eruptions, decay of vegetation, winds and sand of dust storms. Man-made pollutants e.g., CO_2, NO_x, SO_2, CO_2, hydrocarbons, particulates etc are also released into the atmosphere. These have surpassed the pollutants contributed by nature thousand-fold. The magnitude of the problem of air-pollution has increased alarmingly due to population explosion, industrialization, urbanization, automobiles and others for greater comfort. The pollutants travel through the air, disperse and may interact with other substances in the atmosphere before they reach a sink, such as an ocean or a human receptor. If the pollutants enter the atmosphere at a faster rate than are absorbed by the natural sinks, then they gradually accumulate in the air. Such a disturbance in the dynamic equilibrium in the atmosphere by the air pollutants released by anthrapogenic activities resulting in considerable accumulation in the atmosphere may affect the very life on earth and its environment. Further, the dilution and dispersion of the gaseous pollutants in the atmosphere depend upon the meteorological conditions prevailing at a given time.

CLASSIFICATION OF AIR-POLLUTANTS

The air pollutants may be classified in different ways as follows:

(*a*) According to origin:

(*i*) Primary pollutants which are directly emitted into the atmosphere and are found as such, e.g., CO, NO_2, SO_2 and hydrocarbons.

(*ii*) Secondary pollutants which are derived from the primary pollutants due to chemical or photo-chemical reactions in the atmosphere, e.g., Ozone, Peroxy- acy1 nitrate (PAN), Photo-chemical smog, etc.

(b) According to chemical composition:

(*i*) Organic pollutants, e.g.,-Hydrocarbons, aldehydes, ketones, amines and alcohols

(*ii*) Inorganic pollutants

Carbon compounds (e.g., CO and carbonates)

Nitrogen compounds (e.g., NO_x and NH_3)

Sulphur compounds (e.g., H_2S, SO_2, SO_3 and H_2SO_4)

Halogen compounds (e.g., HF, HCI and metallic fluorides)

Oxidising agents (e.g., O_3)

Inorganic particles (e.g., fly ash, silica, asbestos and dusts from transport, mining, metallurgical and other industrial activities).

(c) According to state of matter :

(*i*) Gaseous pollutants which get mixed with the air and do not normally settle out, e.g., CO, NO_x and SO_2/

(*ii*) Particulate pollutants which comprise of finely divided solids or liquids and often exist in colloidal state as aerosols, e.g.,- smoke, fumes, dust, mist, fog, smog and sprays.

Biochemical effects of some important air pollutants

1. **Oxides of Sulphur (So_x)**

SO_x comprises of SO_2 and SO_3. They are Colourless, heavy water soluble with pungent and irritating odour. SO_x pollution is due to volcanic activity, combustion of fuels, Coal fired power stations, transportation, Refineries, metallurgical operations, chemical plants and other natural and human activities. In atmosphere, oxidation of SO_2 in to SO_3 by photolytic and Catalytic processes (in presence of O_3, NO_x or hydrocarbons) giving rise to the formation of photochemical smog. In humid conditions of the atmosphere SO_3 reacts with water vapours to produce droplets of H_2SO_4 aerosols, give rise to the so called "Acid Rain".

$$SO_2 + O_3 \ \lambda \rightarrow \ SO_3 + O_2$$

$$SO_3 + H_2O \ \lambda \rightarrow \ H_2SO_4$$

Biochemical effects: Absorbs quickly and irritates the upper respiratory tract. Reacts with cellular constituent chemicals e.g., enzymes. The H_2SO_4 formed lowers pH, impairs enzymatic functions and destroys various functional molecules. Leads to bronchial spasms, breathlessness, impaired pulmonary function via airway resistance, impaired lung clearance and increased susceptibility for infection.

2. **Oxides of Nitrogen (NOx)**

Characteristics: NOx mostly comprises of NO, NO_2 and N_2O. NO is colourless gas and is slightly soluble in water. NO_2 is reddish brown gas, somewhat water-soluble, oxidizing agent, can react with water to form HNO_3 which is a powerful oxidizing agent and capable of reacting with almost all metals and many organic compounds. NO_2 can travel into the respiratory system. It is also involved in the formation of ozone in the atmosphere.

Biochemical effects: Oxidises cellular lipids, Forms bonds with haemoglobin and reduces the efficiency of oxygen transport. Disrupts some cellular enzyme systems. Higher levels and prolonged exposures may cause pulmonary fibrosis, inflammation of lung tissues and may eventually lead to death. Causes nitric acid mediated effects some of which are similar to that of H_2SO_4 described above. NO can form addition compound with haemoglobin, if it enters the blood stream.

3. **Carbon monoxide (CO)**

Characteristics: Colourless, odourless, toxic gas, slightly water soluble but still is extremely dangerous because it has a greater affinity for haemoglobin than that of O_2.

Biochemical effects : It competitively inhibits combination of O_2 and haemoglobin. It attacks haemoglobin and displaces O_2 to form carboxyhaemoglobin, and thus reducing the oxygen carrying capacity of blood.

(Under normal conditions) : O_2 + Hb $\lambda \rightarrow$ O_2Hb (oxyhaemoglobin)

(In presence of CO) : O_2Hb + CO $\lambda \rightarrow$ CO Hb + O_2 (Carboxyhaemoglobin)

The immediate response to CO-poisoning is loss of judgment, which is responsible for many automobile accidents. Further exposure to higher levels of CO leads to various metabolic disorders such as asphyxiation and causes death. CO-poisoning can be cured by providing fresh O_2' which reverses the above reaction.

4. Ozone (O_3) and other photochemical oxidants such as peroxyacetyl nitrate (PAN) present in photochemical smog.

Characteristics : Ozone is a pale blue gas, fairly water soluble, unstable, sweetish odour. Very reactive oxidizing agent capable of combining with many organic compounds in cells and tissues as well as with rubber and other materials. O_3 and PAN are harmful to plants, animals and humans.

Biochemical effects : Oxidize cellular constituents. PAN and ozone toxicity is produced via generation of free radicals. The free radicals produced may damage DNA and thus alter cellular genetic integrity too. The toxic effects of ozone are manifested after inhalation and absorption in the lungs causing accumulation of fluids in the lungs (pulmonary edema), damaging lung capillaries and mortality if continued or high level exposures occur. Both O_3 and photochemical smog cause irritation of the eyes and respiratory tract. The free radicals produced by O_3 and other photochemical oxidants attack the sulphydryl groups on the enzymes and also inactivate enzymes like isocitric dehydrogenase, malic dehydrogenase and glucose-6-phosphate dehydrogenase, which are so much involved and essential for citric acid cycle and generation of cellular energy. O_3 may also inhibit the activity of some enzymes involved in synthesis of cellulose and lipids in plants. Among the sulphur containing amino-acids, cysteine is strongly attacked by PAN.

5. Hydrocarbons (and other volatile organic compounds)

Characteristics: Very reactive. React with many kinds of compounds yielding many kinds of products. Volatile hydrocarbons and other organic compounds participate in atmospheric reactions generating ozone.

Biochemical effects : Some of these compounds can react with the constituents of the cells. Carcinogenic hydrocarbons like benzopyrene can react with DNA causing mutations and cancer.

6. Particulate Matter

Characteristics : Solid particles or liquid droplets including fumes, smoke, dust and aerosols. Solid particular can adsorb various chemicals.

Biochemical effects : Effects vary with the nature of the particles. Carbon particles and other particles cause scarring of lungs via complex walling off and fibrogenic reactions leading to a disease condition known as "pneumoconiosis". Particles carrying absorbed mutagens lead to damage of DNA in the lungs and elsewhere.

Pulmonary fibrosis in asbestos mine workers, black lung disease in Coal miners and Emphysema in urban populations are attributed to the particulate pollution. Particulates may accelerate corrosion of metals and cause damage to paints and sculptures. They may absorb solar radiations and reduce visibility.

7. Heavy metals and other principal pollutants:

Some inportant pollutants including heavy metals are given in table no. 5.1. with their source and effects.

Table 5.1. Sources of some important air pollutants and their effect on human beings and animals

Pollutant	*Major Sources*	*Typical Effects*
Carbon monoxide (CO)	Incomplete combustion of fuels, automobile exhausts, jet engine emissions, blast fumaces, mines and tobacco smoking.	Toxicity, blood poisoning, increased proneness to accidents, CNS impairment. CO combines with haemoglobin, forming carboxy haemoglobin, which is useless for respiratory purposes and hence leads to death.
Sulphur dioxide (SO_2)	Combustion of coal, combustion of petroleum products, burning of refuse, petroleum industry, oil refining, power house, sulphuric acid plants, metallurgical operations and from domestic burning of fuels.	Increased breathing rate and feeling of air-starvation, suffocation, aggravation of ashma and chronic bronchitis, impairment of pulmonary functions, respiratory irritation, sensory irritation, irritation throat and eyes.
Oxides of Nitrogen (NO)	Automobile exhausts, coal-fired and gas-fired fumaces boilers, power stations, explosives industry, fertilizer industry, manufacture of HNO_3, combustion of wood and refuse.	Respiratory irritation, headache, bronchitis, pulmonary emphysema, impairment of lung defences, edema of lungs, lachrymatory effect, loss of appetite, corrosion of teeth.
Hydrogen sulphide (H_2S)	Coke ovens, kraft paper mills, petroleum industry, oil refining, viscose rayon manufacturing plants, manufacture of dyes, tanning industry and sewage treatment plants.	Headaches, conjunctivitis sleeplessness, pain in the eyes irritation of respiratory tract, respiratory paralysis, asphxiation malodorous. In high concentrations, it may lead to blockages of oxygen transfer, poisoning cell enzymes and damaging nerve tissues.
Chlorine (Cl_2)	Accidental breakage of chlorine cylinders, electrolysis of brine, bleaching of cotton pulp and other process industries using chlorine.	Irritation of eyes, nose and throat, toxicity, respiratory irritation, lachrymatory effects. In large doses it may cause edema, pneumonitis, emphysema and bronchitis.
Hydrogen fluoride (HF)	Glass fibre manufacture, chemical industry, fertilizer industry, aluminium industry, ceramic industry, phosphate rock processing.	Irritation, respiratory diseases, fluorosis of bones, mouling of teeth.
Carbon dioxide (CO_2)	Combustion of fuels, automobile exhausts, jet engine emissions.	Toxic in large quantities, Hypoxia.
Hydrocarbons	Organic chemical industries, petroleum refineries, automobile exhausts, rubber manufacture.	Some hydrocarbons have carcinogenic effects, lachrymatory effect
Oxidants (e.g.O_3)	Photochemical reactions in atmosphere involving organic materials and NO_2 etc. Reactions induced by silent electrical discharge and intense u.v. radiations in the atmosphere.	Irritation of lungs, eyes and respiratory tract. Accumulation of fluids in lungs and damage to lung capillaries. These biochemical effects of O_3 mostly arise from generation of free radicals which attack the-SH groups present in the enzymes.

Dusts	Asbestos factories, mining activities, power stations, metallurgical industries, ceramic industry, factory stacks, glass industry, cement industry, foundries.	Respiratory diseases, toxicity from metallic dusts, silicosis and asbestosis from the specific dusts. Asbestos dust cause pulmonary fibrosis, pleural calcification and lung cancer.
Ammonia (NH_3)	Chemical industries, coke oven refineries, stocks yards, fuel incineration.	Damage to respiratory tracts and eyes, corrosive to mucous membranes.
Formaldehyde (HCHO)	Waste incineration, automobile exhausts, combustion of fuels, photochemical reactions.	Irritation to eyes, skin and respiratory tract.
Hydrochloric acid (HCl)	By-product from chlorination of organic compounds, combustion of gasoline in which ethylene dichloride is present, burning of coal, paper and chlorinated plastics.	Inflammation and ulceration of upper respiratory tract, clouding of cornea, coughing and choking while inhalation.
Radioactive gases and dusts	Radioactive gases and suspended dusts from natural and artificial radioisotope sources.	Somatic effects such as leukaemia and other types of cancer, cataracts and reduction in life expectancy. Genetic effects such as mutations in human gametes.
Arsenic (As)	Arsenic containing fungicides, pesticides and herbicides, metal smelters, by-product of mining activities, chemical wastes.	Inhalation, ingestion or absorption through skin can cause mild bronchitis, nasal irritation or dermatitis. Carcinogenic activity also is suspected. Attack-SH groups of enzymes, coagulate proteins.
Beryllium(Be)	Coal, nuclear power and space industries, production of fluorescent lamps, motor fuels and other industrial use.	Damage to skin and mucous membranes, pulmonary damage, perhaps carcinogenic.
Boron (B)	Boron producing units, production and use of petroleum fuel and additives, burning coal and industrial wastes, detergent formulations.	Ingestion or inhalation as dust causes irritation and inflammation. Boron hydrides can damage CNS and may result in death.
Cadmium (Cd)	Cadmium producing industries, electroplating, welding. By-product from refining off Pb, Zn and Cu, fertilizer industry, pesticide manufacture, cadmium nickel batteries, nuclear fission plants, production of TEL used as additive in petrol.	Inhalation of fumes and vapours causes kidney damage, bronchitis, gastric and intestinal disorders, cancer, disorder of heart, liver and brain. Chronic and acute poisoning may result. Renal dysfunction, anaemia, hypertension, bone-marrow disorder and cancer.
Chromium (Cr)	Metallurgical and chemical industries, processes using chromate compounds, cement and asbestos units.	Toxic to body tissues, can cause irritation, dermatitis, ulceration of skin, perforation of nasal septum. Carcinogenic action suspected.
Lead (Pb)	Automobile emissions, lead smelters, burning of coal or oil, lead arsenate pesticides, smoking mining and plumbing.	Absorption through gastrointestinal and respiratory tract and deposition in mucous membranes, cause liver and kidney damage, gastro-intestinal damage, mental retardation in children, abnormalities in fertility and pregnancy.

Manganese (Mn)	Ferromanganese production, organo-manganese fuel additives, welding rods, incineration of manganese containing substances.	Poisoning of CNS, absorption, ingestion, inhalation, or skin contact may cause manganic pneumonia.
Nickel (Ni)	Metallurgical industries using nickel, combustion of fuels containing nickel additives, burning of coal and oil, electroplating units using nickel salts, incineration of nickel containing substances vanaspati manufacture.	Respiratory disorders, dermatitis, cancer of lungs and sinus.
Mercury (Hg)	Mining and refining of mercury, organic mercurials used in pesticides, laboratories using mercury.	Inhalation of mercury vapours may cause toxic effects and protoplasmic poisoning. Organo-mercurials are highly toxic and may cause irreversible damage to nervous system and brain.

Effects of Air Pollutants on Man and his Environment

(I) **Damage to materials.** The materials that may be affected by air pollutants include metals; building materials, rubbers, elastomers, paper, textiles, leather, dyes, glass, enamels and surface coatings. The types of possible damage to these materials by air pollutants include corrosion, abrasion, deposition, direct chemical attack and indirect chemical attack. The intensity of damage depends upon factors such as moisture, temperature, sunlight, air movement and of course the nature and concentration of the pollutants.

(2) **Damage to Vegetation.** Air pollutants, such as sulphur dioxide, HF, particulate fluorides, smog, oxidants like ozone, ethylene (from automobiles), NO_x, chlorine and herbicide and weedicide sprays exert toxic effects on vegetation. The damage usually manifests in the form of visual injury such as chlorotic marking, banding, silvering or bronzing of the underside of the leaf. Retardation of plant growth may also occur in some cases. The extent of damage to a plant depends upon the nature and concentration of the pollutant, time of exposure, soil and plant condition, stage of growth, relative humidity and the extent of sunlight.

(3) **Damage to farm animals.** Arsenic, lead and fluorides are the main pollutants which cause damage to livestock. These air-borne contaminants accumulate in vegetation and forage and poison the animals when they eat the contaminated vegetation.

Arsenic occurs as an impurity in coal and many ores. It is also used in insecticides. Livestock near smelting and other industrial operations suffer arsenic poisoning with symptoms like salivation, thirst, liver necrosis, inflammation of depression of central nervous system.

Lead is emitted from metallurgical smelters coke-ovens and coal combustion operations, lead arsenate sprays and automobile exhausts. Lead poisoning occurs in horses and other animals with symptoms such as depression, lethargy, gastritis, paralysis and breathing troubles.

Cattle and sheep are particularly susceptible to fluorine toxicity which may cause fluorosis of teeth and bones.

(4) **Darkening of sky and reduction in visibility**. Sky darkening may be caused by heavy smoke and fog or by dust storms. The reduction in visibility may be due to smoke, fog and industrial fumes which contain particulates in the size range of 0.4 to 0.9μ m that scatter light. The intensity of these effects depends upon the particle size, the angle of the sun, aerosol density, thickness of the affected air mass and meteorological factors such as inversion height, wind speed and humidity.

(5) **Effect on human health and human activities.** The effects of air pollution on humans, animals and vegetation has already been discussed in earlier sections. Air pollution can effect the

health of workers within the industrial premises, causing absenteeism, sickness and drop in production. Industrial hygiene measures are being taken by many industrial managements to combat these occupational disease. However, apart from the effects on industrial workers, air pollution also affects larger segments of general population. The notorious London smog of 1952, which lasted for 5 days causing 4000 deaths, is an example. Epidemiological and toxicological studies indicate a link between air pollution and respiratory conditions like chronic bronchitis, bronchial asthma, pulmonary emphysema and lung cancer. Irritation of nose, eyes and throat and bad odours due to air pollutants cause annoyance, allergy and health hazards.

Air pollution may cause sickness, absenteeism among workers and general lethargy, which naturally result in decrease in efficiency in all facets of human activity.

MEASURES TO CHECK AIR-POLLUTION

It is not easy to control/check air pollution at resonable cost, because it is not so simple. Our every life style/amentities of modern life is facing for air pollution. But we can check it or prevent by careful planning for industries, better design, operation of equipments and general awareness to do this. The following are the general methods of air pollution control :

1. Controlling the air pollution at source.
2. Site selection/Zoning
3. Controlling air pollution by devices/equipments/process modification.
4. Air pollution control by growing vegetation.
5. Air pollution control by Fuel selection and utilization.

1. Controlling of Air Pollution at Source

This is the best to check air pollution at source. This can be achieved by :

(i) Modifying the process in such a way that pollutants do not form at all beyond the permissible limits.

(ii) Before release the pollutants, they should be reduced to telerable levels by methods equipments to destroy, alter, trap or so.

To control or minimised air pollution at source, the following steps should be strictly follow :

(a) This step can also be done in two ways, Ist we should select the raw material in such a way to release minimum pollutants. The supplements may also be used if needed. Secondly use suitable fuels avoiding sulphur fuels. Non-essential ingredients are removed before processing of the raw material.

(b) Air pollution can easily be checked by using modified procedure or new process. Timely it should be monitored.

(c) Equipment alternations such as the use of vented tanks should be avoid. Use floating root tanks. For industry, new furnaces, modified equipments should be used.

(d) By using modified equipment, the pollutants are stored at one place, form where they should removed timely.

2. Site Selection/Zoning

To instal the industry, site selection is important, which results in the production of single source of pollution. Control measure based on the knowledge of the mechanics of the atmosphere is called "Zoning". While setting the factories the meterological and micro-meterological conditions should be considered. Other factors such as facilities for as material supply, transport labour and market for products are also important for selecting the site of industry.

For improvement of the people's health. Zoning should be done properly. The industries running on electrical power and causing, no nuisence may be sitted near residential area. But opposite to this, they may be located far away from residential area.

3. Controlling of Air Pollution by Devices/Equipments/Process Modifications

Large number of factories/industries release various types of gases, along with particulates which are measure source of air pollution. In order to prevent these pollutants two types of methods are used.

(A) Methods/Equipments used to Control gaseous Pollutants

For gaseous pollutants, following methods are generally used :

(i) Absorption (ii) Adsorption

(iii) Combustion (iv) Cold trapping or condensers

(v) Others.

But the first three are in common use.

(i) Absorption : Scrubbers are mostly used for the removal of gaseous pollutants. They have suitable liquid as absorbent to remove or modify one or more of the pollutants present in the stream. Through scrubbers gaseous effluents are passed. The efficiency of gas absorption depend upon the following factors :

(a) Chemical Activity of the gas pollutant. (b) The extent of the surface for contact.

(c) The contact time. (d) The concentration of the absobing medium.

This technique is used for removal of NOx, H_2S, SO_2, SO_3 fluorides etc.

(ii) Adsorption : Here, the gaseous effluents are passed through porous solid adsorbent taken in suitable containers. The efficiency of adsorption depends upon the surface area per unit weight of the absorbent. Constituents of the gas effluents are held at the interface of the adsorbent by chemisorption. When the effluents have higher concentration of Nox, Sox etc. the gases can be recovered economically and used for the manufacture of acids, *i.e.*, HNO_3 & H_2SO_4 etc.

Some important adsorbent/adsorbent are given in table :

Table : Absorbents and Adsorbents for Some Gaseous Pollutants

Pollutants	Absorbent	Adsorbent
Nox	H_2O & aq. HNO_3	Silica gel, commercial zeolites
HF	H_2O & NaOH	Porous pellets of NaF, limestone.
H_2S	Ethanol amines, $NaOH^+$ Phenol, Sodium Alanine, Soda Ash, Sodium thioarsenate, Tripotassium Phosphate.	Iron-Oxide
SO_2	Water, Alkaline H_2O, Sulphites of Ba, Ca or Na, Dimenthyl aniline, Aluminium Sulphate.	Dolomite, Alkaline alumina

The Sox can be removed from the boiler/furnaces by injecting CaO (Pulverised lime stone). Thus the emission of SOx is prevented, because CaO formed $CaSO_3$ & $CaSO_4$ with SOx.

(iii) Combustion : The flame combustion or catalytic combustion of organic gaseous pollutants convert them in to H_2O & CO_2. Flame combustion include fume incinerators, steam injection while catalytic combustion is resorted where lower temperature is needed.

(B) Methods/Equipments used to Control Particulate Emission

The particulate collection devices are based on the size, shape, properties of the particulate, which are generally originate from stationary and mobile sources. The various methods are :

(i) Filtration (ii) Mechanical

(iii) Precipitators (iv) Scrubbers.

(i) Filtration : Different type of filters are generally used, *i.e.*, Fibrous, deep bed, cloth bag filters. The particulate matter is passed through filters, particles are trapped & collected in filters. Cloth & Nylon filters are used up to 80-90ºC while silicone & asbestos covered glass, cloth filters can be used up to 250-350ºC. After filtration gas devoid of the particles are discharged out.

(ii) Mechanical : It includes the two mechanisms :

(a) Gravity settling in which the velocity of the horizontal carrier gas is reduced so that particles settle by gravitational force.

(b) Sudden change of direction of the gas flow causes the particles to separate.

Gravity settling chambers consists of a chamber in which dust is separated from the gas but reducing the velocity of the gas. As a result dust particles settle down in the chamber, coarse particles are thus removed. These chambers are capable of removing only the large particles of size, 25-30 μm diameter.

(iii) Precipitators : These are generally electrostatic in nature hence called electrostatic precipitators. Particles with diameter as small as 0.0001 cm. can be removed by passing the stack gas through electrostatic precipitator.

It works on the principle that when particulates move through a region of high electric potential, they become charged and then they are attracted to an oppositely charged area, where they are collected and removed. It consists of a series of plates which are charged to high voltage alternatively + and –. Particles approaching given plate tend of aquire its charge. They are then attracted to the surface of the next plate, from which they fall into the hopper below. Thus particles pick up charge as they pass between the plates and are precipitated on plates of opposite charge. The potential across the plates is around 50,000 V. The high voltage in the wires produces billions of electrons and bombard the gas molecules, which become positively and negatively charged. The positive ions return to the negative end electrode and gain electrons while the negative ions combines with the dust particles and make them negatively charged. The negatively charged dust particle collect as the positively charged plates. Electrical attraction becomes weak when dust layer becomes 6 mm thick.

(iv) Scrubbers : In this device the particles are wash out of the gas flow by a water spray. The object is to transfer suspended particulate matter in the gas to the scrubbing liquid which can be readily removed by the gas cleaning device. This leaves the gas clean to pass onwards to the process for which it is being used or alternatively to be discharged to the atmosphere. The most common scrubbers are :

(i) Cyclonic scrubbers

(ii) Venturi scrubbers

In cyclonic scrubbers the aerosol is introduced in a centrifugal manner. Water is sprayed at the entrance of gas and plates are provided to remove the moisture from the gas after the removal of the dust. This is followed by a control equipment such as gravity settling chamber or cyclones. Cyclonic scrubbers have an efficiency of 90% and it can remove dust particles of 5 μm size. It is capable of cleaning about 2000 litres of gas per minute.

In venturi scrubber consists of venture throat though which dirty gas is passed at a rate of about 3400-12600 mm per minute. Water added in the direction of low so that the water enters at the throat. Efficiency, of this is also 99% and is capable of cleaning even very fine particles. It can clean 4000 litres of gas per minute.

(4) Controlling of Air Pollution by Growing Vegetation

Planting of trees is very helpful in reducing air pollution due to dry flyash and coal dust. Trees should be planted all around the source in order to reduce the spreading of air pollution from pollutants coming out from industry/source. Cultivation of pollution resistant species is the best possibility of reducing air pollution. The odours of gases can be absorbed by passing them through beds of activated charcoal of sand or soil. The odours can also be controlled by the oxidation of the compounds using chlorine, ozone & Hydrogen peroxide as oxidising agent.

WATER POLLUTION

INTRODUCTION

Water is essential for the survival of any form of life. On an average, a human being consumes about 2 litres of water everyday. Water accounts for about 70% of the weight of a human body. About 80% of the earth's surface (i.e., 80% of the total 50,000 million hectares in area) is covered by water. Out of the estimated 1,011 million km^3 of the total water present on earth, only 33,400 m^3 of water is available for drinking, agriculture, domestic and industrial consumption. The rest of the water is locked up in oceans as salt water, polar ice-caps and glaciers and underground. Owing to increasing industrialization on one hand and exploding population on the other, the demands of water supply have been increasing tremendously. Moreover, considerable part of this limited quantity of water is polluted by sewage, industrial wastes and a wide array of synthetic chemicals. The menace of water-borne diseases and epidemics still threatens the well-being of population, particularly in under-developed and developing countries. Thus, the quality as well as the quantity of clean water supply is of vital significance for the welfare of mankind. An analysis conducted in 1982 revealed that about 70% of all the available water in our Country is polluted.

Municipal water is mainly used for drinking purposes and for cleaning washing and other domestic purposes. The water that is fit for drinking purposes is called potable water.

Characteristics of Potable Water

1. It should be colourless, odourless and tasteless.
2. It should be free from turbidity and other suspended impurities.
3. It should be free from germs, bacteria and other pathogenic organisms.
4. It should not contain toxic dissolved impurities, such as heavy metals, pesticides, etc.
5. It should have a pH in the range 7-8.5.

6. It should be moderately soft, having hardness preferably in the range 50-100 PPm. Its hardness should not be above 150 PPm.
7. It should be aesthetically pleasant.
8. It should not be corrosive to the pipelines and should not cause any incrustations in the pipes.
9. It should not stain clothes.

Table 5.1. Standards (maximum permissible limits) for drinking water as recommended by World Health Organisation (WHO).

Parameters	*Level WHO standard*
pH	6.5-9.2
BOD	6
COD	10
Arsenic	0.05 PPm
Calcium	100 PPm
Cadmium	0.01 PPm
Chromium	0.05 PPm
Ammonia	0.5 PPm
Copper	1.5 PPm
Iron	1.0 PPm
Lead	0.1 PPm
Mercury	0.001 PPm
Magnesium	150 PPm
Manganese	0.5 PPm
Chloride	250 PPm
Cyanide	0.05 PPm
Nitrate + Nitrite	45 PPm
Polyaromatic hydrocarbons (PAH)	0.2 PPm
Selenium	0.01 PPm

Water Pollutants and their Sources

The various types of water pollutants are:-

(a) **Oxygen-demanding wastes.** These include domestic and animal sewage, bio-degradable organic compounds and industrial wastes from food-processing plants, meat-packing plants, slaughter-houses, paper and pulp mills, tanneries etc., as well as agricultural run-off. All these wastes undergo degradation and decomposition by bacterial activity in presence of dissolved oxygen (D.O.). This results in rapid depletion of D.O. from the water, which is harmful to aquatic organisms. The optimum D.O. in natural waters is 4-6 ppm, which is essential for supporting aquatic life. Any decrease in this D.O. value is an index of pollution.

(b) **Disease-causing wastes.** These include pathogenic microorganisms which may enter the water along with sewage and other wastes and may cause tremendous damage to public health. These microbes, comprising mainly of viruses and bacteria, can cause dangerous water-borne diseases such as cholera, typhoid, dysentry, polio and infectious hepatitis in humans. Hence, disinfection is the primary step in water pollution control.

(c) **Synthetic Organic Compounds**. These are the man-made materials such as synthetic pesticides, synthetic detergents (syndets), food additives, pharmaceuticals, insecticides, paints, synthetic fibres, elastomers, solvents, plasticizers, plastics and other industrial chemicals. These chemicals may enter the hydrosphere either by spillage during transport and use or by intentional or accidental release of wastes from their manufacturing establishments. Most of these chemicals are potentially toxic to plants, animals and humans.

(d) **Sewage and agricultural run-off.** Sewage and run-off from agricultural lands supply plant nutrients, which may stimulate the growth of algae and other aquatic weeds in the receiving water body. This unwieldy plant-growth results in the degradation of the value of the water body, intended for recreational and other uses.

(e) **Oil.** Oil pollution may take place because of oil spills from cargo oil tankers on the seas, losses during off-shore exploration and production of oil, accidental fires in ships and oil tankers, accidental or intentional oil slicks (as in the Gulf War between Iraq and U.S.-led allied forces in the year 1991) and leakage from oil pipe-lines, crossing waterways and reservoirs. Oil pollution results in reduction of light transmission through surface waters, thereby reducing photo-synthesis by marine plants. Oil pollution in Seas has been increasing due to the increase in oil based technologies, massive oil shipments, accidental oil spillages etc.

(2) **Inorganic Pollutants**

Inorganic pollutants comprise of mineral acids, inorganic salts, finely divided metals or metal compounds, trace elements, cyanides, sulphates, nitrates, organometallic compounds and complexes of metals with organics present in natural waters. The metal-organic interactions involve natural organic species, such as fulvic acids and synthetic organic species, such as EDTA. The heavy metals such as Hg, Cd and Lead, metalloids such as As, Sb and Se are most toxic. The water pollution by heavy metals occurs mostly due to street dust, domestic sewage and industrial effluents. Polyphosphates from detergents are also water pollutants.

(3) **Suspended solids and sediments**

Sediments are mostly contributed by soil erosion by natural processes, agricultural development, strip mining and construction activities. Suspended solids in water mainly comprise of silt, sand and minerals eroded from the land. Soil erosion by water, wind and other natural forces are very significant for tropical countries like India. It is estimated that 5.37 million Tonnes of NPK fertilizers are washed away in to the sea. Sediments and suspended particles exchange cations with the surrounding aquatic medium and act as repositories for trace metals such as, Cu, CO, Ni, Mn, Cr, and Mo.

(4) **Radioactive Materials**

The radioactive water pollutants may originate from the following anthropogenic activities:

(a) Mining and processing of ores, e.g., Uranium tailings.

(b) Increasing use of radioactive isotopes in research, agricultural, industrial and medical (diagnostic as well as therapeutic) applications, e.g., I^{131}, P^{32}, Co^{60}, Ca^{45}, S^{35}, C^{14}, Rb^{86}, Ir^{132} and Cs^{137}

(c) Radioactive materials from nuclear power plants and nuclear reactors, e.g., Sr^{90}, Cs^{137}, Pu^{248}, Am^{241}.

(d) Radioactive materials from testing and use of nuclear weaponry, e.g., Sr^{90}, Cs^{137}

The radioactive isotopes found in water include Sr^{90}, I^{131}, Cs^{137}, Cs^{141}, Co^{60}, Mn^{54}, Fe^{55}, Pu^{239}, Ba^{140}, K^{40}, Ra^{226}.

(5) **HEAT.** Considerable thermal pollution results from thermal power plants, particularly the nuclear-power-based electricity generating plants. In such industries, where the water is used as a coolant, the waste hot water is returned to the original water bodies. Hence the temperature of the water body increases. This rise in temperature decreases the DO content of water, which adversely affects the aquatic life.

EFFECTS

Some important effects of various types of water pollutants are as:-

(i) Tannery effluents contain several constituents which are deleterious, irrespective of the fact that where they are discharged viz., into river, stream, sewer, land or sea.

(ii) It imparts persistent dull brown colour to the receiving water causing aesthetic and other problems described earlier.

(iii) Highly repulsive odour is imparted to the receiving water. The dissolved constituents like proteins are purifiable.

(iv) The acidic or alkaline effluents are corrosive to concrete and metal pipes.

(v) Excess NaCl in the effluent is also corrosive and renders the receiving water unsuitable for irrigation.

(vi) The effluents may contain pathogenic bacteria.

(vii) The dissolved chromium present is toxic to fish and aquatic life and thus affects the natural self-purification property of the stream.

(viii) The suspended solids such as hair, flesh, $CaCO_3$,' etc. interfere with aeration and photosynthetic activities of the aquatic flora.

(ix) If the wastewater is discharged into sewer, the suspended impurities such as $CaCO_3$, hairs etc. may choke the sewerage pipes. The sulphides present in the wastewater cause "crown corrosion" to the concrete structures, etc.

(x) The chromium and sulfides present in the waste water being toxic to microorganisms disrupt the biological treatment operation such as trickling filtration. The suspended lime etc. also interfere with the biological activities in the sewage treatment plants.

(xi) The presence of excessive salt and Cr in the wastewaters may deteriorate the quality of the ground water in the affected areas.

(xii) High amounts of fluoride (over 1000 mg/1) present in phosphatic fertilizer effluent enrich the fluoride content of the receiving waters causing dental and skeletal fluorosis to humans, abnormal calcification of bones in animals and adverse effects on plants. Presence of Cr, cyanide, ammonia are harmful to aquatic life.

(xiii) Volatile substances such as alcohols, aldehydes, ethers and gasoline may cause explosion in sewers.

(xiv) Suspended solids such as silt and coal may injure the gills of the fish and cause asphyxiation.

(xv) Suspended solids may also cause bad odour and tastes and also may promote conditions favourable for growth of pathogenic bacteria.

(xvi) Radioactive isotopes are toxic to life-forms. For instance, Sr^{90}, which emanates from testing of nuclear weapons, accumulates in bones and teeth and causes serious disorders in human beings. The maximum permissible level of Sr^{90} in water is 10 pico curies per liter (1 pico curie = 10^{12} curie).]

CONTROL OF WATER POLLUTION

The control of water pollution is difficult, but we may try for its prevention and minimisations. Industrialised and Developing countries spend a handsome amount of their GNP (Gross National Product) on pollution control measures, but the problem is going to be worsening day by day. The main worries before water pollution control are:-

1. How pure the water should be? 2. How to prohibit the effluents and discharge in to water? 3. To what extent water quality be improved? 4. How to create public opinion against

water pollution? Therefore, we should adopt the respective safety measures to achieve acceptable water quality at the least cost. Some of these are :-

1. Scientific techniques are necessary to be adopted for the environmental control of catchment areas of rivers, lakes, ponds or streams.
2. Industrial plants should be based on recycling operations.
3. The possible reuse or recycle of treated sewage effluents and industrial wastes should be emphasized and encouraged.
4. Instead of throwing wastes in to water, the recycling should be done for better use. Gobar gas plant, composting, manufacture of hardboard, paper etc. such examples where respective waste can be used.
5. Minimum, appropriate quantity and concentration of Fertilizers, pesticide & insecticides should be used, because excess will cause pollution.
6. There should be propaganda for water pollution control, on radio, TV, Newspapers etc. because public awareness is a must.
7. Treatment plants should be constructed and Govt should also help by funding for domestic, sewage and industrial effluents.
8. Local authorities, Industrialists, Govt officials, with public participation should co-ordinate to finds ways to control water pollution.
9. Water resources should be used in the best possible economic way.
10. To conduct seminars and training courses for helping those, who are directly or indirectly engaged in water management and water pollution control.
11. Govt. should encourage people to participate in research programmes like disposal of sewage and industrial effluents.
12. Destruction of forest should be discouraged our goal should be –"Conservation of forests" and "Plant more trees".
13. Techniques like adsorption, electro dialysis, ion exchange and reverse osmosis etc can be used for the removal of water pollutants.
14. Plants should be developed to recover metal from metal bearing waste water.

SOIL POLLUTION

INTRODUCTION

Soil is a very important constituents of the lithosphere. The word "Soil" is derived from a Latin word "Solum" which means earthy material in which growth of plants takes place. "Soil" may be broadly defined as the weathered layer of the earth's crust with living organisms and their products of decay. It is a complex physio-biological system containing water, mineral salts, nutrients and dissolved oxygen. It provides anchorage to plants. The study of Soil science is called pedology or edaphology.

Earth's crust comprises of rocks and the loose material derived from these rocks, the disintegration of which results into the formation of gravel, sand, silt and clay. About 99% by weight of earth's crust is constituted by about 12 elements only, whereas the other elements exist in traces. The basic composition of the earth's crust is shown in table (5.2).

Table 5.2: Elementary composition of the earth's crust.

Elements present	*Percentage by weight*
Oxygen	49.85
Silicon	26.03
Aluminium	7.28

Iron	4.12
Calcium	3.18
Sodium	2.33
Potassium	2.33
Magnesium	2.11
Hydrogen	0.97
Titanium	0.41
Chlorine	0.20
Carbon	0.19
Others	1.00

The earth's crust basically consists of the following three rock types.

1. **Igneous rocks** : These are formed by cooling and solidification of molten rock material called 'Magma'. Ex:- Basalt and Diorite.

2. **Sedimentary rocks** : These are developed as a result of gradual accumulation, consolidation and hardening of products of weathering of mineral materials brought about by wind or waters. These rocks are characterized by the presence of distinct sedimentary layers. Ex:- Lime stone, sand stone and shale.

3. **Metamorphic rocks** : These are formed as result of metamorphosis of igneous and sedimentary rocks under the influence of high pressure and intense heat. Ex:-Quartzite, Slate, Marble and Schist.

Out of about 2000 minerals known, nearly 99% of the earth surface is made of only about 10 to 12 minerals. The following four mineral groups account for over 90% mass of terrestrial rocks.

Feldspars	58%
Pyroxenes	16%
Quartz	13%
Mica	4%

The common minerals found in the earth crust are shown in Table (5.3).

Table 5.3 : Common minerals found in earth crust and their composition.

Name of the mineral	Chemical composition
Feldspars	$K_2Al_2Si_6O_{18}$, $NaAlSi_3O_6$, $CaAl_2Si_3O_8$
Pyroxenes	(Mg, Fe) SiO_3
Quartz	SiO_2
Micas	$KAl_2(AlSi_3O).(OH)_2$
K Mg Fe-Al Silicates	
Olivine and serpentine	$(Mg.Fe).SiO_4$
Amphiboles	$(Mg.Fe).\ (Si_4O_{10}.(OH)_2$
Calcite	$CaCO_3$
Magnetite	$MgCO_3$
Dolomite	$CaCO_3.MgCO_3$
Oxides of Iron	
(Haemetite, Magnetite, Limnite)	$Fe_2\ O_3$, Fe_3O_4, $FeO(OH).XH_2O$
Montmorillonite	$(Ca\ MgO).Al_2O.5SiO_2.5H_2O$
Kaoline	$Al_2O_3.2SiO_2.2H_2O$

Importance of Soil to the Biosphere

Soil plays a vital role in determining the quality and composition of the biosphere which develops over it. The multiferious functions of soil are as follows:

(i) Soil provides mechanical support to the plants.

(ii) Owing to the porosity and water-holding capacity of soil, it serves as a reservoir of water and supplies water to the plants through their roots even when the land surface is dry.

(iii) The ion-exchange capacity of the soil ensures availability and supply of micro-and macro-nutrients for the growth of plants, animals and microbes. It also helps in preventing excessive leaching of nutrient ions, while maintaining proper pH.

(iv) The colloidal components of soils which comprises of clay micelles and humus particles (less than 0.002mm) tightly adsorb a number of nutrient ions and supply them evenly to the plants.

(v) Soil contains organotrophic bacteria, nitrifying bacteria, nitrogen-fixing bacteria. Fungi, protozoans and other microbes which help in decomposition and mineralization of organic matter and regeneration of nutrients.

Major types of soil in India

The major types of soil occurring in different parts of our country are given in Table (5.4). below:

Table 5.4 : Major types of Soil in Our Country.

Type of Soil		Remarks
1. Alluvial Soil	-	Occur in the great-northern plains of India and deltas of rivers in peninsular India.
(a) Khadar	-	Very fine and new alluviam. Very fertile.
(b) Bangar	-	Relatively coarse and old alluviam. Relatively less fertile.
2. Mountain or hill soils and forest soils	-	Rich in organic matter (humes), occur in eastern and western ghats hills in Central India and northern hilly regions of Himalyas.
3. Black soils	-	Originate from volcanic rocks. Highly fertile clay soils. Occur in parts of M.P. Tamilnadu, Gujarat and Deccan trap regions of Maharashtra.
4. Red soils	-	Relatively less fertile. Deficient in organic and nitrogenous matter. Occur in plateaus of Kerala, Karnataka, A.P., Tamilnadu, Orissa and Southern parts of Bihar.
5. Lateritic soils	-	These are poorly fertile but can support pastures and scrub forests. These soils occur in regions of tropical rainy climate such as western ghats, chota Nagpur plateau, assam, Orissa, A.P., Tamilnadu and Kerala.
6. Desert soils	-	These are arid sandy soils with low moisture and low humus content. They occur mostly in Rann of Catch, Southern Punjab, Western Rajasthan and Haryana.

Plants as pollution indicators

Some plant species have been used as indicators of pollution of soil, air and water. Certain plant species such as chara utricularia and Wolffa were found to grow well in polluted water. Species like agrotis, festuca, anthoxanthum and impatiens are used as metallic tolerance plant indicators for Cu, Pb, Zn and Cd respectively. Presence of diatoms indicate sewage pollution.

Escherichia coli bacteria indicate water pollution. Leaf cabbage indicate accumulation of polycyclic hydrocarbons in the soil. The growth of lichens was found to decrease when the soil is polluted. Soil pollution inhibits plant growth and reduce productivity. Plant growth is inhibited or destroyed in areas near smelters.

Sources of Soil Pollution

Soil pollution differs from water pollution or air pollution, because the pollutants remain in direct contact with the soil for relatively longer periods and hence alter the chemical and biological properties of the soil. The hazardous chemical can also enter the human food chain from soil or water plants.

The major sources of metallic contamination of soils include mining, smelting, sludge, fertilizers, pesticides, composted town refuse etc. Metals such as Cd, Pb, Hg, Ni, Mo, Ni, Cr etc. are toxic to plant and animal life.

Indiscriminate dumping of industrial wastes and municipal wastes leads to the leaching and/or seepage of toxic substances into the soil and pollution of ground water. Further due to some modern agricultural practices, obnoxious pesticides, fungicides, insecticides, biocides, bacteriocides, etc. contaminate land. Direct pollution of soil by dangerious pathogenic organisms is also important.

Fly ash generated from thermal power plants, industrial waste discharged into streams or dumped into the surrounding land, mining wastes, non-biodegradable organic pollutants, industrial sludges such as flue gas desulphurization sludge, heavy metal sludge sets, cause serious water and soil pollution problems.

Commercial and domestic urban wastes consisting of cried sewage sludge as well as garbage and rubbish materials such as plastics metal cans, glasses, street sweepings, waste paper, fibres, rubber etc contribute to soil pollution.

Human and animal excretia, farm wastes, soil conditioners, soil fumigants, radioactive wastes, etc, also cause soil pollution.

EFFECTS OF SOIL POLLUTANTS

"*Soil Pollution*" was originally defined as the contamination of the soil system by considerable quantities of chemical or other substances, resulting in the reduction of its fertility or productivity with respect to the qualitative and quantitative yield of the crops. However, if some of the contaminants are such that if they are taken up by the plants (with or without any detrimental effects on them), and enter into the food chain and impart detrimental or toxic effect on the consumers, then that also should be treated as soil pollution.

Soil pollution is receiving greater and greater attention due to its direct impact on public health. The major effects of various types of pollutants are given below:

(a) Effects of modern agricultural practices:

Synthetic Fertilizers : Synthetic fertilizers are employed to increase the soil fertility and crop productivity. These fertilizers concentrate the essential nutrients in layer of top soil. However, the soil enriched by chemical fertilizers cannot support the microbial flora which are so essential to enrich the humus that helps in plant growth. Excessive and indiscriminate use of chemical fertilizers may result in the following undesirable effects:

(i) Wheat, maize, corn, etc. grown on soils fertilized with NPK fertilizers may result in considerable reduction in protein content of the crop.

(ii) Excessive use of nitrogenous fertilizers leads to the accumulation of nitrates in the soil which may contaminate the ground water. Nitrate concentrations exceeding 90 ppm in drinking water may lead to diarrhoea, blue Jaundice (Cyanosis) in children, "methemoglobinemia" (or blue baby syndrome) in infants. Further, the nitrates and nitrites entering the human body may be eventually converted nitroso amines and other nitroso-compounds which are suspected to cause stomach cancer. Surveys in Rajasthan and other parts of the country indicated much higher nitrate levels than the permissible 45 ppm levels.

(iii) Vegetation growth in nitrate-rich soils may exert toxic effects in cattle.

(iv) Excessive use of chemical fertilizers may enter the water bodies and contribute to "eutrophication". (Eutrophication is the excessive growth of algae and aquatic plants to undesirable levels).

(v) Excessive use of chemical fertilizers may reduce the ability of plants to fix nitrogen.

(vi) Excessive quantities of potassium fertilizers in soils may reduce the quantities of valuable ascorbic acid (vitamin C) and carotene in fruits and vegetables grown in such soils.

(vii) The large-sized fruits and vegetables grown in highly fertilized soils may be more vulnerable to attacks by pests and insects.

Pesticides

As per the reports of the World Health Organization (WHO), about 50,000 people in developing countries are poisoned and about (5,000) people die because of improper use of pesticides and other chemicals in modern agricultural practices.

Pesticides pose potential hazard to animals, humans and aquatic life. They also cause deleterious effect on soil fertility and crop productivity. Pesticides applied to crops are retained in the soil in considerable quantities. They enter into cyclic environmental processes such as absorption by soil, leaching by water, etc, and contaminate both lithosphere and biosphere. Pesticides including herbicides, fungicides and rodenticides, are persistant pollutants. Owing to interactions between lithosphere and biosphere, pesticides may enter the food chain and pose serious health hazards. Some of them undergo metabolic transformation and bio-degradation. The degradation products of some of the pesticides are more dangerous than their respective parent compounds. Some of the pesticides residues are carcinogenic while their metabolic products too are toxic. The rate of degradation of pesticides depend upon their properties and structural characteristics.

The following types of pesticides are commonly used:

(a) Chlorinated hydrocarbons (eg. DDT, Aldrin, Dieldrin, Lindane, BHC etc.)

(b) Carbamate compounds (eg. Carboryl or Sevin, Zectrion etc.)

(c) Organo-Phosphorous compounds (eg. Methyl or ethyl parathion, melathion, guthion etc.)

(d) Inorganic compounds (eg. As_2O_3, PbO_2, $NiCl_2$, $CuSO_4$ etc)

(e) Miscellaneous compounds (eg, Organic mercurials, 2.4D; 2.4,5T etc)

Some of the adverse effects of pesticides are given below:

(i) Some arsenic pesticides may render the soil permanently infertile.

(ii) Pesticide residues in soil may be taken up by plants and cause phyto-toxicity. They may enter the aquatic environment and enter the food chain.

(iii) Pesticides such as, endrin, dieldrin, DDT, hepachlor etc. may seep through the soil and contaminate groundwater and surface waters. They may eventually contaminate drinking water supplies.

(iv) Fruit, vegetables, rice, wheat, barley, maize etc. are known to contain considerable quantities of toxic pesticide residue such as of DDT, BHC and other organochloro pesticides.

(v) Polychlorinated biphenyls (PCB) having half-life periods of about 25 years in soil are among the most hazardous soil pollutants. They may accumulate in soil and plants and when they eventually enter the animal or human body, they may cause severe health disorders including eye damage, skin problems, nervous disorders, foetus deformities and liver or stomach cancer.

(vi) Irrigated water from pesticide contaminated soils may evaporate and spread the toxic pesticide vapours in the atmosphere.

(vii) DDT can enter the food chain and accumulate in human fats and may lead to disorders such as impotency.

(viii) Persistant pesticides can damage human tissues and interfere with the normal metabolic activities by disturbing enzymatic functioning.

(ix) Chlorinated pesticides and herbicides are hazardous soil pollutants which can affect the soil texture and damage the ecosystem.

(x) Herbicides such as dioxan may cause congential birth defects in offsprings.

(xi) Hunting birds feeding on grains contaminated with DDT are threatened of extinction.

(xii) Organophosphate pesticides may cause muscular disabilities, tremors and dizziness.

(xiii) Excessive use of synthetic pesticides may lead to defoliation of forests and adverse effect on fauna and flora.

(xiv) Farm animals drinking stagnant water in fields sprayed by pesticides developed toxic symptoms and some mortalities were reported.

(xv) Farmers and farm workers are particularly prone to pesticide poisoning because of greater exposure while handling and spraying.

(xvi) Accidental spillages and leakages in pesticide manufacturing industries cause disastrous effects on the people residing in nearby areas due to pollution of air, water and soil. The Bhopal tragedy on 3rd December, 1994 is a lingering example.

(xvii) Contaminated soils may act as potential carriers of pathogenic bacteria and other dangerous organisms which may endanger human health.

(xviii) Volatile pesticides may cause pollution of air in the surrounding areas.

(b) **Effects of Industrial Effluents:**

Solid, liquid and gaseous chemicals from various industries as such paper and pulp, iron and steel, fertilizers, dyes, automobiles, pesticides, tanneries, coal-based thermal power plants etc. contain a variety of pollutants such as toxic heavy metals, solvents, detergents,plastics, suspended particulates and refractory/non-bio-degradable/recalcitrant chemicals. If they are not properly treated at source, they give rise to water, air-and soil pollution. Fly ash resulting from coal-based thermal power plants is one of the alarming and continuously increasing source of soil-pollution leading to degradation of soil, apart from water-and air-pollution in the nearby areas. Some trade wastes such as tannery wastes may contain pathogenic bacteria.

Indiscriminate dumping of untreated or inadequately treated domestic, mining and industrial wastes on land is an important source of soil pollution.

Fall-out of gaseous and particulate air-pollutants from mining and smelting operations, smoke-stacks, etc. is a major source of soil pollution in nearby areas.

(c) **Effects of urban wastes**:

Millions tones of urban waste are produced every year from critically polluted cities. The inadequately treated or untreated sewage sludge not only pose serious health hazards but also pollute soil and decrease its fertility and productivity. Other waste materials such as rubbish, used plastic bags, garbage, sludge, dead animals, waste medicines, hospital wastes, skins, tyres, shoes, cans, etc. also cause land and soil pollution. Some solid wastes may cause clogging of ground water filters. Suspended matter present in sewage can act as a blanket on the soil and interfere with its productivity.

Apart from the above major sources, radioactive waste dumped in the soil from natural and man-made sources, soil erosion due to deforestation, unplanned irrigation and unscientific agricultural practices are also result in land and soil pollution.

Control of Soil Pollution

As discussed in earlier sections, the major sources of soil pollution are the domestic wastes, industrial wastes and agricultural wastes including those toxic chemicals (eg. Pesticides) arising from modern agricultural practices. The various approaches to control soil pollution are as follows :

(1) Implementing stringent and pro-active population control programmes.

(2) Launching extensive afforestation and community forestry programmes.

(3) Implementing deterrent measures against deforestation.

(4) Formulation of stringent pollution control legislation and effective implementation with powerful administrative machinery.

(5) Imparting informal and formal public awareness programmes to educate people at large regarding the health hazards and undesirable effects due to environmental pollution. Mass media, educational institutions and voluntary agencies should be involved to achieve these objectives.

(6) Banning the use of highly toxic and resistant synthetic chemical pesticides or atleast regulating/restricting their use only for special purposes under thorough monitoring.

(7) Encouraging the use of bio-pesticides in place of toxic chemical pesticides.

(8) Conservation of soil to prevent the loss of precious top soil from erosion and to maintain it in a fertile state for agricultural purposes.

(9) Transforming intensive agriculture into a sustainable system by measures such as

(a) maintaining a healthy soil community in order to regenerate soil fertility by providing organic manures, increasing fallow periods avoiding excessive use of chemical fertilizers and pesticides.

(b) Infusing bio-diversity in agriculture by sowing mixed crops, crop rotation etc.

(10) Effective treatment of domestic sewage by suitable biological and chemical methods and adopting modern methods of sludge disposal.

(11) Municipal wastes have to be properly collected by segregation treated and disposed scientifically in land fills. Recycling and re-use of materials should be done wherever possible, such as recycling of glass, plastics, paper and production of bio-gas.

(12) Industrial wastes have to be properly treated at source, by segregation of wastes and/or adopting integrated waste treatment methods. Proper care should be taken in treating heavy metal wastes and other obnoxious waste materials.

(13) Security land-fills have to be constructed for permanent disposal of hazardous and recalcitrant industrial wastes.

(14) Sponsoring more intensive R & D efforts on bio-fertilizers, bio-pesticides utilization of wastes by recovery, reuse and recycling processes, and safer treatment and disposal of hazardous waste.

(15) Enforcing environmental audit for industries and promoting ecolabelled products.

(16) Avoiding excessive use of chemical fertilizers and insecticides and providing more organic manures to the fields and thereby maintaining healthy biota. This in turn regenerates soil fertility. A soil rich in organic mater also helps in controlling soil erosion.

MARINE POLLUTION

INTRODUCTION

Seas are the unlimited source of water for man. Secondly they are the main source of food and earnings for persons living in coastal areas. When the marine water is polluted it effects the animals and other food chain components. Researches shows that many marine animals secrets the medicinal chemicals which are useful to mankind and other living organisms. When water will be polluted it will effect the animals present in seas. Generally drainage from rivers, industries, human activities from coastline area, disposal of radioactive wastes and toxic materials, leakages from ships are main source of marine pollution. When ships and other animals take such polluted water get affected. When man take them dies. Leaked oil other pollutants from ship spread over in sea, get absorbed on sediments. It effects the marine life. This is due to dumping of wastes material from outside, which is harmful too, in ocean affect their ecosystem. This is known as marine pollution.

Sources of marine Pollution

The main sources of marine pollution are:-

1. Rivers are the main source of marine pollution. They carry wastes in their drainage and joins the sea/ocean. The drainage include sewage sludge, industrial effluents, detergents, agrochemicals, plastics, metal scraps etc.
2. Catchment area-like India and other countries too, many big cities and industries are situated along the coast line. Every large amount of wastes from hotels, wastes effluents mixed with detergents, sewage from corporations and industries, other wastes from human activities are mixed in sea water.
3. Ships which carry toxic substances, lubricating oil, paints heavy oils, fuels, automotive materials and other chemicals from one place to another, some times by accident or by leakages pollute the marine water.
4. Testing of atomic weapons, space aircrafts, missiles (generally developed country do this) and other radioactive wastes when dumped in seas, causes heavy loss to aquatic biota.
5. Harmful effluents from nuclear power stations or from other scientific organizations like BARC in India, chemical industries, fertilizers, Pesticide and insecticide industries when mixed in marine water causes harmful effects to marine life.
6. Marine pollution also caused by oil drilling in seas, tourism activities and heat released from industries. etc.

Effects of marine pollution:

The major effects of marine pollution are as follows:-

1. Oil is most dangerous pollutant when afloat on sea or mixed with water a great threat to marine life specially fish, birds, invertebrates and algae. Thousands of birds killed every year because once they oiled, seldom survived despite efforts to clean themselves.
2. Oil of sea also effects sensitive flora and fauna, phytoplankton, zooplankton and other animals. In Alaska, Brittany (France), Elbe (Germany) thousands of birds died by oil spillage.
3. Plastic or plastic materials when dumped into sea by commercial ships or from drainage, animal take it through their food in stomach. It causes ulcer and reduces hunger.
4. Marine pollution effects the food chain in seas. Serious diseases like cancer are the caused when affected animals are taken by man from ocean.
5. Detergents, either from cleaning up the spills or from drainage, also responsible for high mortality of marine life.
6. Heavy metals (like lead and mercury), factory materials, mineral oils, acids and other biocides are also measure threat to marine life when mixed with sea water.

Apart from these major effects, there is a heavy loss of economy after getting polluting animals and chemicals from marines.

Control of marine pollution:

The control of marine pollution can be studied in following two steps:-

1. **Steps already in operation**

   (i) Port authorities are alert and introduced antipollutant measures by creating pollution cell. But deeper in sea coastal guards are doing this job.

   (ii) Various research organizations, institutions are working in this field to check the marine pollution.

   (iii) In most of the countries (India too), the monitoring and survey in operation to control the marine pollution.

   (iv) Authorities are taken care of effective measures to check the oil leakage from ships and tankers.

   (v) Urban and coastline corporations are trying to check the dumping of wastes from human activities & Municipal etc. solid waste management is helping to recycle or reuse.

Suggesting steps to control marine pollution

(1) Dumping of oil ballest, hazardous and toxic substances, gases from radioactive labs into sea, should be banned or should be properly treated before dumping.

(2) Drainage, sewage sludge and effluents from industries should not be discharged in to rivers which joins sea.

(3) Developmental activities on coastal areas should be minimised.

(4) Toxic pollutants from industries and treatment plants should not be discharged into sea.

(5) Ships and ports should have certain facilities for reducing pollution.

(6) Certain biological and other methods should be followed to restore species diversification and ecobalance in the water body to prevent pollution.

(7) Effective measures should be developed to check the leakage in ships and oil tankers.

(8) Nuclear explosions and other nuclear activities in sea should be minimized.

(9) Wastes from municipal, industries, sewage and thermal power stations should be recycled for reutilization. Such plants should be developed. Some are in operation.

(10) We should develop awareness in people to reduce the amount of waste in their daily life.

(11) Drilling should not be allowed in coastal areas.

NOISE (SOUND) POLLUTION

INTRODUCTION :

The term 'noise' may be defined as an unwanted sound at a wrong time and a wrong place. Although noise is undesirable. It could be meaningful or meaningless. A meaningful noise is generally meant for inviting attention or expecting a consequent response such as the cry of a baby or a screaming of a person for help. On the contrary, in irresponsible or meaningless noise is disturbing and annoying. Whether a given sound is wanted or unwanted may depend upon the person involved, the loudness, the rhythm, and the length of time for which one is exposed to it. A sound may be music to one person but noise to another; acceptable when soft, rhythmic or for short time, but unacceptable when loud, random or prolonged; reasonable when made by himself but unreasonable when it is made by other. However, prolonged and loud sound is generally considered as noise which is mostly caused because of industries, vehicles, aeroplanes etc. Sound is a special kind of wave action which is usually transmitted through air in the form of presence waves. These waves are received by hearing approaches of animals, including man transformed in to electrical impulses in the ear and carried to the brain which enables us to hear.

Sound has several physical properties among which " Frequency" and "Intensity" are the most relevant for the present discussion. Sound frequency is the rate at which compression waves arrive at or pass a fixed point. "Pitch" is the human perception of sound frequency (and also intensity to some extent). Sound intensity is the acoustical power (i.e. the energy delivered by sound) per unit area. "Loudness" is the human perception of the sound intensity (and also frequency to some extent). "Hertz´(Hz) or cycles per second is a measure of sound frequency. Human beings can hear only sounds ranging from 20 Hz to 20,000 Hz. The range of frequencies of human speech is 200 to 3000Hz, which is best heard by humans. Thus we can hear only sounds of certain frequencies and even among them, we are more sensitive to some than to others. Sounds too high in frequency (above 20,000 Hz) is called "ultrasound" and that which is too low in frequency (below 20 Hz) is called "infrasound".

The response of ear to sound is proportional to the logarithm of its intensity or pressure. The loudness of two sounds is judged subjectively by the ear by the ratio of their intensities or pressures. The loudness is expressed in terms of a unit called "decibel" ("deci" comes from the Latin word for ten, and a "bel" is "the" logarithm of ratio" of any two acoustical or electrical intensities. In terms of sound, a "decibel" (dB) is ten times the logarithm of the ratio of two sound intensities, one being the intensity of any sound of interest and the other being a reference sound (I).

$$\text{Decibel (dB)} = 10 \log \frac{\text{Sound intensity measured}}{\text{Reference sound intensity}}$$

In the United States, the Reference sound intensity is 10^{12} watts per square meter, which is the sound intensity barely audible to human beings. This zero dB is the threshold of hearing. Thus, a sound which is 100.000 times louder (more intense) than the reference level would be called a 50-dB sound (because the logarithm of 100,000 is 5 and 10 times 5 is 50). Similarly, sound with 10 times the intensity of the reference level would be a 10 dB sound. A normal conversation is done at 60 dB sound levels. A jet plane during its takeoff produces a 150 dB sound. The decibel scale for some familiar sound sources is given in Table 5.5. The psychological and physical harm at different decibel levels is shown in Table 5.6.

Table 5.5. The Decibel Scale

Sound Source	*Decibel, dB*
Launching of space rocket	170
Jet plane at take off	150
Threshold of pain	140
Pneumatic riveter	130
Running motor cycle	118
Jet fly over at 150 m	115
Rock band	111
Jet fly over at about 300	103
Farm tractor	98
Motor cycle at 25 ft	90
Heavy city traffic	85
Alarm clock	80
Average city traffic	70

Normal conversation	60
Business office or light city traffic	50
Living room	45
Library	35
Broadcasting studio or A quiet room at night	20
Rustling of leaves	10
Threshold of hearing	0

Table 5.6. Psychological and Physical Effects at Different Decibel Levels

135 dB	..	Painful
110 dB	..	Discomfort
88 dB	..	Hearing impairment on prolonged exposure
80 dB	..	Annoying
65 dB	..	Intrusive

Effects of Noise

(a) Physiological Effects:

The acute effects caused by noise depend upon the pressure and frequency. At high levels of about 150 dB, immediate permanent hearing impairment may be caused. At sound levels in the range of 120-150 dB, effects on respiratory system, dizziness, disorientation, loss of physical control, other physiological changes resulting from stress, nausea and vomiting may be caused. Even sounds of the order of 70 dB can have measurable physiological effects, although they may not result in any immediate impairment.

Loud sounds can cause an increased secretion of many hormones of the pituitary gland e.g., adrenocorticotropic hormone (ACTH). ACTH in turn stimulates the adrenal gland, which secretes several other hormones. Through a variety of influences, these hormones in turn trigger various effects such as (a) enhancement of the sensitivity of the body to adrenalin, (b) increase of blood-sugar levels (c) suppression of immune system and (d) decreasing the efficiency of liver to detoxify blood.

(b) Psychological effects

Although there is little specific evidence regarding the onset of mental or nervous illness caused by noise, some reports are available to indicate temporary effects such as deterioration in concentration and even mental disorientation at high noise levels.

Loud continuous noise reduces the working efficiency, interferes with communication, increases the frequency of errors which may, at times, cause accidents. Noise reduces the mental capability. Noise has psychological effects on humans ranging from mild distress to complete unhinging.

Noise interferes with deep sleep and interrupts sleep. Because sleep is important to emotional stability, noise may contribute to distress and emotional disturbance. Noise also aggravates any existing psychological conditions and mental illness.

(c) **Hearing Loss**

Prolonged exposures to loud noise can cause temporary or permanent loss of hearing. People working in noisy places such as industrial establishments, factories etc. Often suffer from temporary loss of hearing. The ciliary cells in the inner ear are inactivated or numbed and the threshold of hearing of the subject is raised. If the loudness of noise is moderate or the duration of exposure is short, the damage is only temporary. The auditory system recovers itself when the exposure ceases. In Audiometric tests, the phenomenon is referred to as **Temporary Threshold Shifts** or **TTS**.

Longer exposures to louder noises may cause permanent shift in the threshold of hearing of an individual. The individual in such cases suffers from partial but permanent loss of hearing. He is no longer able to hear low sounds which are audible to normal persons. This is caused by slow and chronic damage to ciliary cells in the inner ear. Still, medical science is of little help in such cases.

Very loud, sudden and impulsive noises, such as a bomb blast, are capable of causing acute damage to auditory system and an abrupt loss of hearing. With or without involvement of inner ear, it is the middle ear which is affected in most of the cases. High intensity sound waves damage the ear drums and may disrupt the delicate bony chain which carry sensation from ear drums to the inner ear. Very fine surgical techniques have been developed to restore the hearing ability where only middle ear is involved.

(d) **Other health effects of noise pollution**: Loud noise is nuisance which affects sleep, concentration and work or performance of an individual. Work which needs a high degree of skill and precision is considerably affected. It may cause headache, irritability and fatigue. It is interesting to note that our optical system is considerably affected by noise pollution. Dilation of pupils, impairment of night vision, decrease in colour perception ability are some of the effects caused by exposure to loud noise for long durations.

Noise affects our cardiovascular system also. Loud noises tend to decrease the output of blood from heart, cause arterial blood pressure to fluctuate and smaller blood vessels of the body constrict reducing the flow of blood to the organs concerned. Heart beat rate is affected. Changes in breathing amplitude have been reported due to sudden and impulsive noises. Eosinophilia, hyperglyaemia, hypokalaemia and hypoglycaemia may also be caused by changes in blood circulation and other body fluids due to noise pollution (Kryter, 1970).

(E) Prevention and Control of Noise Pollution

Loud noise is the form of pollution which often causes much public concern. Therefore, necessary steps have to be taken to control the nuisance. Some of these are :

1. **Reduction of noise at the source of its origin:** Often a little precaution can reduce much of the nuisance caused by loud noise. This can be achieved by replacement of noisy rattling devices or machines with quieter ones. Noise level can be reduced effectively by replacement of noisy and rattling parts, providing better cushioning to check the vibrations, proper oiling and greasing to ensure smooth running and using effective silencers etc.

2. **Application of sound proofing techniques to muffle down loud noises:** Sound waves are absorbed by porous material such as perforated sheets and other objects. Just as putting cotton plugs in the ears reduces noise level for the individual concerned, sound barriers placed around the source of origin of loud noises drastically reduce the intensity of sound on the other side of the obstacle. For example, little of loud noise produced in picture halls and auditoria escapes out because of effective sound proofing and acoustic techniques are applied for the purpose. The same can be done for industrial units also.

3. **Keeping residential localities free of noisy industries, busy highways, aerodromes etc.**

Residential localities should be established away from noisy industries, busy highways, aerodromes or else these noisy establishments should be developed away from quiet residential areas. Industrial units can be displaced to some industrial area whereas by passes may be developed to divert busy railway tracks and highways away from domestic establishments. Only that part of traffic should be allowed to get into a residential area which is barely necessary. This shall curb much of the nuisance caused by noise pollution.

4. **Enactment of strict legislation and its effective compliance**: In most of the countries including our own, legal framework against noise pollution has been developed. However, in most of the cases little efforts are made to enforce these rules and regulations effectively. If we ensure

only effective compliance of these rules much of the nuisance of noise pollution shall automatically be curtailed.

5. **Noise control methods in industrial plants**

Excessive noise is produced from various types of machines, petrol and diesel engines, electric motors, construction site equipment, pumps and pumping systems, compressed air systems, hydraulic systems, air distribution system, industrial fans, etc. It is always advantageous, economical and effective to identify the noise sources and noise problems right in the design and erection stages and incorporate the necessary noise control measures rather than attending to the problems at a later stage.

The various noise sources in industrial plants and the methods available for noise control are summarized in Table 5.7: given below:

Table 5.7: Typical Noise Sources in Industrial Plants and Methods of Noise Control

Equipment	*Noise Source*	*Methods for noise control*
Bulldozer, crane, compactor, excavator, dumper, shovel, scraper, etc.	Engine.	Fitting of more efficient silencer or exhauster, closing the enclosure panels, if fitted.
Rotary drills, diamond drilling and boring	Drive motor and bit.	Use of machines inside acoustic covers.
Riveters.	Impact on rivet.	Enclose the working area in acoustic screen.
Pumps	Engine pulsing.	Enclosure in acoustic screen, allowing for engine cooling and exhaust, use of antivibration mounting, flexible couplings and hoses, maintaining adequate inter pressure.
Motors	Cooling fans	Intake muffler, unidirectional fan.
	Cooling systems.	Absorbent duct liners
	Electrical motors.	Enclosure
Engines.	Air intake and exhaust.	Use of mufflers
	Cooling fan.	Enclosing intake and discharge lines, use of quieter fan
Vibrating screen.	-	Stiffening and damping.
Pneumatic equipment	Air and steam vents.	Use of mufflers: quieter valves
	Air jets	Use of mufflers, limiting of air velocities, improving orifice design.
	Ducts.	Lagging, adsorbent lining, limiting of air velocity.
Furnaces	Combustion	Use of acoustical plenum: seals around control rods.
	Ducts	Lagging, use of mufflers.
Turbines	-	Use of enclosure, use of intake and outlet mufflers.

Compressors.	Discharge piping and expansion joint.	Lagging and use of in-line muffler.
	Air intake and exhaust.	Use of mufflers
	Intake piping and suction drum	Lagging.
	Speed changers.	Enclosure or constrained damping on case.

Approaches for Noise Control

The following four approaches are available for noise control :

(1) Modifying some of the present practices and procedures in order to minimize the noise.

Ex:Reducing automobile traffic, outlaying sirens, discouraging stereos without headsets, using glue instead of rivets, etc.

(2) Shielding the sources of noise generation.

Ex: Use of sound-absorbing motor mountings, better installation, better design, use of motor enclosures, use of vibration damping or absorbing materials in automobiles and dishwashers, etc.

(3) Shielding the noise receiver.

Ex: Using earplug, control booths, etc.

(4) Shifting noisy sources and things away from people

Ex: Isolating airports, industrial complexes, etc.

Obviously, some of the above measures can be implemented successfully only if they are mandatory.

THERMAL POLLUTION

The term Thermal Pollution has been used to indicate the detrimental effects of heated effluents discharged by various power plants. It denotes the impairment of quality and deterioration of aquatic and terrestrial environment. Various Industrial plants like thermal, atomic, nuclear, coal fired plants oil field generators and mills utilize water for cooling purposes.

The heated effluents are discharged at a temperature 8 to 10^0 C higher than the temperature of intake waters, which reduces the concentration of D.O. (Dissolved Oxygen).

Thermal Pollution: It can be defined as:

1. The warning up of an aquatic system to the point where desirable organisms are adversely affected.

2. Addition of excess of undesirable heat to water that makes it harmful to man, animal, plant or aquatic life or other wise causes significant dangers to the normal activities of aquatic communities in water.

3. Heated effluents either from natural or man made sources, contaminated with water supplies, may be harmful to life because of their toxicity, reduction in Dissolved Oxygen (D.O.), aesthetically unsuitable and spread diseases.

4. It reduces the number of aquatic species and destroys the balance of life in streams as is evidenced by the biological indices of community and diversity.

5. It is a by-product of rapid and unplanned industrial progress and over population.

SOURCES OF THERMAL POLLUTION

The accelerated pace of development, rapid industrialization and extensive population density have increased demand of thermal power plants. Human activities, today, are constantly adding pollutants to air and water at an alarming rate. The following sources contribute to thermal pollution:

1. **Nuclear Power Plants**: Nuclear power plants, including drainage from hospitals, institutes, nuclear experiments and explosions, discharge a lot have unutilized heat and trapped radionuclides into nearby water streams. Emissions from nuclear reactors and processing instruments are also responsible for increasing the temperature of water bodies. Heated effluents from power plants are discharged at 10°C higher than the coolant receptor and severely affect the aquatic flora and fauna.

2. **Coal-fired Power Plants:** Some thermal power plants ultimately discharges effluent having temperature difference of 15°C between effluent and water body. The Thermal power plants utilize coal as fuel and they constitute the major source of thermal pollutants. The heated coils are cooled with water from nearby like or river and discharge the hot water back to the receptor water body and thereby increasing the temperature of the nearby water. The heated effluent decreases the dissolved Oxygen content of water. It results into killing of fish and other marine organisms.

3. **Industrial Effluents:** Industries generating electricity, like coal as fuel and Nuclear powered thermal plants, require huge amounts of cooling water for heat removal. Other industries like textiles, paper and pulp as well as sugar also release heat in water but to a much lesser extent. The heat from the turbo-generators installed in industries have temperature of effluent as 5°C to 9°C more than the normal temperature of stream. To cope with the increased demand of electricity and rapid industrialization the number of installations are raised which results in discharge of more volume of water/heated effluent and above the receptor water body temperature.

4. **Hydro-electric Power:** The generation of hydroelectric power, sometimes, results in negative loading in water systems. Apart from electric power industries, various factories with cooling contribute to thermal loading. It has been reported that about 18% more heat is given to cooling ponds in nuclear power plants than any other plant of equivalent size.

5. **Domestic Sewage**: Domestic sewage is commonly discharged into rivers, lakes and canals with or without waste treatment. The municipal sewage normally has a higher temperature than receiving water. The discharged water not only raises the stream temperature to a measurable extent but also creates numerous deleterious effects on aquatic biota. The organic matter present in the sewage utilizes the dissolved oxygen present in the surface water for oxidation. With the increase in the temperature of the water, the D.O. content decreases and the demand of oxygen increases. Hence, the anaerobic conditions will result in the release of foul and offensive gases. The marine life dependent upon the D.O. will die out and the quality of water is also adversely affected.

EFFECTS OF THERMAL POLLUTION: The various effects of the thermal pollution are:-

1. **Reduction in Dissolved Oxygen**: Concentration of dissolved oxygen decreases with increase in temperature of water. For example, the D.O. content is 14.6 ppm in water at a temperature of 32°F and 6.6 ppm at 64°F. Thus cold-water fish, which requires about 6 ppm to survive, would not tolerate the high water temperatures. If they remained in the area they would die of oxygen starvation. Since the aquatic biota live acrobically, so a healthy stream should have an adequate supply of dissolved oxygen.

2. **Change in Water Properties:** A rise in temperature changes the physical and chemical properties of water. The vapour pressure increases sharply, while the viscosity of water decreases. The decrease in density, viscosity and solubility of gases increases the settling speed of suspended particles, which seriously affect the food supply of aquatic organisms.

3. **Increase-in Toxicity:** The rising temperature increases the toxicity of the poison present in water. A 10°C rise in temperature doubles the toxic effect of Potassium cyanide, while an 80°C rise in temperature triples the toxic effect of O-Xylene causing massive mortality of fish.

4. **Interference with biological Activities:** Temperature is considered to be of vital importance to physiology, metabolism and biochemical process in controlling respiratory rates,

digestion, excretion and overall development of aquatic organisms. The temperature changes totally disrupt the entire ecosystem. Sharp changes in temperature are often destructive. Because, the life of aquatic animals involves several chemical reactions and the rate of these reactions vary according to changes in temperature.

5. **Interference with Reproduction:** In fishes, several activities like nest building, spawning, hatching, migration and reproduction etc. depend on some optimum temperature. For instance, the maximum temperature at which lake trout will spawn successfully is 8.9°C. The warm water not only disturbs spawning but also destroys the laid eggs.

6. **Variations in Reproductive Rate:** The increase in temperature triggers deposition of eggs by female. The triggering is particularly dramatic in estuarine fish, which spawn in four hours after the water temperature reaches critical level.

7. **Changes in Metabolic Rate:** Fishes show a marked rise in basal rate of metabolism with temperature to the lethal point. The respiratory rate, oxygen demand, food uptake and swimming speed in fishes increase.

8. **Increased Vulnerability to disease:** Activities of several pathogenic microorganisms are accelerated by higher temperature. Hot water causes bacterial disease in certain fishes such that they fail to develop eggs above critical temperature.

9. **Invasion to destructive organisms:** Thermal pollutants may permit the invasion of organisms that are tolerant to warm waters and highly destructive e.g. invasion of ship worms into New Jersey's Oyster Creek.

10. **Undesirable Changes in Algae Population:** The life in an ecosystem is greatly influenced by the algal growth. Excess nutrients from the washout waters from farmlands, thermal plants cause an excessive algal growth with consequent acceleration of eutrophic and other undesirable changes.

11. **Destruction of Organisms in Cold Water:** The volume of water required for cooling purposes from a stream is enormous. Unfortunately many of plankton, small fish, insect larvae that are sucked into the condenser along with cooling water are killed by the thermal shock, increased pressure and water viscosity.

12. **Biochemical Oxygen Demand**: When the temperature of stream carrying biodegradable organic matter rises, the intensified action of aquatic organisms causes B.O.D. to be accomplished at a lower temperature. When the temperature of stream carrying biodegradable organic matter rises fish death may occur due to synergistic action, which is caused due to accelerated chemical or biochemical action.

13. **Effect on Marine Life:** Temperature plays an important role in affecting the physiology, metabolism, growth and development of marine animals. Sea organisms are poilkilothermic i.e. their body temperature varies with the surrounding water. Some marine creatures cannot tolerate wide changes of temperature, so they die at higher temperature.

14. **Effect on Bacteria:** Due to the heated discharges from the industries and plants (industrial), the bacteria are severely damaged. The effect includes coagulation of body Protein, melting of cell fats, toxic action of metabolic products etc.

CONTROL OF THERMAL POLLUTION

Heat must be removed from the condenser cooling waters prior to their disposal into water bodies. The major principles involved in the process of heat loss are:-

1. Conduction
2. Convection

3. Radiation
4. Evaporation

The following methods can be adopted to control high temperature caused by thermal discharges:

(1) **COOLING PONDS:** The cooling towers are beneficially used in dissipation of heat as shown in fig.5.1.

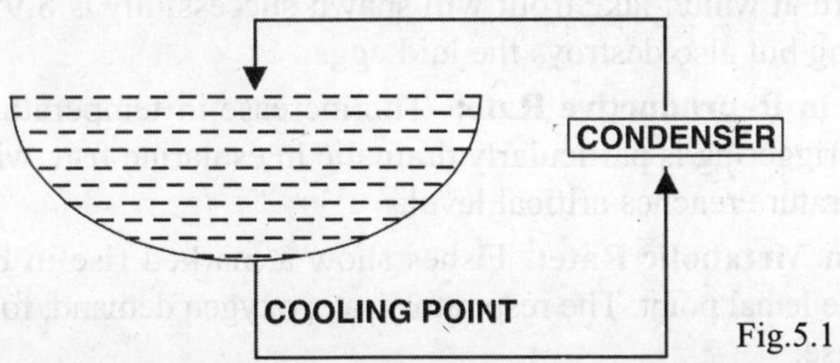

Fig.5.1

The water from the condensers is stored in the earth like ponds where natural evaporation brings down the temperature. The water is re-circulated again. Another method for installation of cooling ponds is shown in fig.5.2.

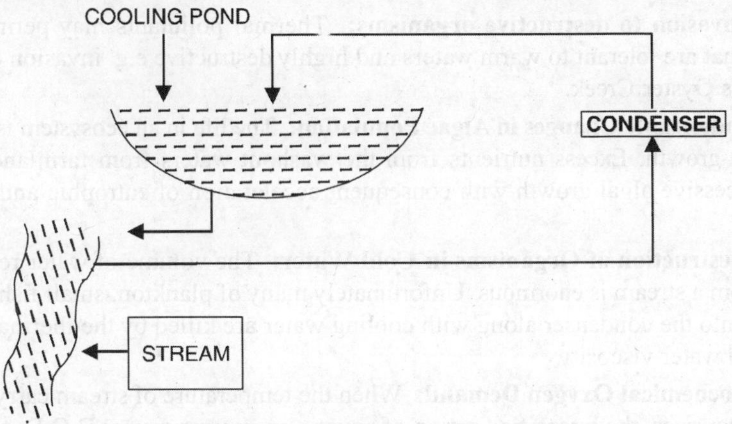

Fig.5.2

(2) **SPRAY PONDS**: In spray ponds, the water is sprayed in the cooling ponds with the help of spray nozzles to convert it into fine droplets which provide more surface area to facilitate efficient heat transfer to atmosphere.

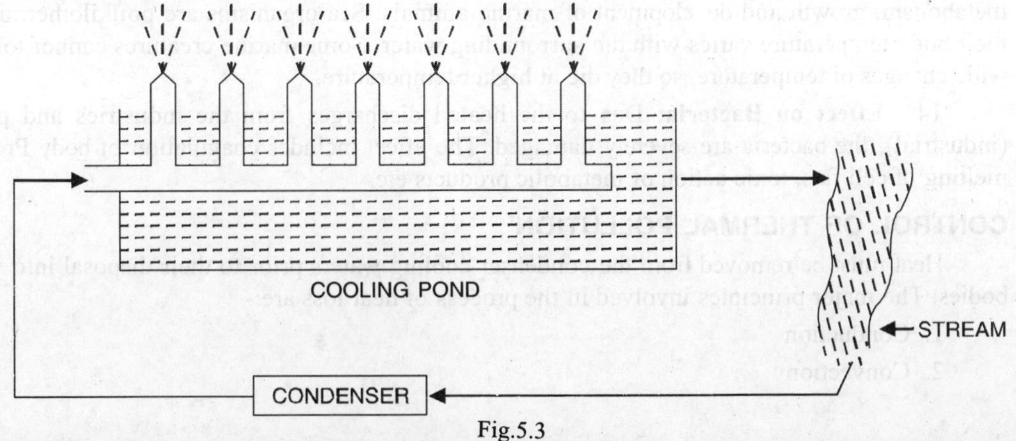

Fig.5.3

(3) **COOLING TOWERS:**

Wet Cooling Towers: In wet cooling towers, the heated water is brought in direct contact with continuously flowing air. The evaporation brings down the temperature. To increase the surface area of contact, the water is broken down into droplets by use of spray nozzles or by splashing it on the packing or baffles in the cooling towers.

(4) To handle large quantities of heated effluents, large tanks or reservoirs should be constructed to retain the water for a little longer time. When water cool down to a tolerable temperature, it may be released.

(5) The heated effluents discharged from the chemical industries and thermal power plants can be put in to certain beneficial uses like green house, frost protection during colds, aquaculture, heating the buildings etc.

NUCLEAR HAZARDS

Hazard mean dangerous to human being or so by external source. This external source is from environment. When our environment is polluted then no one can escape from the pollution hazard. It has become a part of our life. A number of atoms possess the ability to limit radiations and thereby cause radioactive pollution. Radiations originate from instability of the nuclei of an atom which loses sub-nuclear particles and energy to acquire a stable state i.e radioactivity. It is the state of nuclei which is responsible for the phenomenon. Neutrons and protons constitute the nucleus while electrons revolve round the nucleus (units outer orbits). When the member of protons are equal to number of electrons, the chemical properties shall remain the same. Neutrons and protons constitute the mass while electrons constitute change to the element. Thus *Radioactive element* is defined to be the collection radioactive mass with the same change of the nucleus. The radioactive atom has the same change of the nucleus and the same mass is called *Radioactive isotope*. The *Radioactivity* of a radioactive substance is expressed by the number of nuclear transformations in unit time.

A radio isotope is characterized by the following properties:-

(*i*) Half life period. (*ii*) Mode of decays.
(*iii*) Energy of radiations (*iv*) Definite energy state

Radiation is the emission of rays and particles or release of energy from the source (atom). There are two types of radiations ionizing and non-ionizing radiations. These radiations destroy the organic molecules of which the body cells are composed. If ion pairs enter into a living protoplasm, they damage it and the damage is proportional to the number of ion-pairs absorbed. The following types of radiations are given out when an element transmutates or decays.

(i) **Emission of alpha (α) particles**. Alpha particles are nothing but Helium nuclei. Emission of alpha particle will change into elements of lower atomic number. These are deflected by electric and magnetic fields. They are to show moving, strongly ionizing, weakly penetrating and stopped by 80 mm of air.

(ii) **Emission of Beta particles**:- **(β)** Emission of Beta particle changes into another element with a higher atomic number. Beta particles are high velocity electrons. Strongly deflected in electric and magnetic fields. The penetrating power of Beta particles varies with the energy of particles.

(iii) **Emission of Gama rays (γ).** These are high energy electromagnetic radiations. Can penetrate several cm. of Lead sheet depending upon the energy. These are undeflected in magnetic fields.

Radioactive decay is a spontaneous process arising from nuclear instability.

Sources of Radio Active Pollution

The two main sources of radioactive pollution are, natural and manmade.

NATURAL SOURCES

The natural sources of radioactivity are considered mainly of the cosmic radiation received from the space, and the naturally occurring radioisotopes present in the environment and those contained within the body of the organisms. The cosmic radiations are of extraterrestrial origin, which probably arise from the sun or even beyond it. They are consisted of particles of very high energy, primarily of protons and some heavy nuclei. These cosmic particles collide with the gas molecules of the upper atmosphere bringing about intense ionization in gases accompanied with the formation of secondary cosmic rays composed mainly of neutrons, mesons, and gamma rays. Eventually a complex mixture of particles reaches the earth as cosmic rays. These particles also form substantial quantities of 3_H and 14_C in the atmosphere.

Another source of natural radiation is the presence of radionuclides in the lithosphere, hydrosphere and atmosphere. All the elements above atomic number of 82 (Lead) are radioactive in nature and emit a variable quantity of radiations. The most abundant naturally occurring radionuclides on the earth are Uranium, Thorium and Potassium-40. Soils, rocks and even building material contain small quantities of 40_k; and Uranium and its daughter elements.

Man-Made Source

Man causes radioactive pollution by testing of nuclear weapons, establishment of nuclear power plants, mining and refining of plutonium, and thorium, and preparation of radioactive isotope.

1. Nuclear weapons

Testing of nuclear arms comprises:

(a) The use of Uranium 235 and Plutonium 239 for fission.

(b) Hydrogen or lithium as fusion material.

Atomic explosions are uncontrolled chain reactions. They give rise to very large neutron flux conditions that cause other materials in the surrounding environment to become radioactive.

Huge clouds of fine radioactive particles and gases are thrown up in the environment and are carried away to distant areas by the agency of wind. Gradually they settle down on earth as fall out or are brought down by rain.

2. Atomic Reactors and Nuclear Fuel

The most common fuel used for fission in the nuclear power plants are uranium, thorium and plutonium. Uranium undergoes several processes, right from its mining to its inception into the reactors. The spent materials obtained from the reactors, after the energy has been utilized, are reprocessed to recover unburnt uranium, plutonium and some other important isotopes, which can be used in medicine or for some other useful purposes. The whole operation from the mining of the fuel to its final disposal is called "nuclear fuel cycle".

At almost all stages of the nuclear fuel cycle, liquid, gaseous and solid radioactive wastes are released having a tremendous potential to contaminate the environment and hence, a great care is to be taken for the environmental safety during the nuclear operations.

3. Radioactive Isotopes

Radioactive isotopes such as 125_I, 14_C and 32_P and their compounds find wide usage in scientific research institutions contain varying amounts of radioactive materials. When this waste water reaches the different water sources such as rivers, streams, lakes etc. through the sewers they cause water pollution. Radioactive iodine and phosphorus also enter the food chain through water and may finally reach man through fish etc.

4. Other Sources

During different medical treatments, varying concentrations of radiations enter the human body for instance, X-rays are common for detecting skeletal disorders, and therapy for cancer patients often includes radium and other isotope radiations.

It has been reported that about 240 million dental and medical X-rays are taken annually and that 15 million tests using radioactive materials as tracers in the human body are also made.

A common type of ionizing radiation is X-ray which is produced by radiographic equipment. X-ray therapy equipment, dental X-rays that can operate at a voltage about 10 KV produce more penetrating radiation and may be more hazardous, if not properly shielded.

Damages to a Biological System

Most of the damages caused by radioactive pollutants stem from their capacity to produce high energy radiations, which are very harmful to a living system. There are two main modes in which radioactive pollution can be dangerous to a biological system.

(i) Damages caused by radiations from outside source.

(ii) Damages caused by radiations from sources inside the body.

Damages caused by Radiations at different levels

(i) Damages at Molecular level

Damages to macromolecules such as enzymes, DNA, RNA etc. through ionization cross-linkages within and between two affected molecules.

(ii) Damages at sub-cellular level

Damages to cell-membranes nuclei, chromosomes such as fragmentation, mitochondria etc.

(iii) Damages at cellular level

Inhibition of cell division death, decay and transformation to malignant state.

(iv) Damages to Tissues and Organs

Disruption of such systems as central nervous system, loss of sight, inactivation of bone marrow activity resulting in blood cancer malignancy and ulceration of intestinal tract.

(v) Damages to an Individual and whole population

Death or shortening of life due to radiations changes in characteristics due to mutations. In human beings exposure of radiations results in little visible effects in early stages. But after 12-24 hours injury symptoms manifest themselves. This includes reddening of skin, anemia, anorexia. Vomiting, and diarrhoea and with heavy doses, blister formation, pigmentation of skin, burning sensation all over the body, loss of sight etc. It must be noted that for all this there is no cure available. Once a person is exposed to radiation he has to bear its consequences. Medical aid can do little.

HAZARDS ASSOCIATED WITH RADIO ACTIVE POLLUTION

Radio-active pollutants are not like other pollutants which are sooner or later converted into harmless material and degraded into simpler constituents to be recycled into the ecosystem and used again. It is not the element itself, but the instability of its nuclei which is responsible for damages caused by these pollutants. As long as radiations continue, these wastes are dangerous for the living beings. After the emission of radiations nuclei attain stable state and behave like any other element in the environment or the biosphere. Life on earth's crust could evolve only when the nuclear activity ceased, atoms acquired stable state and radiations were reduced to the level which could be tolerated by living beings. Major hazards associated with radio-active pollution can be summed up as follows:

(1) No physical, chemical or biological process can influence the process of radio active emissions. The unstable nuclei have to decay and acquire a stable state.

(2) A number of radio-active isotopes have a very long half-life. Thorium-232 ($_{90}Th^{232}$) takes 14,0000,000,00 years to lose half of its radio-activity. Half of Uranium-235 ($_{92}U^{235}$) takes 710,000 years to disintegrate. Half of Neptunium-237 ($_{93}Np^{237}$) decays in 21,00,000 years. This makes these radio-active wastes almost a permanent hazard for the biosphere.

(3) Most of the radiations have a high penetrating power. Thick sheets of steel, cement concrete walls etc, can not contain them. They can easily penetrate to deep seated organs and cause injury.

(4) Nucleic acids (DNA and RNA) effectively absorb these radiations. Even low level radiations which do not cause any visible damage are completely absorbed by nuclear material which causes carcinogenic, mutagenic and teratogenic effects.

(5) A biological system is unable to distinguish between a radio-active and a normal isotope of an element as their physical and chemical properties are similar. Radio-active isotopes are therefore, absorbed and incorporated within the bodies of living organisms as normal isotopes are. This lodges a radio-active source within the body of the organism itself.

(6) Like any other element radio-active isotopes are also absorbed, accumulated and bio-magnified thousands of times. Thus the entire food chain becomes contaminated. Organisms at higher trophic levels may, therefore, receive a highly concentrated source of radio-active material through their food supply.

(7) There is no other way to dispose off these hazardous wastes except to store them for thousands or millions of years away from living beings. This is too long a period on human scale of time. Even the safest burial places for radio-active wastes, which represent the best of human efforts, have shown signs of leakage. At present it appears very difficult, though not impossible to store radio-active wastes away from the biosphere for such long periods.

(8) In spite of all these hazards, nuclear reactors and tests are still continuing and an increasingly large amount of radio-active wastes is accumulating every day while no solution to the problem of their safe disposal is in sight till date.

The *harmful effect of radiation* upon human beings is due to its ability to wise and ultimately destroy the organic molecules of which body cells are composed of. The damage depend upon, the energy and the type of radiation. The energy is expressed in **Rads** (i.e. absorption of 100 ergs or 10^8 joules of energy per gram of tissue). The total biological effect of radiation expressed in **Rems.**

Number of rems = n × Number of rads

Where n = 1 for β, γ and x-rays.

= 10 for α radiations or high energy neutron.

Control of Radioactive Pollution

Control of natural radioactive pollution may not be possible. Out of all the sources, only artificial radioactivity is the scope of intervention, wherever controls can be thought of. Radioactive pollution can be controlled by strict enforcement of the following safety measures.

All low or high level wastes have tremendous capacity to pollute the environment. As low level wastes are often produced in large quantities, their containment is not possible. They are visually subjected to a treatment for removal of radioactivity and then discharged in the water bodies or on land in usual way. High level wastes, on the other hand, cannot be disposed of freely in the environment, but have to be concentrated, contained and stored out of the reach of human's environment.

The radioactive wastes concerned with water pollution are usually in liquid or solid state. These different kinds of wastes pose various problems, as disposal techniques suitable for one kind may be risky for other. All techniques however have a single goal that radioactive constituents of wastes are not allowed to cause harm to organisms and in particular humans.

SOLID WASTE MANAGEMENT

Any material that is thrown away or discarded as useless and unwanted by human or from animal activities is considered as solid waste. In earlier period, the disposal of solid waste was simple but now a days it is a great challenge. The management of waste is the fundamental concern of the activities encompassed in solid waste management. The purpose of the study of solid wastes is to –

(i) Identify the various types of solid wastes and their sources.
(ii) Examine the composition of wastes.
(iii) Consider the elements involved in their management.

The activities involved with the management of solid wastes from the point of generation to final disposal have been grouped in to six Functional Elements –

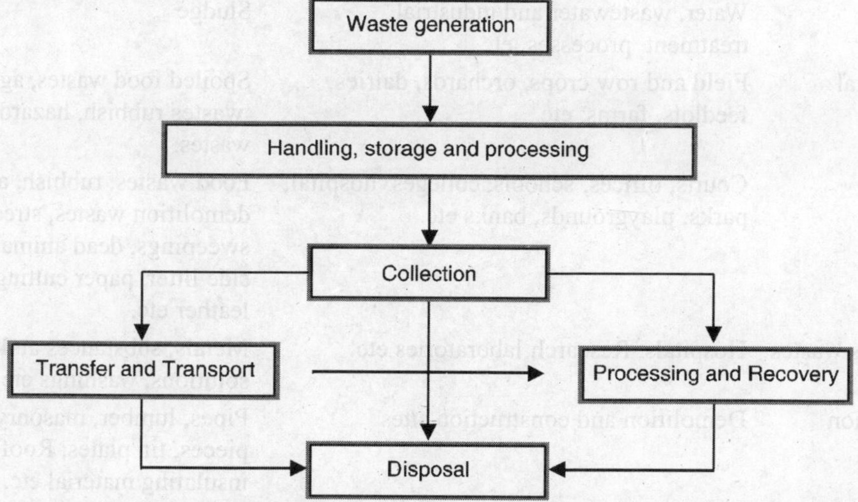

The total quantum of solid waste generated in an area depends upon its population and urbanization. Solid wastes generation is directly related with income. Higher the income greater is the waste generation.

Sources of solid wastes

Sources of solid wastes can be clarified in to following categories:-

1. Residential
2. Commercial
3. Municipal
4. Industrial
5. Open areas
6. Treatment plants
7. Agriculture
8. Hazardous wastes
9. Construction sites

Details of the above is given in this table:-

Table 5.8 Sources of solid wastes

Source	*Location—Wastes are generated*	*Types of solid waste*
Residential	Single-family and Multi-family Houses, low-medium and high rise apartments, etc.	Food wastes, rubbish, ashes, special wastes.
Commercial	Restaurants, Markets, Stores, hotels Institutions, office, Workshops etc.	Food wastes, rubbish, Ashes, demolition and construction wastes, special wastes.
Industrial	Construction, Fabrication, Light and heavy manufacturing, chemical plants, Mining, power plants, deduction, etc.,	Food wastes, rubbish, ashes, demolition and construction wastes, special wastes, hazardous wastes.
Open areas	Sheets, parks vacant lands, playgrounds, beaches, etc.	Special waters, rubbish.
Treatment	Water, wastewater and industrial treatment processes, etc,	Sludge
Agricultural	Field and row crops, orchards, dairies, feedlots, farms, etc.	Spoiled food wastes, agricultural wastes rubbish, hazardous wastes.
Municipal	Courts, offices, schools, colleges, hospital, parks, playgrounds, banks etc.	Food wastes, rubbish, ashes, demolition wastes, street sweepings, dead animals, road side litter, paper cuttings, glass, leather etc.
Hazardous wastes	Hospitals, Research laboratories etc.	Metals, substances and their solutions, washings etc.
Construction	Demolition and construction sites	Pipes, lumber, masonry brick pieces, tin plates, Roofing and insulating material etc.

TYPES OF SOLID WASTES:

Garbage: Food wastes are the animal, fruit, or vegetable residues resulting from handling, preparation, cooking, and eating of foods. It is also known as garbage.

Rubbish: Rubbish consists of combustible and non-combustible solid wastes of households, institutions, commercial activities, etc, excluding food wastes or other highly pursuable materials.

Ex, Combustible-Paper, Cardboard, Leather etc.,

Non-Combustible -Aluminium cans, tin cans, glass etc.

Ashes and Residues:

Materials remaining from the burning of wood, coal coke and other combustible wastes are categorized as ashes and residues.

Demolition and Construction wastes:

Wastes from buildings and other structures are classified as demolition wastes. Wastes from the construction, remodeling, and repairing of individual residences, commercial buildings, and other structures are classified as construction waste.

Special Wastes:

Wastes such as street sweepings, roadside litter, catch basin debris, dead animals and abandoned vehicles are classified as special wastes.

Agricultural Wastes

Wastes and residues resulting from diverse agricultural activities-such as the planting and harvesting of roe, field, and tree and vine crops, the production of milk, the production of animal for slaughter, and the operation of teed lots are collectively called agricultural wastes.

Hazardous wastes:

Chemical, biological, flammable, explosives, or radioactive waste that are harmful to human, plant or animal life are classified as hazardous wastes.

Collection of Solid Wastes:

Collection of solid wastes in urban areas is difficult and complex because the generation of residential and commercial-industrial solid wastes is a diffuse process that takes place in every home, every apartment building, and every commercial and industrial facility as well as in the streets, parks, and even the vacant areas of every community. The mushroom like development of suburbs all over the country has further complicated the collection task.

As the generation patterns become more diffuse and the total quantity of waste increases, the logistic problems associated with collection become more complex. Although these problems have always exist to some degree, they have now become more critical because of the high cost of fuel and labour. Of the total amount of money spent for the collection, transportation, and disposal of solid waste in 1975, approximately 60 to 80 percent was spent on the collection phase. This fact is important because small percentage improvement in the collection operation can effect a significant saving in the overall cost.

Effects of Solid Wastes

The accumulation of waste at any place is a bad and risky situation. Varieties of micro-organisations like bacteria, fungi, viruses, worms etc creep in to the accumulated waste and start its decomposition. Later on they grow and increase in number.

Various types of germs develop in the waste. They reach us through air, water and food. Most of the infectious diseases like cholera, diarrhoea, dehydration etc. spread in these ways. Air pollution, water pollution and soil pollution are caused due to the accumulation of different types of wastes.

Harmful fumes from industries and other waste effects eyes, skin, historical moments etc. Asbestos particles from Asbestos Industry causes Asbestosis. Accumulation of heavy metal particles cause serious health hazards. Mercury can cause Mina Mata disease.

Wastes material when accumulated here and there disturbs the drainage system. Decomposing wastes reach underground and contaminate underground water and soil.

Improper disposal of Municipal wastes and throwing the household wastes here and there effects the community and themselves. This produces foul smell and breeds various types of insects. Industrial solid wastes are the sources of toxic metals and hazardous wastes, which effects the soil and water. Wastes some times caused the fire in farm and forests, which produce dioxins, furans and polychlorinated biphenyls. These causes the serious ailments like cancer and other chronic diseases. Wastes like cans, pesticides, plastics, batteries, cleaning solvents, radioactive matters, paper, scraps etc. which can be recycles, cause serious effects on mankind in one or other way. Animals are also effected by taking poisonous waste and polythenes.

Management of solid waste

Waste management is the collection, transport, processing or disposal of waste materials so as to reduce their effects on local environment and community. Because it can not be stopped absolutely.

Methods of solid waste disposal. There are following methods:-

(a) **Physical removal.**- It is generally done by manual activities like, collection of wastes and sorting out in to reusable, decomposable and non decomposable. Then disposal becomes easy. Dustbins should be used in homes, offices and dispose accordingly i.e. to kabadi or for reuse, recycle. Some Municipals are also doing such jobs.

(b) **Dumping**- Transfer of solid waste from place of collection to the site of disposal is called dumping. Corporations and Municipal bodies collect and dump them on some suitable and safe site located far away from human habitation.

(c) **Compaction and Bailing**- The solid wastes are often spread on a plane and hand surface and later pressed by bulldozer. This is called compaction. These compacted layers are rolled and piled. This is called bailing. Now such compacted and bailed solid wastes are dumped for decomposition.

(2) 3R or Reduce, Reuse and Recycle of solid waste

(A) Reduce of waste material- We should reduce the household waste by using maximum part of the goods. Before throwing out side, we should select the parts for Reuse/Recycle. When we purchase the things, avoid polythene and heavy packages.

Hazardous waste can be controlled by reduction at source. We should suggest friends, relatives to save all clean papers and other various means to save paper. Gaseous wastes are generally removed through combustion, absorptions and adsorption techniques. There should be proper co-operation and co-ordination among individuals, local bodies and Govt Institutions for proper waste management in an area. Reduced demand for any metallic product will decrease the mining of their metal and cause loss production of waste. Thus, every individual has a responsibility of creating less waste and managing its properly.

(B) Reuse of waste materials. After selecting the waste (which can be reused) use after the proper treatment. We should not use, cups, plates, utensils, napkins etc of paper. If they are of permanent nature, therefore, they can be reused after washing. Plastic bags, wrap, foils, rotten articles should not be used. We should use refillable lighters, containers, and other usable items. We should discourage use and throw policy. Sell or donate goods instead of throwing them out. Furniture, clothes and other repairable articles should be reused after repair instead of throwing. We should develop quality of borrow, share and rent in ourselves.

One should take lesson from poors, villagers who reuse their materials to the maximum due to their financial conditions. We should utilize paper of their optimum use.

(C) **Recycling of waste materials**. Sewar and other drainage systems are associated with sewage treatment devices that centralize toxic effects of sewage before releasing it to the local water systems. Principal operations of solid waste disposal incorporate composting, senitary, land filling, thermal process or incineration.

(i) **Sewage treatment.** It is done through following steps:-

(a) The sewage is sent through setting chambers, where lime is mixed with it. Thus it becomes neutralized and most of the sediment is removed.

(b) Neutralized sewage is passed through Upflow Anaerobic Sludge Blanket (UASB). Here, decomposable material is decomposed through bacterial activities in absence of oxygen. After that water is passed through aeration tanks where air and bacteria are mixed.

(c) Dissolved substances are removed by processes like chlorination, evaporation, exchange technique and absorption.

The treated water is used accordingly.

(ii) *Pulverisation*:- The volume of solid waste is reduced through grinding or smashing for easy handling to transport and disposal.

(iii) *Composting* :-The process of making manure of decomposable waste with the help of microbial activities is called as composting. It is of two types Aerobic i.e. in presence of air and Anaerobic i.e. in absence of air. For this different size pits are dug in the ground and all the biodegradable solid, semi solid wastes are dumped and fully filled pits are covered with a layer of soil. Water is added time to time. The average time for composition is 1-6 months.

Sanitary Landfilling: In this process solid wastes are scientifically filled in to low lands. As all this wastes can not be recycled or burnt, these will be always a need for land fill. In sanitary landfills, garbage and other waste is spread out in their layers, compacted and covered with clay or plastic foam. The process of filling is done in such a way that wastes can not create any type of hazard to public health.

(V) **Thermal process :** Burning of solid waste under controlled conditions is called as thermal process. The heat produced in this process may be utilized. It is carried out in both the presence and absence of air.

Burning in presence of air is called INCINERATION when in absence is done, it is called PYROLYSIS. Incineration of waste is considered to be an unsound practice, because-

- It destroys most of the waste.
- It creates toxic gases and ash, which can harm local populations.
- Release of Dioxins after burning of mixed wastes is hazardous.

We should not throw or dump waste in open or burn illegally, because it effects the environment. Recycle and reuse are very important because limited resources of environment are the properties of coming generations. Therefore, they must be used properly and judiciously in order to keep some of them safe for those generations.

ROLE OF AN INDIVIDUAL IN PREVENTION OF POLLUTION

People say, one and one becomes eleven, eleven and eleven becomes one thousand one hundred eleven and so on .... Then, why an individual can not do for pollution or so. People go with us, if one can start some meaningful project. Mahatma Gandhi was one, Sunder Lal Bahuguna is one, but we can have knowledge of their movements. Gandhi ji said – The soil is capable to fulfil the complete demands of people but can not fulfil the excess desire of people. Soil is mother of a man, then it is the duty of every individual to miscible with that rather destroy. We should not over exploit the nature. We have to control our necessities.

Pollution and poverty are complimentary to each other. Some time people forced to go with the path of pollution. Illiteracy is another factor in the prevention of pollution. Unlimited desires, selfishness, urbanization, industrialisation, deforestation, to increase the life style etc. are some of big factors which are main cause of pollution. For that the need is to understand first, apply then be a lesson for others. Gandhiji was not an environmentalist but was for the concept of sustainable development.

In short, an individual can do as following safety measures to prevent the pollution -.

(1) One should start first in the field of environmental awareness to protect the pollution.

(2) We should go place to place to teach the lesson of awareness and prepare volunteers.

(3) Give the message to save environment through papers, magazines, T.V. and radio.

(4) To promote for plantation and conservation of forest.

(5) To organize seminars, on the subject related to pollution.

(6) One should go in rural areas during festivals, functions, local gatherings, and religious occasions to convince people for prevention of pollution.

(7) Awareness is very effective in childhood, hence we should go to schools, organize rallies to teach the lesson of environment.

(8) World forest day, world environmental day and other such function should be organized for general awareness. On these functions, Govt. should also take interest in this regard but we should not depend on Govt.

(9) Population growth should be reduced.

(10) We should use and promote mass transport system. If possible go on foot or use bicycle for short distances.

(11) We should not use materials containing CFC eg Refrigerators, Cups, etc.

(12) We should discourage the use of more fertilizers insecticides and pesticides but should encourage the use of bio fertilizers.

FLOOD

Floods have ravaged portions of India from time immemorial. Though floods are one of the very few well recorded natural phenomena, the catastrophic damages caused by them attracted focused attention in recent decades. With increasing population pressure and accelerated economic development, the adverse effects of floods are being increasingly felt now. Floods cause great distress whenever they damage crops and property and endanger lives. The term Flood is generally defined as a relatively high flow or stage in a river and the inundation of low land which might result therefrom. In a broader sense the term flood is used to convey all their outfalls into main rivers, outflow due to jamming or blocking of rivers by landslides and inadequate drainage to carry away surface water speedily. Coastal floodings are also covered.

In India vast stretches of land are submerged under water and other adverse effects are caused, such as destruction or damage to houses, property, bridges roads and other means of communication lives lost etc. year after year. The disastrous floods of 1954 and the immediate succeeding years resulted in the initiation of organized and coordinated flood management efforts to mitigate the problem.

CAUSAL PHENOMENA AND CHARACTERISTICS

Flood are natural phenomena characteristic of all rivers. As is known, the rainfall in India is largely dependant on the monsoons and cyclonic depressions. Most of the rainfall is received during the southwest monsoon season during which heavy spells of rain are often experienced in the catchment over the period of a few days at a time. It could therefore be said that high rainfall coupled with inadequate channel capacity leads to flooding. Choking of rivers beds by natural causes or artificial obstructions aggravate the problem.

Flood damages are the combined result of natural phenomena of the floods coupled with the human activity in the flood plains. The fertile river silt has promoted large-scale settlements and cultivation of lands near the riverbanks and adjacent areas or even in the river bed region. The social and economic activities of the people increase. While these activities are going on in one hand, on the other the river continue to experience varying magnitudes and intensities of floods which cause damages, sometimes in disastrous proportions. In a way flood damage is the price paid for the human occupation and exploitation of the flood plain of the river. Even single events could result in a heavy toll of death as also property loss.

As we noticed, the basic cause of flooding is the high rainfall. Apart from that, the size of the catchment also usually governs the character of the flooding. On large rivers with big basins, such as the Ganga or the Brahmaputra, the riverflow in the lower reaches is relatively slow to change; in contrast to this, flash floods, most commonly associated with small catchment lead to very high build up as also lowering. They record very little time between the start of the flood and the peak discharge. Coastal floods are associated with tropical cyclones, storms surges and tidal conditions.

The general characteristics of floods are generally on the lines discussed so far but it must be noted that floodings are the complex results of interaction off a number of connected phenomena and that the flooding characteristic of each river is different from another. They cannot be easily classified even in types or groups.

In this unit, our interest in floods and flooding is not in the scientific phenomenon as such but the damages and economic disruptions caused by them. If there would have been no occupation of the riverfront or economic activities nearby, high floods might come as also subside without mankind being affected or bothered much. We, however, are concerned with flood losses. Flood losses may be defined as the destruction or impairment, partial or complete, off the value of goods and services or of health, resulting from the action of flood waters and the silt and debris they carry.

India is one of the highly flood prone countries of the world. Flood damage statistics, compiled from reports from the State Governments indicate that on an average (based on data for 1953 1990) about eight million ha. on land are affected by floods in India, involving about thirty three million people. In a high flood year, the figures will be many times more. Our neighbour Bangladesh also suffers seriously from floods. The floods of 1988 which caused high losses in India also caused serious flood problems there, affecting 45 million people and crop damage on two million ha. of land.

VULNERABILITY

From the earlier days mankind has learnt to live with nature. As people settled in environs with fertile soils and by the side of waterfronts, for raising food or on strategic considerations such as trade, commerce, communication or defence, they also realized that these regions that sustained them are also disasterprone. They soon learnt that lessons and started taking precautions so as to reduce their risks. The evidence noted in the form of houses build by silt, on the banks of major rivers are of this nature. In course of time the population pressures increased and the vigilance of the people also slackened. Thus mankind's vulnerability started increasing.

An extreme natural phenomenon capable of causing disaster (leading to loss of lives or damage to property) is known as a natural hazard. The process of identifying the probability of occurence of a natural hazard of a given intensity at a specific location, based on an analysis of natural processes and site conditions is termed Hazard assessment. Vulnerability indicate the conditions (physical, socioeconomic/political) which increase the community's susceptibility to disaster or which adversely affect its ability to respond to events. It thus gives an idea of the expected degree of damage to a construction or an economic activity when exposed to a natural hazard of a given intensity. Risks are the probable losses in a given area or to an infrastructure system caused when the hazard materializes.

The type and degree of flooding is influenced by many factors. The principal factors can be classified to fall under three groups.

(1) climatological

(2) hydrological and environmental conditions

(3) local geomorphology off the flood plain

In addition, coastal flooding also depends on the coastal configuration and tidal conditions.

ADVERSE EFFECTS OF FLOODS

All over the world, and throughout history, natural disasters have imposed human suffering and extracted heavy toll of losses. Recent instances have revealed that it is not merely the developing countries that have so suffered. The loss in some of the highly developed Nations is mind boggling notwithstanding the high standards of construction and extensive protection measures that they had undertaken.

Apart from the casualties, injuries and disablement, many sections of the population get affected by the floods. Cropped area gets submerged, eroded and strewn with sand leading to loss of crop production and consequential disruptions. Many houses are destroyed completely; others are damaged. Damage and loss to public and private utilities and industrial disruptions occur. Breakdown of economic activities occurs with corresponding loss of wealth.

As we noted briefly earlier, the statistics of flood losses reported by the State Governments, compiled by the Central Water Commission for the period 1953 to 1990 shows that on an average eight million ha. of land are affected involving some thirty three million people. Over a hundred thousand heads of cattle and more than one thousand and five hundred people are lost. The average annual loss or damage to houses, public utilities and crops was Rs.940 Crores. The extent of damage varies from year to year. In years of high floods the loss is many times the average figure.

The statistics compiled suffer from one disabilities and many suggestions for better compilation of flood damages have been also offered. Moreover damage figures compiled by interested parties or even the Govt. for other purposes may not indicate the precise picture of losses. However the broad figures as indicated above serve the purpose of indicating the order of losses. In any case the exact assessment of the comprehensive loss to the economy of the Nation or to the individuals is a near impossible task.

PREPAREDNESS

Disaster preparedness could be defined as the detailed planning for the prompt and efficient response immediately as soon as the anticipated event materializes. This effort to be very comprehensive inclusive of public education and awareness compaign ahead, provisions for the issuance of timely warnings, development of orderly evacuation plans, and preparations for providing the evacuees with food, clothing and shelter on emergency basis. The moment the disaster strikes will also mark the start of the emergency response period. The immediate onsite responses are spontaneous actions of local residents but their effectiveness could be improved by advance training. The speed and efficiency of the community reaction to save lives and mitigate suffering and losses is determined by adequate planning, training and rehearsals.

In the context of floods, it is well known that floods damage human settlements, necessitate evacuation to safer areas, damage crops and disrupt farming, wash away infrastructure items like irrigation, communication etc. and make land unusable. Disaster preparedness should also deal with all these aspects and other connected matters.

The National Flood Commission (1980) set up by the Government of India made a comprehensive study of the flood management scene in India and had made many valuable recommendations or flood management including flood disaster and cyclone disaster mitigation steps needed. The Government of India and the various State Govts. are also engaged in identifying and implementing the many steps needed to be taken in different parts of India to take care of local conditions. These steps include those on flood disaster preparedness.

The United Nations General Assembly designated the decade of the 1990s as the International Decade for Natural Disaster Reduction. The various activities taken up under this programme have focused attention on the many disaster preparedness measures.

EARTHQUAKE

Earthquakes are considered to be one of the most dangerous and destructive natural hazards. The commencement of this phenomenon is usually sudden with little or no warning. It is not yet possible to predict earthquakes and to make preparation against damages and collapse of buildings and other man-made structures. Actually earthquake consists of a sudden shaking (vibrations) of ground caused by disturbances in the earth's crust. An earthquake generates a set of horizontal and vertical vibrations of the ground which are random in character.

Earthquakes may be defined as a natural phenomenon which tends to create panic due to the trembling vibrations of sudden undulation of a portion of earth's crust caused by splitting of a mass of rock (Tectonic) or by volcanic or other disturbances.

This unit provides a general discussion about earthquakes. For clear understanding, we will first explain the general characteristics of earthquakes. Besides this precursors: instrumental and non-instrumental and vulnerability of the different regions of the country will be discussed to analyse the impact and effect of earthquake. Lastly nature of damage caused by earthquakes will be briefly described.

GENERAL CHARACTERISTICS

Impact of Earthquakes is sudden with little or no warning. However, following a major Earthquake, the after-shocks may give warning of a further earthquake. On some occasions, an earthquake may be preceded by a less intense tremors or foreshocks.

- It is not yet possible to predict magnitude, time and place of occurrence of an earthquake.
- The onset is usually sudden.
- Earthquake prone areas are generally well identified and well known on the basis of geological features and past occurrences of earthquakes.
- Major effects arise mainly from ground movement and fracture or slippage of rocks underground. The obvious effects include damage (usually very severe) to buildings and infrastructures alongwith considerable casualties.
- On the average about 18000 people die each year due to this disaster throughout the world.
- About 200 large magnitude earthquake (M>6.0) occur in a decade.
- The world's earthquake problem seems to be increasing with the increased population, high rise buildings and crowded cities.

The exact spot underneath the surface of the earth at which earthquake originates is known as "focus" while the point lying vertically above the focus is defined as "epicenter" of the earthquake. The seismic shocks originating at a depth of about 50 km. or less below the surface are termed as shallow focus earthquakes; otherwise these are known as deep focus earthquakes.

The energy released from the focus, due to elastic rebound of rocks is transmitted in all directions in the form of Earth's crust leading to earthquakes.

The power (energy) of an earthquake is reckoned in terms of its "magnitude" which is measured on an open-ended Richter Scale from 1 to 9. But it is not a linear scale and not even, a logarithmic scale. This will be clearly understood from the following Table 5.9 which gives the equivalence of earthquake magnitude (on Richter Scale) and energy released by the explosion of a certain mass of TNT which is the well known measure of explosive power in any blast.

Table 5.9

Magnitude of Earthquake(on Richter Scale)	*Approximate TNT Equivalent*
1.0	170 gms.
3.0	180 Kg (180×10^3 gm)
6.0 (like Latur, 1993)	5700 tonne (570 × 107 gm)
8.5 (like Assam 1897 & 1950)	28 700000 tonne (287×10^{11} gm)

From the above, it should be clear that the energy released by an earthquake (and hence the destruction) increases enormously as the magnitude on Richter Scale rises. Another way to appreciate the enormous destruction potential of an 8.5 magnitude earthquake is to know that the energy released is approximate equal to 10,000 Hiroshima type Atom Bombs.

The primary waves (or P-waves) are transmitted due to longitudinal vibrations set up within the earth. These waves have the velocity of the order of several kilometers per second and cause the preliminary tremors on the surface of the earth. These waves create an effect of horizontal pull and push and are also called pull and push waves.

The secondary (or S-waves) on the other hand are transmitted due to transverse vibrations. These are known as surface or slow waves. Even though the amplitude and size are small compared to other waves, these are the most destructive since they create vertical up and down movements in the ground surface as against horizontal oscillation due to longitudinal waves.

While the "magnitude" of an earthquake defines the energy released by the event the "intensity" of the earthquake will depend on the particular place where it is measured. Obviously the intensity will decrease as the distance from the epicenter increases.

Earthquakes are graded in two ways based upon the magnitude or intensity of an earthquake. The magnitude of earthquake is measured on the Richter Scale which has been explained above. It is calculated based on the amplitude of the waves generated by the earthquake.

The intensity is calculated for a particular location and is dependent on the distance of that location from the epicenter.

PRE-CURSORS INSTRUMENTAL AND NON-INSTRUMENTAL

We have already stated that it is not yet possible to predict earthquakes. However, sometimes there are some indication that would indicate that perhaps an earthquake would occur. Such indications are called "precursors". There could be either instrumental, i.e., those that are measured by instruments or non-instrumental, i.e. those which can only be perceived and not measured. Needless to say, the non-instrumental precursors are more subjective.

Some of the generally recognized precursors are listed below:

Table 5.10

Instrumental Precursors	*Non-Instrumental Precursors*
(a) Changes in Velocities of P & S Waves	(a) Sudden rise or fall of water level in wells and lakes in a Country like India. With numerous villages and each village with several wells, this precursor of rise fall of water level is easy to notice. Similarly, changes in water level in lakes can be observed.
(b) Fore-shocks & after shocks.	(b) Mud and sand shows up in surface waters.
(c) Statistical pattern of shocks.	(c) Changes in flows of natural springs.
(d) Uplift or subsidence of ground	(d) Increase in salinity of water.
(e) Changes in gravity	(e) Advance and retreat of seas.
(f) Faults, displacements in Earth	(f) Unusual behavior of animals.
(g) Tilt and strain of underground rock formations.	
(h) Changes in elective resistance of rocks	
(i) Changes in earth's magnetic field.	
(j) Emission of Radon Gas from the ground	
(k) Unusual sounds from inside earth	

VULNERABILITY

Disasters result from vulnerable societies being exposed to a hazard. There can be physical vulnerability, social vulnerability and economic vulnerability on account of an earthquake disaster.

Physical vulnerability relates to buildings, infrastructure and agriculture. The vulnerability of buildings is dependent on their sets, shape, materials used, construction techniques, maintenance and proximity of buildings to others. The weightage attached to each factor will vary according to the characteristics of the particular earthquake.

Infrastructure may be considered in three broad groups: transport systems (roads, railways. bridges, airports, port facilities): utilities (water, sewerage and electricity); telecommunications; dams and flood protection embarkments.

Vulnerability analysis is especially concerned with the risk faced by critical facilities (sometimes termed "life-lines") which are vital to the functioning of societies in disaster situations especially such as in case of earthquakes. These facilities include hospitals, dispensaries and emergency services. Special consideration is given also to protect heritage buildings of great cultural and historical importance.

Social Vulnerability

Records of past earthquake disasters suggest that the following groups of people are particularly at risk and require special attention:

- Single parent families:
- Women, particularly when pregnant or lactating.
- Mentally and physically handicapped people:
- Children; and
- The elderly.

Poor people are less concerned with infrequent hazards. If there are groups whose livelihoods are at risk, living or working in densely populated areas, with low perceptions of risk and without institutional support, the cumulative effect would be high social vulnerability.

ECONOMIC VULNERABILITY

It measures the risk of hazards causing losses to economic assets and processes. It focuses only valuating the direct loss potential (i.e. damage or destruction of physical and social infrastructure and its repair or replacement cost, as well as crop damage and losses to the means of production); indirect loss potential (i.e. the impact on cost production, employment, vital services and income-earning activities); and secondary effects (epidemics, inflation, income disparities and isolation of outlying areas). With the insights provided by economic vulnerability analysis, it is possible to estimate direct and indirect losses and to design ways and means to mitigate them in relation to the estimated costs of relief/recovery actions and mitigation measures required.

IMPACT AND EFFECTS

In general terms, typical impacts and effects of earthquake disasters tend to be :

- Loss of Life.
- Injury
- Damage to and destruction of property.
- Damage to and destruction of subsistence and cash crops.
- Disruption of production.
- Disruption of lifestyle.
- Loss of livelihood.
- Disruption to essential services.
- Damage to national infrastructure and disruption to administrative and organizational systems.
- National economic loss.
- Sociological and psychological after-effects.

The following problem areas need particular attention in case of Earthquake disasters:

- Severe and extensive damage, creating the need for urgent counter measures, especially search and rescue, and medical assistance.
- Difficulty of access and movement.
- Widespread loss of or damage to infrastructure, essential services and life support systems.
- Recovery requirements (i.g., restoration and rebuilding) may be very extensive and costly.
- Occurrence of earthquakes in areas where such events are rather rare may cause problems due to lack of public awareness.

NATURE OF DAMAGE

Damages due to earthquakes are the related terms and depends upon various factors listed below:

(a) Nature of earthquake.

(b) Geological and soil conditions

(c) Quality of construction.

(d) Sociological factors.

Essential services such as water-mains, drainage systems, and electrical transmission lines are seriously damaged. Broken water-main cause flooding of the area and leave no water for drinking or for fire-fighting. The sparking of high tension over-head electric cables cause fires, setting ablaze whatever combustible material is in the vicinity. Leaks from cooking gas cylinders or supply lines also cause fires.

Disrupted drainage lines spread noxious fluids and give rise to diseases and epidemics.

Geological faults in the Earth's crust become activated and accentuate displacement of the ground, producing gaping fissures in which human beings and animals are known to have been engulfed. Telephone and telegraph poles fall down and the services go out of order. Communication are seriously hampered or altogether stopped. Railway lines are twisted out of shape and rail communication to and from the affected area is broken off. In some cases the only access to the affected area is by helicopter.

Large dams in the vicinity may be affected, and in some cases may even burst and cause floods. On the coast, huge wanes called tsunamis lash the shore and bring down houses and other structures and dislocate fishing and navigation.

In the Makran Coast Earthquake of 26th November, 1945, four new islands had come up through the huckling of two sea-floor. The islands were roughly circular in shape, 100 to 200 meters in diameter and rose to some 10 to 20 meters above the sea level.

Such creation of islands is a rare phenomenon but does occurs due to some earthquakes. They were composed of loose sand and clay and were being eroded fast due to waves and tides.

CYCLONE

Cyclones are one of the most disastrous natural hazards in the tropics and are responsible for deaths and destruction more than any other natural calamities. Cyclones bring with them extremely violent winds, heavy rain causing floods and storm tides causing coastal inundation.

Cyclones form over the warm ocean waters (sea surface temperature of the order of 26°C or 27°C) little away from the equator within the belt of 30°N and 30°S. In our area, cyclones form in the Bay of Bengal and the Arabian sea. As they move westward or northwestward, those forming in the Bay of Bengal come to the Indian territory while those forming in the Arabian Sea generally go away from India but sometimes they turn around to hit Gujarat.

CHARACTERISTICS

Tropical cyclones are large, rotating, atmospheric, phenomena extending horizontally from 150-1000 Km and vertically from surface to 12-14 Km. These are intense low pressure areas with a spiral shape. Fierce winds spiraling anti-clockwise in the northern hemisphere blow around the cyclone centre. Cyclones generally move 300-500 Km in 24 hours over the ocean. The severest category of cyclones have wind speeds of 115 kmph or more and are classified as Severe Cyclonic Storm with a core of Hurricane winds.

Cyclones develop from areas of low atmospheric pressure and go through the stages of depression and deep depression before attaining the category of cyclone. Each category is recognized on the basis of windspeed as indicated below:

Table 5.11

	Categories	Wind Speed
1.	Low Pressure Area	<30 kmph
2.	Depression	30 to 55 kmph
3.	Deep Depression	55-65 kmph
4.	Cyclonic Storm	66-90 kmph
5.	Severe Cyclonic Storm	90-115 kmph
6.	Severe Cyclonic Storm with a core of Hurricane wind	>115 kmph

A well developed cyclone consists of a central region of light winds known as its "Eye". The eye has average radius of about 20 to 30 km, but it can be 40 to 50 km in large cyclones. The eye is an almost cloud-free zone and it is surrounded by a ring of very strong winds extending on an average up to 30 to 50 km beyond the centre. This area is known as zone of maximum wind. Surrounding this region, winds spiral in the counterclockwise direction in the northern hemisphere, extend outward to large distances, with speeds gradually decreasing as one moves further away from the centre.

On an average, about 5-6 cyclones form in the Bay of Bengal and the Arabian Sea every year, out of which 2-3 may be severe. More cyclones form in the Bay of Bengal than in the Arabian Sea. The ratio is 4:1. Tropical cyclones in these seas generally form between 5°N and 20°N. There are two distinct seasons of cyclones in our area. One is from May to June (Pre-monsoon) and the other is from October to mid-December (Post-monsoon). May, June, October and November are known for severe cyclonic storms.

Almost the entire east coast is vulnerable to cyclones with varying frequency and intensity. In the west coast, the North West coast (coast north of Mumbai) is more vulnerable as compared to southwest coast (South of Mumbai).

WARNINGS

Cyclone warnings are provided through six cyclone warning centres located at Calcutta, Bhubaneswar, Visakhapatnam, Madras, Bombay and Ahmedabad. These centres have their distinct area wise responsibilities covering both the east and west coasts of India and the oceanic areas of the Bay of Bengal and the Arabian Sea, including Andaman and Nicobar and Lakshadweep. Cyclone warnings are issued to the All India Radio (AIR) and the Doordarshan for broadcast/telecast in different languages. Cyclone warnings are also given to control room and Crisis Management Group in the Ministry of Agriculture, Government of India, who are finally responsible for coordinating various activities of Centre and Government and other agencies in respect of cyclone warnings.

Cyclones are tracked with the help of INSAT, powerful cyclone detection radars and conventional meteorological observations including weather reports from ships. At present cyclone detection radars are installed at (i) Calcutta, (ii) Paradip, (iii) Visakhapatnam, (iv) Machhlipatnam, (v) Chennai, (vi) Karaikal on the east coast; and (vii) Goa, (viii) Cochin, (ix) Mumbai and (x) Bhuj along the west coast. Present cyclone surveillance system in India is such that no cyclone in the region will go undetected at any time of its life cycle.

The important components of cyclone warnings are the forecast of future path and intensity of a cyclone and the associated hazardous weather. For the preparation of future position (path) of tropical cyclones and for estimation of storm surges, modern computer based techniques are used in addition to conventional methods. Intensity forecasts are made by using satellite techniques.

Cyclone warnings are provided in two stages. In its first stage, a "Cyclone Alert" is issued 48 hours before the anticipated time of commencement of adverse weather along the coast in the 2nd stage, a "Cyclone Warning" is issued 24 hours before the cyclone's anticipated landfall warnings for the ports and fisheries start much earlier. Ports are warned day and night through a specially designed port warning system. Informatory messages on cyclone are issued to All India Radio and Doordarshan much earlier, as soon as a tropical cyclone is detected in the Bay of Bengal or in the Arabian Sea.

Cyclone warnings are disseminated through the following means:

- Telegrams with highest priority
- Telecast through Doordarshan
- Broadcast through AIR

- Bulletins to the press
- Broadcast through Department of Telecommunications, Coastal Radio Stations for ships in the high seas and coastal areas, and
- INSAT based Disaster Warning System.

In addition to above, cyclone warnings are disseminated through teleprinters, telex, facsimile and telephones wherever such facilities exist with the recipients.

PREPAREDNESS

The preparedness means measures which enable government organizations, communities and individuals to respond rapidly and effectively to disaster situations. The preparedness measures include the formulation of viable disaster mitigation plans.

The preparedness actions have to be planned ahead of disaster. It would consist of a plan of action to be implemented on the receipt of the Cyclone Alert message from Cyclone Warning Centre. A cyclone alert is issued generally 48 hours before the possibility of the area being affected by strong winds, heavy rain and storm surges. The Action Plan would indicate how evacuation of people would be effected and the places where they could be evacuated to. The identification of strong buildings which would withstand the fury of the storm is an important segment of preparedness action plan. The safe storage of non-perishable food and other essential needs, adequate collection of stocks of drinking water and medicines, have to be made. Most of the maritime states have prepared Cyclone Disaster Preparedness handbooks or manuals, where action plans of various organizations have been indicated in the case of cyclone threat. It is desirable that as an essential component of preparedness, the action points indicated in the manuals are rehearsed at the beginning of the cyclone season.

To deal with cyclone situation a contingency plan has been evolved by the Ministry of Agriculture, who is the nodal agency at the Centre to co-ordinate the activities of various Central departments and the affected State/States to cope up with the natural disaster in general.

Training programmes for the disaster management officials and Non-Government Organisations (NGOs) are arranged by several management and public institutions in India.

RISK REDUCTION PROCESSES

The prevention of tropical cyclone formation is not within the realm of possibility. However, the loss of human lives and destruction of properties can be minimized by adopting prescribed short and long term measures for risk reduction. While cyclone warning system is the most important constituent of short term risk reduction measures against cyclone disaster, the risk assessment of tropical cyclone falls under long term measures.

As prevention of formation of tropical cyclone is not in the realm of possibility, some structural and non-structural preventive measures of long term nature can be undertaken to mitigate the suffering of cyclone affected people. Structural measures like construction of cyclone shelters, embankments, dykes, reservoirs and coastal afforestation are some of the long-term risk reduction measures for cyclone disasters. Creation of proper awareness, training and education of people in the vulnerable communities, introduction of insurance can be some of the non-structural measures.

EFFECTS

Severe tropical cyclones are responsible for large casualties and considerable damage to property and agricultural crop. The destruction is confined in the coastal districts and the maximum destruction being within 100 km from the centre of the cyclones and on the right side of the storm track. Principal dangers from a cyclones are: (i) very strong winds, (ii) torrential rain, and (iii) high storm tides. Most casualties are caused by coastal inundation by storm tides. Maximum penetration of storm surges varies from 10 to 20 km inland from the coast. Heavy rainfall and floods come nest in order of devastation. They are often responsible for much loss of life and damage to property. Death and destruction directly due to winds are relatively less. The collapse of buildings, falling trees, flying debris, electrocution, aircraft accidents and disease from contaminated food and water in the post-cyclone period also contribute to loss of life and destruction of property.

Floods generated by cyclone rainfall are more destructive than winds. The rainfall is the heaviest around wall cloud zones. Rainfall of the order of 20 or 30 cm per day is very common.

Tropical cyclone's worst killer comes from the ocean, viz., the storm surge. Over the deep oceans wave generated hurricane force winds may reach height of 50 feet or more. Below the storm centre, the ocean surface is drawn upward by 30 cms or so above normal due to the reduced atmospheric pressure in the centre. As the storm crosses the continental shelf and moves coastward, the mean water level increases. This abnormal rise in sea level caused by cyclone is known as storm surge. The surge is generated due to interaction of air, sea and land. The cyclone provides the driving force in the form of very high horizontal atmosphere pressure gradient and very strong surface winds. As a result the sea level rises and continues to rise as cyclone moves over shallower water, and reaches a maximum on the coast near the point of landfall (Point of crossing coast). Surge is maximum in the right forward sector of the cyclone and about 50-100 km from the centre coinciding with the zone of maximum wind. Winds in this sector is from ocean to land.

Due to significant improvement in cyclone warning system and adequate and timely steps taken by the government and other agencies, the loss of human lives is in the decreasing trend, although, loss of properties shows an increasing trend. The increase in the loss of properties is due to increased activity but unplanned human activities and non-engineered construction along the coast also contribute to the damage suffered by property. In support of the above statements we present some data on recent cyclones in the table 5.12 below. It may be seen that although the November 1977 and May 1990 cyclones, which occurred in the same coastal area of Andhra Pradesh and had the peak wind speeds of the same order, yet the loss of human lives in the case of the 1990 cyclone was much less in comparison to that of 1977 cyclone but the economic losses were many times more in the 1990 cyclone.

Table-5.12

Cyclone	*Peak(m/Sec)*	*Human Loss*	*Loss of (Millions Rupees)*	*Month/year*
Chirala	70	10.000	3500	Of November 1977
Machhlipatnam	58	700	1700	Cyclone of May 1979
Sriharikota	58	604	4000	Cyclone of November 1984
Machhlipatnam	65	967	22.480	Cyclone of May 1990

LANDSLIDES

Often it is not realized that a large part of India consists of mountainous terrain. In the north, there is the extensive Himalayan mountain system extending all along from the west to the east. Its lofty peaks rise to more than 8000 metres height. The middle ranges of the Himalayas are about 5000 metres high on the average while the foothills rise to about 6000 metres. The Himalayas abound in glaciers and are the origin of many rivers and streams. There is abundant rainfall and snowfall often accompanied by strong winds.

The peninsular region of India starts from the Vindhyachal ranges and consists of the Deccan Plateau which slopes eastwards. On its edges, this great plateau is bound by the mountain ranges of the Eastern Ghats and the Western Ghats. The Nilgiri mountains are in the southern parts of the plateau. The west-central parts of the country have the ranges of the Aravali mountains.

Many of these mountains systems are relatively new (in the geological sense) and are still growing such as the Himalayas. The rock systems are therefore fragile.

Given these special geological and geographical features and combined with the heavy rainfall system of the two monsoons (the summer monsoon and the winter monsoon) and also the not so rare occurrence of earthquakes, it is but natural that the mountainous areas of India are vulnerable to the hazards of landslides. In the snowy regions of the Himalayas, snow avlanches are the additional dreaded disasters.

LANDSLIDES IN INDIA

Landslides affect the remotely located, often isolated, small communities in villages or hamlets in the mountain regions of the country where external assistance takes time to reach in times of emergency when the normally difficult terrain and tracks may become almost impossible to negotiate. Many a times, even the information about the occurrence of such events and the damage done takes days to reach the district and state headquarters. Because of these reasons, landslides and snow avalanches assume the status of major natural disasters even though the affected area and population may be rather small.

Areas Struck with Frequency and Intensity

Landslides are a frequency and recurring phenomenon in the various hill ranges of India from Kerala to the Himalayas. Areas prone to landslides also include the Eastern and Western Ghats, the Nilgiris, the Vindhyachals, the mountains in the northeastern States and the great Himalayan range. The incidence of landslides in these regions is a recurring feature especially during and after spells of heavy rains. As the geological history of the rocks and the rainfall regime have strong bearing on the incidence of landslides, there are variations in the occurrence of landslides in different parts of the country as is indicated in Table 5.13 given below.

Table 5.13 : Incidence of landslides in India

Region	*Incidence of Landslides*
Himalayas	High to very high
Northeastern Hills	High
Western Ghats and the Nilgiris	Moderate to high
Eastern Ghats	Low
Vindhyachals	Low

Kind and Magnitude of Damage

There is no doubt that anything that comes in the ways of a landslide will suffer severe damage and may even by totally buried or wiped out. Anything located on top of a landslide will also not survive when the rock or mud slips out from below it.

Landslides : More often, the major landslides are combinations of rockslide and rockfall. They all involve movement or mass (soil, debris or rock). The process of movement of mass may vary from slow soil creep to abrupt and sudden rockfall. Landslides, also known as landslips, range from low angle and rather slow slides to sudden vertical falls.

Based on the type of movement, relative rate of movement and kind of material involved, landslides can be designated into 5 kinds as follows:

- Slump with earthflow
- Debris slide
- Debris fall

- Rock slide
- Rock fall

Landslides, being more widespread in different mountainous or hilly regions of the country (as against snow avalanches which are confined to the snowy of the Himalayas), cause damage which is more varied and more widespread. Increased population, spurt in quarrying, mining and construction activities near unstable hill slopes, ill conceived developmental activities in the vulnerable hilly areas, have resulted in more landslides and greater damages. Apart from the catastrophic damages suffered by communities living on or near unstable hill slopes as their houses along with persons and property may be destroyed by a landslide, the most crippling damages due to landslides are suffered by (i) roads and (ii) productive soil. Damage to roads leads to considerable inconvenience and economic loss. The disappearance land and the cultivable top soil takes away the agricultural potential of the affected area thus depriving them of their already livelihood seriously.

Landslides are also known to result in blocking of streams or overflowing of lakes thus causing flash floods because large volumes of debris falling in a lake or reservoir cause its water to overflow or the temporarily blocked stream may suddenly release the huge quantity of impounded water to cause a devastating flash flood downstream.

Relief and Rehabilitation

Essentially, the relief steps comprise the following:

(1) Search and Rescue
(2) Medical assistance to the injured
(3) Disposal of the dead
(4) Food and water
(5) Emergency shelter for the homeless
(6) Opening up access roads if blocked; and restoration of communication channels
(7) Psychological counseling of the survivors who have lost their close relatives
(8) Repair of houses and facilities
(9) Assistance (technical and financial) to restart economic activity to restore regular work and income
(10) Reconstruction through proper planning.

Measures for the rehabilitation of a community affected by landslide or snow avalanche will depend very much on the extent of the damage done by the disastrous event.

If the damage has not been severe, the rehabilitation will take the form of (a) short-term relief to restart life and (b) taking long-term measures so that any future landslide or snow available does not hurt the community at all or at least, not as much.

We have already discussed the relief steps in the preceding section . As regards the long-term measures, these will comprise the following:

(1) Reducing the hazard proneness of the site through engineering measures such as strengthening or modifying the slopes, removing fragile and unstable portions, securing snow accumulations by snow fences, snow nets or by cribbing, and improvement of drainage.
(2) Stopping indiscriminate quarrying and mining in mountain areas.
(3) Afforestation of zones prone to landslides so that trees and vegetation provide a binding force to prevent slippage of debris, rock, and snow.

(4) Creation of a voluntary, community based preparedness system of watch, monitoring and alert. This will not only be useful in times of a disaster but will provide enough self confidence (and thereby self reliance) which is an essential objective of an effective rehabilitation programme.

(5) Provision of assistance for economic rehabilitation by arranging work, employment loans, and grants.

In the extreme case of severe damage to a community by a landslide or snow avalanche, the site may be rendered totally unusable. In that case, rehabilitation takes the form of relocation and reconstruction. In such an event, the new site should be carefully chosen so as to minimize vulnerability and risks.

QUESTIONS

Short answer type questions:–

1. What is environmental pollution?
2. State, how air pollutants are classified ?
3. What are the characteristics of potable water ?
4. How water pollutants are classified ?
5. What is soil pollution ?
6. How pesticides contribute to soil pollution ? Explain.
7. Differentiate noise with sound.
8. Explain hearing loss.
9. What are types of solid wastes ?
10. What are -3R.
11. Discuss the role of an individual in prevention of pollution.
12. Describe the environmental effects of Bhopal gas tragedy.
13. Write control measures of water pollution.
14. Write ill effects of noise pollution.
15. What is decibel ?

Long answer type questions :–

1. Explain pollution. Give an account of various kinds of pollution.
2. Name the different types of air pollutants. Explain the characteristics and biochemical effects of air pollutants.
3. How air pollutants effects man and his environment ?
4. What are water pollutants ? Discuss their effects and to control them.
5. What are various sources of soil pollution ? Discuss how soil pollution is controlled ?
6. What are the main sources of Marine pollution ? Give the effects and methods to control marine pollution.
7. Describe noise pollution. Differentiate between thresold pain and thresold of hearing.
8. Write an essay on "radioactive pollution".
9. Write short note on–
   (*i*) Thermal pollution. (*ii*) Solid waste management. (*iii*) Floods
   (*iv*) Earthquake (*v*) Cyclone (*vi*) Bhopal Gas tragedy.
10. Write an essay on disaster management.
11. Write an essay on air pollution.
12. Write about the role of an individual in prevention of pollution.
13. Write a detail account of disaster management.
14. Explain various causes of water pollution.

UNIT
6

Social Issues and the Environment

Undoubtedly man is related with environment and he is solely dependent on nature. But since few years, over exploitation of natural resources disturbed the environment. Environment effects our life style, culture, social components etc. Technological development also have great impacts on social and natural resources. Development does not mean the increase in GNP(Gross National Product) but it should be visualized in a holistic manner, where it brings benefits to all. Though the natural environment undergoes continual change but man also produced some changes like domestification of animals, introduction of agriculture, development of industries etc. Increased population and higher consumption per head greatly effects the environment. The success of environmentally sound development depends on proper understanding of social needs, opportunities and of environmental characteristics

FROM UNSUSTAINABLE TO SUSTAINABLE DEVELOPMENT

More and more natural resources were consumed in the process of satisfying the rapidly growing needs of the habitat. Every development activity has some impact on the environment. For meeting the needs, the human can not live without the developmental activities. Consequently, there is need to harmonize developmental activities in such a way that environment should not polluted at least. Unsustainable development means the development of a few privileged nations both in science and technology. Such developments are at the cost of our life supporting systems like air, water, soil and over exploitation of our natural resources which may lead to the collapse of the inter- related systems of the earth. If growth continues in the same way, very soon we will be facing a "doom's day".(Meadow et al 1972)

To be sustainable, development must process both economical and ecological sustainability. The concept of sustainable, development has received much recognition after the Stockholm Declaration 1972. The Brundtland (1987) has defined that sustainable development is development that meets the needs of the present without compromising the ability of the future generations to meet their own needs. The earth summit held at Rio-de-Janeiro in 1992 put the world on the path of sustainable development which aim at meeting the needs of the present without compromising the ability of future generations to meet their own needs. The Rio declaration has taken cognizance of the fact that in order to achieve sustainable development, eradication of poverty is indispensable and thus development process and environment protection must go on simultaneously.

Freedom and sustainable development are mutually exclusive ideas. Freedom encourages people to do what they want to do, sustainable development dictates what people may and may not do. Freedom empowers people to control Govt., sustainable development empowers Govt. to control people.

There are two aspects of sustainable development :-

(i) *Inter-generational equity* – This emphasizes that we should stop over-exploitation of resources, reduce waste discharge and emissions and maintaining an ecological balance. It expects to hand over a safe healthy and resourceful environment to the future generations.

(ii) *Intra–generational equity* – This emphasizes that technological development should support economic growth of the poor countries so as to reduce the weather gas within and between the nations.

Measures for sustainable development – There are following major measures for sustainable development:-

1. *To promote environmental education and awareness* – From childhood, we should develop a feeling of belongingness to earth. This can be possible by introducing environment as a subject in education from primary stage. Media can also be helpful in developing such feelings. The transformations of policy making process from one that reflects and is controlled by the will of the people, to a process controlled by the selected elite who have found ways so impose policies fashioned by the international community to achieve what they believe to be sustainable development.
2. *Three 'R' approach* – Three 'R' means, Reduce, Reuse and Recycle. We should reduce the excessive use of natural resources, but use them again and again instead of passing it on to the waste stream. Recycle the materials to reduce pressure on our existing natural resources.
3. *Appropriate technology* – The technology should use less resources and produce minimum waste. It is over which locally adaptable, eco friendly, resource efficient and culturally suitable.
4. *To utilize resources as per carrying capacity of the environment* – Sustainability of a system depends largely upon the carrying capacity of the system. If carrying capacity of a system crossed, environmental degradation status and continues till it reaches a point of no return. Carrying capacity has two basic components-
   (i) *Supporting Capacity* – It is formed of productive and protective systems.
   (ii) *Assimilative Capacity* – It is formed of the systems which utilize the wastes produced by human activities.

The key to sustainable growth is not to less but to produce differently, offering solutions to a broad range of environmental problems. To achieve sustainability, economic growth can not be based on over exploitation of the resources but must be managed to enhance the resource base.

Urban Problem Related to Energy

Urban areas are developing very fast. In most of cities there is influx of populations from surrounding areas, mostly in search of employment and better living conditions. Therefore, it is difficult to accommodate all the industrial, commercial and residential facilities within limit. As a result, cities are spreading in to sub-urban or to rural areas. Uncontrolled population. irregular development are the main factor for receding facilities in urban areas. Energy is required in every walk of life like industry, transport, defence, agriculture, trade, education, domestic etc. Cities are the main centers of economic growth. Hence, energy is the most important input for development. The energy requirements of urban population are much higher than that of rural ones. Energy problems become more severe due to the limited amount of non-renewable resources of energy. Rapid utilization of fossil fuels produces increased production of wastes which causes environmental pollution. Energy problem day by day becoming serious. People are facing for 'power cut'. Energy demand is higher than production. There are following main causes of energy problems.

1. Increasing use of energy for domestic and commercial purposes (due to increased population and industrialization).
2. Industrial plants using big proportion of energy.
3. Non renewable resources of energy like coal, petroleum and natural gas are decreasing.
4. Increasing of transport means.
5. Decreasing production of Hydro electricity due to insufficient rains.
6. Transmission loss due to defected power distribution system.

There are following steps to solve the energy related problems.

1. To control urbanization.
2. To develop renewable resources of energy like solar radiation, wind power, hydel power, nuclear power, bio mass etc. These are pollution free also.
3. Non renewable energy resources should be used only when no non-conventional source of energy is available.
4. Welcoming the awareness programs to save energy.
5. Effective measures for transition loss and energy theft.

WATER CONSERVATION

Water is needed in almost every sphere of human activity. Without water life is not possible. In many aspects the properties of water are unique. It is called universal solvent. No other liquid can replace it. The global distribution of fresh water on earth's crust including ground water and water present as its vapours in atmosphere.

Table 5.1 Global distribution of fresh water

	Water in Cubic KMS
1. Water in snow caps, Ice sheets, Glaciers etc.	24,000,000
2. Surface ponds, Lakes and reservoirs	280,000
3. Water in Streams and rivers	1,200
4. Water present as soil moisture	85,000
5. Ground Water	60,000,000
Total amount of fresh water on our planet	84,366,200

Water is required for direct consumption or indirectly for washing, cleaning, cooling, transportation or even for waste disposal. Important sectors of human activity, which require water can be grouped as:-

1. Irrigation
2. Industries
3. Live stock management
4. Thermal power generation
5. Domestic requirements
6. Hydroelectric generation, fisheries navigation and recreational activities.

According to 1970 survey, about 3500 cubic kms of water are drawn for human use every year. Agriculture sector is the biggest consumer of fresh water. The amount of water drawn for human use is never used up completely. A large fraction is returned to the surface deposits or stream flows often in a polluted state. Water requirements have greatly increased due to rapid population growth, industrialization and agriculture. The demand for water is likely to exceed its supply by the first or second decade of the next country. The shortage of water shall make many localities barren, devoid of life. Fertile land become deserts. Conservation of water is, therefore an absolute necessity of today. Otherwise the tomorrow will be grion, drier and barren to live through. The following steps should be taken for conservation of water.

1. *Water economy, Re-use and Recycling.* If water meters are installed and charged properly, the consumption of water in domestic establishments, livestock management and industries shall drastically decline. The heated water from thermal power plants, where large amount of water is needed, may be utilized elsewhere after proper cooling. The same is true for many industries, water used once may be used again for another purposes.

2. *Agricultural runoffs from fields.* This can be used to irrigate cropland down the stream, while an efficient use of water with conditions of proper drainage can significantly reduce the agricultural runoffs.

3. *Efficient distribution system.* Water resources are not distributed evenly. Some localities have plenty of water and others have little. Many river basins have plenty of water, which flows down un-used to the sea. Surplus of one basin can be used to make up the deficit at another.

4. *Enhancement of surface storage capacity.* About 27000 Cubic kms of fresh water which rush down to the oceans through stream and rivers are of no use to the mankind. We can store this water in tanks, reservoirs, dams for further use in drier seasons.

5. *Reduce evaporation losses.* Water losses through evaporation and seepage are enormous both from the reservoirs and distribution system. It should be reduced.

6. *Improvement of underground storage capacity.* The fresh water is stored in underground deposits. Every year about 10-15 % of the total precipitation enters the ground water table. These deposits regularly feed streams and rivers during the drier periods. These deposits are cheap and easily obtainable.

7. *Desalination of Sea Water.* A huge store of water exists in our oceans. If the salt content of the sea water is removed, we can use it. This can be done by desalination plants.

8. Afforestation and Reforestation of hill slopes to check loss of water in floods.

9. Artificial rain making and precaution of water pollution.

RAINWATER HARVESTING

Water is an essential natural resource for sustaining life and environment. The available water resources are under tremendous pressure due to the increased demands and time is not far when water, which we have always thought to be available in abundance and free gift of nature, will become a scarce commodity. Conservation and preservation of water resources are urgently required to be done. Water management has always been practiced in our communities since ancient times, but today this has to be done on priority basis. The ministry of Water resources in India is endeavoring to make **rain water harvesting** a part of every day life in our villages and cities as a people's movement, and this will go a long way in the management of ground water as a sustainable resource.

While we have taken impressive strides in advance management of water resources, we also need to learn from the wisdom bestowed by our forefathers. One such wisdom is the concept of rain water harvesting. It is high time that the Government and the people join hands for creating awareness of the importance of rain water harvesting with the main objective of adopting these measures and techniques throughout the country. A judicious mix of ancient knowledge, a modem technology, public and private investment, and above all, people's participation will go a long way in reviving and strengthening water- harvesting practices throughout the nation.

Rainwater harvesting is control or utilization of rainwater close to the point rain reaches earth. It is categorized into domestic rain water harvesting and rain water harvesting for agriculture, erosion control, flood control and aquifer replenishment.

Domestic rain water harvesting, also known as roof water harvesting or roof top rain water harvesting is the technique through which rain water is captured from roof catchments and stored in tanks or reservoirs.

Rain water harvesting systems, both small and large, consists of six basic components.

(a) Catchment area/roof, the surface upon which rain falls.

(b) Gutters and downspouts, the transport channels from catchment surface to storage.

(c) Leaf screens and roof washers, which are systems that remove contaminants and debris.

(d) Cisterns or storages tanks, where rain water is stored.

(e) Water treatment, the filters and equipment as well as additives to settle, filter and disinfect.

The system involves collecting water that falls on zinc, asbestos or tile roof of house during rein storms and conveying it by an aluminum, PVC, wood or plastic drain or collector to a nearby covered storage unit. The rain water yield varies with the size and texture of the catchment area. A smoother, cleaner and more impervious roofing material contributes to better water quality and greater quantity. The most common container now a days is a metal drum or barrel with a capacity of about 200 liters, set up at the foot of the fall pipe. This is enough to supply a family for 5 to 7 days with 6 liters per person per day, with possibility of using more than one storage barrel.

It is interesting to note that annually replenishable resources are assessed as 432 billion cubic meters(BCM) and by adopting water harvesting, an additional 160 BCM shall be available for use. The ground water levels in some areas are falling at the rate of one metre per year and rising in some other areas at the same rate. ***The main causes of fall in ground water levels are:***

(*a*) Overexploitation or excessive pumping either locally or over large areas to meet increasing water demands.

(*b*) Non-availability of other sources of water. Therefore, sole dependence is on ground water.

(*c*) Unreliability of municipal water supplies both in terms of quality and quantity, driving people to their own sources.

(*d*) Misuse of ancient means of water conservation like village ponds, baolis, percolation tanks and therefore, higher pressure of ground water development.

The main effects of overexploitation of ground water resources are:

(*a*) Drastic fall in ground water levels in some areas.

(*b*) Drying up of the wells/bore wells.

(*c*) Enhanced use of energy.

(*d*) Deterioration in ground water quality.

(*e*) Ingress of sea water in coastal areas.

You can capture and recharge 65,000 litres of rain water in Delhi from a 100 Sq. m. size roof top and meet drinking and domestic water requirements of a family of four for 160 days. ***The method and technique include:***

(*a*) Roof top rain water harvesting and its recharge to underground through existing wells or bore wells or by constructing new wells, bore wells, shafts, spreading basins, storm water drains etc.

(*b*) Harnesting runoff in the catchments by constructing structures such as gabions, checkdams, bhandaras, percolation trenches, sub-surface dykes etc.

(*c*) Impounding surplus runoff in the village catchment and water sheds in village ponds and percolation tanks.

(*d*) Recharging treated urban and industrial effluents underground by using it for direct irrigation or through recharge ponds or wells etc.

The main objectives of rain water harvesting are :

(*a*) To restore supplies from the aquifers depleted due to over exploitation.

(*b*) To improve supplies from aquifers lacking adequate recharge.

(*c*) To store excess water for use at subsequent times.

(*d*) To improve physical and chemical quality of ground water.

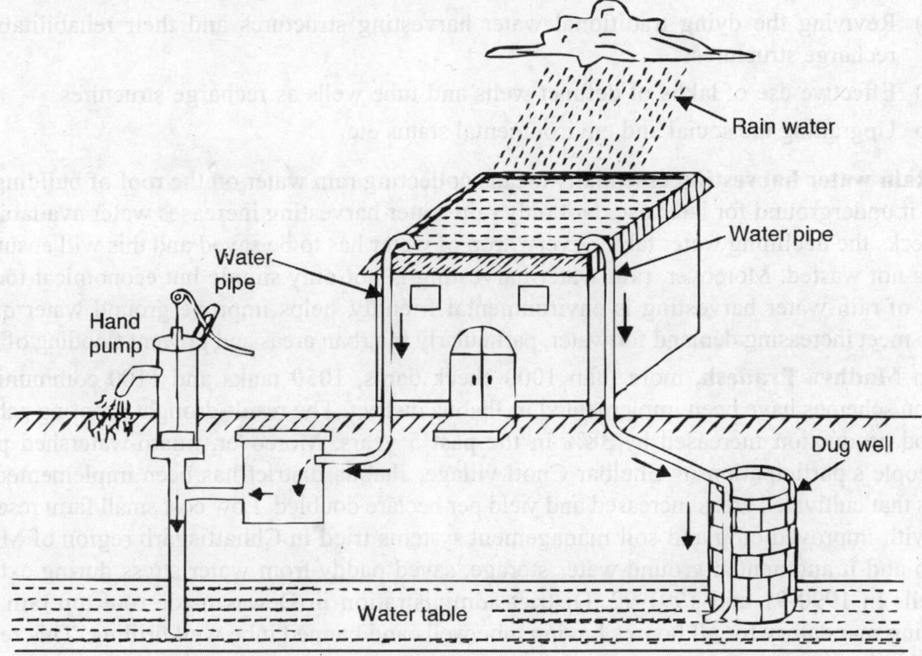

Fig. 6.1. Roof-top rainwater harvesting by recharging (*i*) through hand pump or (*ii*) through abondoned dugwell.

(*e*) To reduce storm water run off and soil erosion.

(*f*) To prevent salinity ingress in coastal areas.

(*g*) To increase hydrostatic pressure to prevent or stop land subsidence.

(*h*) To recycle urban and industrial waste waters etc.

(*i*) To rehabilitate the existing traditional water harvesting structures like village ponds, percolation tanks, baolis, tanks etc.

(*j*) To convert the traditional water harvesting structures into ground water recharge facilities with minor scientific modifications and redesigning.

(*k*) To use the existing defunct wells and bore wells after cleaning and also the operational wells as recharge structures.

The expected advantages of rain water harvesting are :

(*a*) Rise in ground water levels in wells.

(*b*) Increased availability of water from wells.

(*c*) To prevent decline in water levels.

(*d*) Reduction in the use of energy for pumping water and consequently the costs. One metre of water level saves about 0.40 kwh assuming 10 hours of pumping per day for 365 days.

(*e*) Reduction in flood hazards and soil erosion.

(*f*) Improvement in water quality.

(*g*) Arresting sea water ingress.

(*h*) Assuring sustainability of the ground water abstraction sources and consequently the village and town water supply systems.

(*i*) Mitigating the effects of droughts and achieving drought proofing.

(*j*) Reviving the dying traditional water harvesting structures and their rehabilitation as recharge structures.

(*k*) Effective use of lakhs of defunct wells and tube wells as recharge structures.

(*l*) Upgrading the social and environmental status etc.

Rain water harvesting essentially means collecting rain water on the roof of buildings and storing it underground for later use. Not only rain water harvesting increases water availability, it also checks the declining water table. Every drop of water has to be saved and this will ensure that water is not wasted. Moreover, rain water harvesting is not only simple but economical too. The process of rain water harvesting is environmental friendly, helps improve ground water quality, helps to meet increasing demand for water, particularly in urban areas and prevent flooding of roads.

In **Madhya Pradesh,** more than 1000 check dams, 1050 tanks and 1100 community lift irrigation schemes have been implemented in Jhabua district. The result-drought proofing achieved and food production increased by 38% in the past 5 years. Moreover, micro-watershed project with people's participation in Ghelhar Choti village, Jhabua district has been implemented. The result is that cultivated areas increased and yield per hectare doubled. Low cost small farm reservoirs along with improved crop and soil management systems tried in Chhattisgarh region of Madhya Pradesh and it augmented ground water storage, saved paddy from water stress during extended dry spells in 1990-91 and 1991-92. District administration in Dewas made roof top rain water harvesting mandatory for all houses having tube wells and banned tube well drilling. This resulted in improved soil moisture and recharged first-aquifer. This technique is also popular in other states like Maharashtra, Rajasthan, Andhra Pradesh etc.

WATERSHED MANAGEMENT

Watershed is a drainage area on earth's surface from which runoff, resulting from precipitation flows past a single point in to a large stream, a river, a lake or the ocean. It is a geo-hydrological unit and drains at a common point, has been accepted world over as a scientific unit for area development. The watershed can range from a few square kilometer to few thousand square kilometer in size. Damodar Valley Corporation in 1949 adopted first Integrated Watershed Management. Watershed development is the rational utilization of natural resources of soil water and vegetation for increasing and stabilizing the productivity of land on a sustainable basis. The development of watershed will result in increase in sub soil water regime, recharge of wells. The watershed based development approach is undoubtedly an agreeable concept to set the goal. But this demands a massive people's movement to make the village community self reliant.

The watersheds are very often found to be degraded due to uncontrolled, unplanned and unscientific land use activities like over grazing, deforestation, mining, soil erosion, industrialization etc.

Objectives of watershed management. Watershed management is the rational use of land and water resources for optimum production causing minimum damage to the resources. The main objectives of watershed management are as :-

1. To increase agricultural production i.e. increasing the availability of fodder, fuelwood, timber and raw materials for industries.
2. The rational utilization of natural resources like water soil and vegetation.
3. To minimize the risks of floods, droughts and landslide.
4. To Manage the watershed for developmental activities like domestic water supply, irrigation, hydropower generation.
5. To develop the rural areas and their lifestyle.

Under the development of national policy, the watershed management was included in fifth Five Year Plan. Now a days, a number of national watershed development programmes are in progress. Various measures are necessary for watershed management. Some of them are :-

1. Scientific mining and quarrying must be done in the watershed areas because hills loose stability and get disturbed by improper mining.
2. Water harvesting in the watersheds to be used in dry season in low rainfall areas.
3. Afforestation and agro forestry(crop plantation) should be promoted to prevent runoffs loss and soil erosion and increase soil moisture. Woody trees like Eucalyptus and Lencaena should be grown in between crops to reduce the runoffs and loss of fertile soil in high rainfall areas.
4. some mechanical measures like terracing, bunding, bench terracing, contour cropping etc. are used to minimize runoff and soil erosion in the slopy regions of watersheds.
5. To promote soil binding plants like Vitex.
6. People's participation should be ensured including farmers and tribals in the water shed management programmes. This can be done by properly educating people about the campaign or paying some incentives.

The Himalayas are one of the most critical watersheds in the world. Most of the watersheds of our country lie in this region. Successful watershed management has done at Sukhomajri and Panchkula with the active participation of the local people.

Resettlement and rehabilitation of people: Its problems and concerns

Some times for the development of projects like construction of dams, mining, creation of parks etc. and during natural calamities like Earthquake, Landslides, Volcanos, Floods, Droughts, Cyclones, the problems of resettlement and rehabilitation arise. For example recently the Tsunami cyclone affected thousands of families and during construction of Indira Sagar dam in Khandawa district of Madhya Pradesh thousands families were displaced and rehabilitated near Chhanera and other places. This caused permanent loss of the benefits and facilities. This disturbed Socio-economic and ecological base of local community which are generally forest and tribal people. Families are disintegrated and also lost ancestral link between people and the environment. This can easily be seen in old HARSUD where people still go to remember old environment. More than 13000 families are displaced during Tehri dam construction. They rehabilitated in Dehradoon and Haridwar regions. Various types of projects result in the displacement of native people are :-

1. *Displacement due to Dams.* Universe without energy is not imaginable. The most easily accessible and eco-friendly form of renewable energy is hydropower. Water is scarce natural resource and India is blessed with it, Hence it has to judiciously harnessed and managed for welfare of all living beings. India's exploitable hydropower potential is 84044 MW. River Narmada has 3000 MW hydropower potential in the state of Madhya Pradesh. Sardar Sarovar Project (Gujarat),

Hirakund (Orissa), Bhakra Nangal Dam (Punjab) Tehri Dam (Uttaranchal), Indira Sagar project (MP) etc. are some which displaced more than 25 million people.

Case Study

INDIRA SAGAR PROJECT (ISP) - Indira Sagar Project (1000 MW) is in Khandwa district of MP. It is constructed, operate and maintain by Narmada Hydroelectric Development Corporation (NHDC). Narmada is bestowed with rich potential of 29 major, 135 medium and 3000 minor projects. The reservoir of ISP Dam is largest reservoir in India with storage capacity of 12.22 BM^3 of water for irrigation of 2.70 lakh Ha. A total of villages and people affected by this project. They were given plots, transportation grant, shifting facilities, agricultural land and other compensation. A separate township CHANERA was developed for displaced people. Under this project 26000 Ha area is proposed for irrigation, 564 villages to be benefited by irrigation. This will cause the production of 4.00 lakh tonnes of food grains and 10.55 lakh tonnes of other crops every year additionally. Other benefits of the projects are :-

(i) Pisciculture – 1500 tonnes of fish production every year.

(ii) Industrial development – Due to power and irrigation, development of new Industries in Nimar and Malwa region.

(iii) Supply of water to Thermal Power Plant – Water will be supplied to 2000 MW Thermal Power Plant proposed in village Bir.

(iv) Tourism – Resevoir of ISP spread over 913 Sq. km. area will be a boom for tourism development in Madhya Pradesh.

TEHRI DAM – Tehri Dam is on meeting point of River Bhagirathi and Bhilangana. It was controversial since 1991 i.e. from its construction. There was campaign against the project by Tehri people and noted Activist Sunder Lal Bahuguna, the propagator of CHIPKO MOVEMENT. The object of this dam is to produce power and provide irrigation facilities. Its length is 573 meter and 260.50 meter high. It is highest rock fill Dam of India. Its catchment capacity is 34.50 million m^3. Its hydroelectric capacity is 2400 MW. It provide 300 qu.sec drinking water to villages, towns of U.P and Delhi and irrigation facility to 2.70 m Ha. land. The project covered 467 sq. km. area of Tehri, which effected Tehri town and 112 villages. More than 1.25 m people were displaced. It is world's most controversial project. People displaced are called Eco-Refugee because project displaced them Tehri environment. They are rehabilitated in Dehradun and Hardwar. Will they adjust there? It is the question.

SARDAR SAROVAR PROJECT – It is situated in Bharuch district of Gujarat State on Narmada river. This project will submerge 245 villages of MP, Gujarat and Maharashtra and will displace more than one lakh people. It is also a controversial project. The project will also effect 56000 Ha agricultural and 60000 Ha forest land. It is multi-purpose project, and will fulfil the necessity of power drinking water and irrigation. This will result 900 crores of rupees additional in agriculture and 500 crores of rupees in water supply from this project.

(ii) *Displacement due to mining* – Mining is also one of important means in the field of prosperity. Mineral wealth has its own importance. Therefore, it covers large area for this purpose. Due to this developmental activity, thousands of people are displaced. People displaced due to this activity are poorest and mostly tribal. They lost much more than they get from the projects. Maximum persons lost their lives during settlement and transportation.

JHARIA COAL FIELD – a Case Study – It is in Jharkhand. In this projects, underground fires are. Thousands of hectares land effected by this. People are asked to vacate the native place. Since 1976, about 115 crores of rupees have been spent to put out fires. The problem of resettlement and rehabilitation of the people is still there. Estimate for rehabilitation is much more than the availability.

(iii) *Displacement due to National Parks.* To conserve Flora and Fauna (natural resources), some times large forest area is covered under National parks and Sanctuary. It is declared as core area and entry of local dewellers, tribals and villagers (of nearby area) are prohibited. They are deprived of their right of access to the benefits of forests. Therefore, efforts should be made to provide proper rehabilitation and employment to the effected people.

Case Study – (i) Valmiki Tiger Reserve area in west Champaran districts displaced 142 villages of Tharu community tribals.

(ii) **Wayabanad Sanctuary** displaced about 53470 tribal families. Their rehabilitation is still incomplete. The Kerala people are fighting with officials of forest for their cause.

REHABILITATION. The United Nations Universal declaration on Human Rights (Article 25(1) has declared that "Right to housing is basic human right". This suggest better rehabilitation, adequate compensation, job opportunities, civic amenities and religious and cultural benefits. Therefore, National Rehabilitation Policy is needed to honour the human rights of the displaced people. Govt. under Land Acquisition Act 1894 has power to vacate the land from people by giving notice for Govt. use. Therefore, most of the displacements have resulted due to land acquisition by Govt.. There is need of public awareness also in resettlement and rehabilitation plans. In general Govt. and other agencies provide a number of amenities for rehabilitated persons. Every landless person was provided 2 acres agriculture land, Rs 12-18000/- per acre cost of land to displaced person. Rs 20,000 or a plot for residence and 3-5000/- for transportation was also given with certain other compensations.

For displaced persons in case of Indira Sagar Dam, the following compensations for resettlement and rehabilitation were given :-

1. Developed plot or Rs 20,000 for purchase of plot for one family.
2. Rehabilitation grant of Rs 18700/- or 9350/- as per status of PAFs.
3. Transportation grant Rs 5000/- for shifting of one family.
4. Allotted 2-8 hectares of agriculture land per family or land compensation.
5. Attractive compensation for house, trees, wells and other structures.
6. At the plot sites, developed roads, water supply to lights, schools, health centres, worship places, panchayats, community centres, shops etc.
7. R and R work is being executed smoothly.
8. Additionally more benefits were given.
9. Professional training being given to project affected families at I + I Narmada Nagar.
10. Central school (Kendriya Vidyalaya started in June 01)
11. Different socio-economic upliftment programmes such as free medical check up, vaccination, training programmes are being organized.

ENVIRONMENTAL ETHICS

Issues and Possible Solutions

The issues, principles and guidelines relating to human interaction with their environment OR human obligations towards the environment and living beings are called Environmental Ethics or Earth Ethics. Ethics constitute the basic codes of civilized behviour, without which our environment as we know would be impossible. Such rules embody the basic constraints each of us agrees to practice in relationship with others. Ethical codes can be of help in most instances that confront us, but dilemmas do arise in which it seems there are no suitable alternatives.

We can see that our acts will follow what we think i.e. human-centric thinking or earth-centric thinking. The first view urges us to march ahead gloriously to conquer the nature and establish our supremacy over nature through technological innovations, economic growth and development, while second urges us to live on this earth as a part of it, like any other creation of nature and live sustainably. Human beings are over exploiting the natural resources and polluting the environment. These human acts are very dangerous and may lead to environmental crisis. If we want to check the environmental crisis, we will have to transform our thinking and attitude.

In relation to environmental protection or in need of environmental ethics two world views are :-

1. *Eco-centric worldview.* This states that earth resources are limited, and they are not for human beings alone but for all species. So we have to draw our requirements from environment, but not to that extent it degrades the environment. A healthy economy depend upon the healthy environment, therefore, success of mankind depend upon how we cooperate with nature while trying to use the resources of nature.

2. *Anthropocentric World view.* It states that man is the most important species of nature. Earth has an unlimited supply of resources. Most of the industrial societies believe in this view. So the success & healthy economy of mankind depend upon how nicely man derives benefits from nature.

To check the environmental crisis, we must follow the certain environmental ethics for better future. Some of them are -

1. One should love and honour the earth.
2. We should celebrate the turning of the seasons of the earth.
3. Do not waste or exploit the natural resources.
4. To bring about awareness regarding conservation of life support systems.
5. Our should be fair in sharing of resources.
6. We should respectful to plants and animals which provide us food.
7. We should conserve the ecosystem and promote appropriate sustainable development.
8. We should not do any thing at the cost of nature.
9. We should consume the natural resources in moderate amounts so that all may share this treasure.
10. We should concentrate on general awareness regarding environmental ethics from primary education.
11. A healthy environment depends upon a healthy economy

On 5th June 1989 (The world environment day), the eminent environmentalist Dr T. N. Khoshoo gave the concept of the *Dharma of Ecology*, the only way to meet the challenges. There is need to start from an individual to meet this threat to our environment.

CLIMATE CHANGE

Though climate is an average weather of an area or environmental factors of an area. These include quantity of light, temperature, humidity, wind, gases, water etc which average for about 30 yrs. Thus the changes in environmental conditions of an area over long period of time is called **climate change**. These changes effect the agriculture, migration of animals, hydrological cycle, thermal gradient between the poles and equator, wind pattern, distribution of rainfall etc.

The scientific and technological revolution has given multiple facilities to mankind, but at the same time man-made(Anthropogenic) activities are responsible for depletion of resources and upsetting the delicate balance between the various components of the environment. They are, excessive use of fossils fuels, deforestation, desertification, loss of fertility of soil, rapid industrialization, increase of automobiles. Changes in the atmosphere conditions resulting in to serious problems like green house effect, depletion of ozone layer and rise of world temperature etc.

The global change in temperature will not be uniform every where and will fluctuate in different regions. The places at higher latitudes will be warmed up more during late autumn and winter than the places in tropics. Poles may experience 2 to 3 times more warming than the global average, while warming in tropics may be only 50 to 100° C on an average. The increased warming at poles will reduce the thermal gradient between the equator and high latitude regions, decreasing the energy available to the heat engine that drives the global weather machine. This will disturb the global pattern of winds and ocean currents as well as the timing and distribution of rainfall. Shifting of ocean currents may change the climate of iceland and Britain, it may result in cooling at a time when rest of the world warms.

GLOBAL WARMING

The average global temperature is 15°C. The lower most layer of atmosphere i.e., troposphere, traps the heat by a natural process due to the presence of certain gases called **Green house gases**. They are carbon dioxide, ozone, methane Nitrous oxide, Chlorofluorocarbons (CFCs) and water vapours. In absence of these gases the temperature (15°C) would have been - 18°C. Thus warming of the earth's climate owing to the increased concentration of green house gases is called **Green house effect**. Therefore, this effect contributes a temperature rise to the tune of 33°C. These gases act like the glass in the botanical green house trapping the reradiated heat near the earth's surface and warming the planet. These gases along with water vapour and clouds absorbs the infrared radiation, trapping heat near the earth's surface. The two predominant green house gases (1) the water vapours whose level in the troposphere has relatively remained constant is controlled by hydrological cycle while (2) CO_2 whose level has increased is controlled by the global carbon cycle. Other gases whose levels have increased due to human activities are methane, NO and CFCs. Deforestation has also elevated levels of CO_2.

Change the temperature by more than 2°C is disastrous for various ecosystems on the earth including man. Some areas will become inhabitable because of drought or floods following a rise in average sea level.

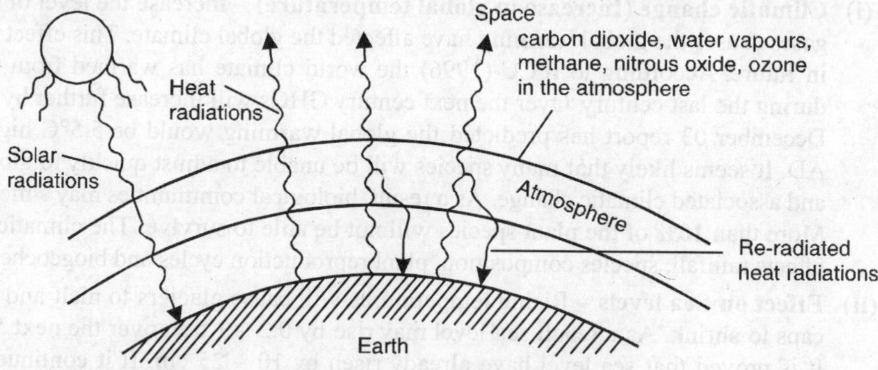

Fig. 6.2 The greenhouse effect

GREEN HOUSE GASES

The green house gases present in the troposphere and resulting in an increase in the temperature of air and the earth are discussed here.

CARBON DIOXIDE (CO_2) – The CO_2 is considered as the most dominant factor responsible for the green house effect. The troposphere contains only 00.0375% CO_2 (by volume) and its amount is controlled by carbon cycle. The four major pools or reservoirs of carbon are fossil fuels, the atmosphere, the biosphere and the oceans. Its concentration in atmosphere has increased from

290 PPM (1860) to 350 PPM (1990) and is expected to be 700 PPM in 2010. It was studied that in 1976 the world consumption of about 5000 million tons of fossil fuels per year was contributing the equivalent of 2.3 PPM of CO_2 to the atmosphere. The net annual increase of CO_2 in the air is 0.77 ppm which indicates that 1.60 ppm of fuel generated CO_2 is being absorbed elsewhere. In last couple of decades the fossil fuels may have contributed more CO_2 to the atmosphere than the terrestrial biosphere. Clearance of forests is another factor for the increase of CO_2.

CHLORO FLUORO CARBONS (CFCs) – The main source of CFCs include leaking air conditioners and refrigerators, evaporation of industrial solvents, production of plastic foams, aerosols, propellants(CFC-11) etc. The concentration of CFCs is rising nearly 5% per year. CFCs trap heat 20,000 times more efficiently than CO_2 and also destroy ozone layer, thus posing a serious twofold environmental problem. CFCs are responsible for 14 – 24% of global warming. It is said (Dickson and Cicerone 1986) that by the end of year 2050 they alone could contribute more than CO_2 to global warming.

METHANE (CH_4) – It is produced in a number of ways including the action of anaerobic bacteria on vegetation, decomposition of organic matter, incomplete combustion of vegetation, natural gas pipeline leaks, burning of biomass during production and uses of oil and natural gas and petroleum oil etc. It is rising approx. 2% every year. It absorbs 20 – 25% times more heat than CO_2.

NITROUS OXIDE (N_2O) – It is released from nylon products, from burning of biomass and fuels (specially coal). From breakdown of fertilizers in soil, livestock wastes and nitrated contaminated ground water, nylon products etc. It is responsible for about 6% of global warming. Besides trapping heat in the troposphere it also depletes ozone in the stratosphere. It absorbs about 250 times more heat than CO_2. The N_2O concentration in atmosphere is 0.3 ppm and is increasing 0.2% annually.

OZONE – It comes mostly from hydrocarbons and nitrogen oxides. It causes irritation to eyes and respiratory organs. It decreases the resistance power to infections and aggrevates illness.

IMPACT OF GLOBAL WARMING

(i) **Climatic change (Increase in global temperature)** – Increase the level of Green house gases causes the global warming have affected the global climate. This effect will increase in future. According to IPCC (1996) the world climate has warmed from 0.3 to 0.6°C during the last century. Over the next century GHGs will increase further by 1°C – 3.5°C December 03 report has predicted the global warming would be 5.5°C higher by 2100 AD. It seems likely that many species will be unable to adjust quickly to global warming and associated climatic change. As a result, biological communities may suffer profoundly. More than 10% of the plant species will not be able to survive. The climatic change also effects rainfall, species composition, plant reproduction cycles and biogeochemical cycles.

(ii) **Effect on Sea levels** – Rising temperatures will cause glaciers to melt and the polar ice caps to shrink. As a result sea level may rise by 0.2 – 1.5 m over the next 50 – 100 yrs. It is proved that sea level have already risen by 10 – 25 cm. If it continues, many low lying areas may be submerged in near future, and it is possible to destroy 20% - 80% of the coastal wetland. Rising sea levels are detrimental to coral reef species, which grow at a precise depth with optimum temperature and water movement. Abnormally high water temperatures in the Pacific Ocean during 1982 – 83 caused the death of symbiotic algae that live inside the coral.

(iii) **Reduction of Biodiversity** – As we have discussed, increased temperatures, inundation of some coastal biological communities and changes in the pattern of distribution of many species over a long period of time are likely to cause reduction in biodiversity in aquatic and terrestrial ecosystems.

(iv) Effect on Agriculture – There are different views regarding the effect of global warming on agriculture. It may be positive or negative. However, the effects of this change will vary for C_3 (i.e. wheat, rice and beans) and C_4 (e.g. maize, millet and sugarcane) plants. As temperature increase with rising CO_2 level, some crops may no longer be grown in certain regions. With rise in temperature soil moisture will decrease and evapo-transpiration and pest growth will increase. This will effect certain crops. With increased CO_2 concentration some plants will show increased photosynthesis, greater root production, increased nitrogen fixation in root nodules which may increase the growth of plants by 30%.

(v) Effect on human health – The global warming will lead to changes in the rainfall pattern in many areas, thereby effecting the distribution of vector borne diseases like malaria filariasis, elephantiasis etc. Warmer temperature and more water stagnation would favour the breeding of mosquitoes, snails and some insects, which are vectors of such diseases. Higher temperature and humidity will increase respiratory and skin diseases. Keeping in view of ill effects of global warming UNEP (United Nations Environmental Programme) is celecrating 5th June as **"World Environmental Day"** every year since 1989.

(vi) Effect on Arctic ecosystems – Global climate change will have profound effects on arctic ecosystems. Tundra is more sensitive to global climate change than most other ecosystems on earth. According to Shaver et al 1992 warmer temperature may increase primary production, thereby increasing Carbon input and soil respiration hence increasing carbon output.

(vii) Ecological disturbance – Global warming increases the desert. It increases temperature in North America, South Africa, Mexico, India and other countries. Changes of hurricanes, cyclones and floods will be more which will damage the lagoons, estuaries and coral reefs. Global warming may cause extinction of more than one million species of animals and plants by 2050 A.D.

MEASURES TO CHECK GLOBAL WARMING

To check the global warming following steps are necessary –

(1) Plant more trees (Afforestation)
(2) Control population growth.
(3) Cut down the current rate of CFCs and fossil fuel.
(4) Use of non-conventional source of energy.
(5) Shift from coal to Natural gas.
(6) To trap and use methane as a fuel.
(7) Reduce beef production.
(8) Efficiently remove CO_2 from smoke
(9) Use photosynthetic algae to remove atmosphere CO_2.
(10) Adopt sustainable agriculture.
(11) Use energy more efficiently.

ACID RAIN

Normal rain water is always acidic because of the fact that CO_2 present in the atmosphere gets dissolved in it forming carbonic (H_2CO_3) acid. Because, the presence of SO_2 (Sulphur Dioxide) and NO_2 (Nitrogen Oxide) gases as pollutants in the atmosphere, the pH of the rain water is further lowered (as low as 2.4). This is known as **Acid rain**.

"Literally acid rain means the presence of excessive acids in rain waters". Acid rain is a mixture of acids mainly H_2SO_4 and HNO_3. H_2SO_4 is a major contributor (60 – 70%) HNO_3 ranks second (30 – 40%) and HCl third.

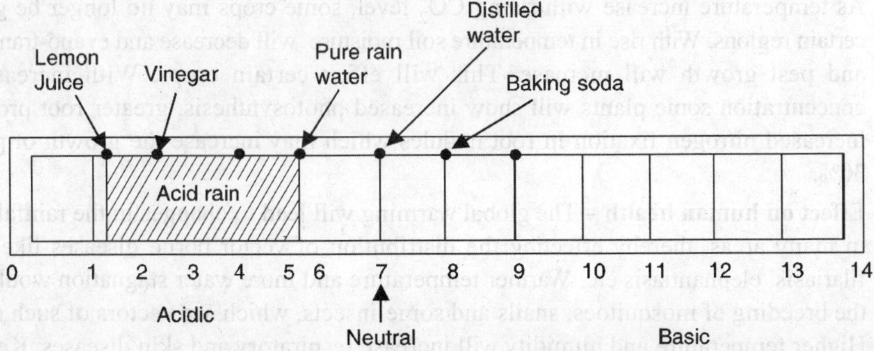

Fig. 6.3. The pH scale of common substances.

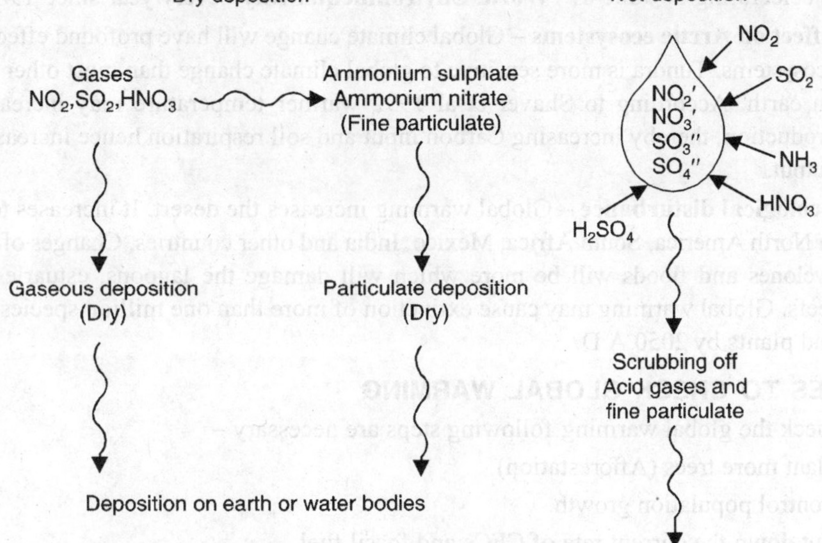

Fig. 6.4. Acid deposition (dry deposition and wet deposition)

Acid rain is a man-made and an environmental problem that knows no boundaries. Increasing acidity in natural waters and soils is becoming a problem all over the world. Acid deposition has degraded lakes, swamps and wet lands to such an extent that organisms can not thrive.

Where does acid come from

Acidification of environment is a man made phenomenon. No doubt that most acid come from human activities i.e. cars, houses, factories, power stations etc. There has always been some acid in rain, coming from volcanoes, swamps and plankton in the oceans, but scientists know that it has increased very sharply over the past 200 years. The acidity is mainly associated with the transport and subsequent deposition of oxides of sulphur, nitrogen and these oxidative products. These oxides are produced by combustion of fossil, fuels, smelters, power plants, automobile exhausts, domestic fires etc. Burning of fossil fuels for power generation contributes to 60 – 70% of the total SO_2 emitted globally. Emission of NO_x from anthropogenic sources ranges between 20 – 90 million tones annually over the globe. Acid rain problem has drastically increased due to industrialization.

Researches show that in Swiss, water is not fit for drinking. Pesticides sprayed on crops evaporate and in stratosphere they react with water vapour. Water contaminated with these toxic substances fall as rain. Due to this pH of Delhi water decreased from 7.0 to 6.1 in 1984. the term acid rain was first coined by Robert Angus Swith in 1952. In 1968, scientists pointed out that rain water was becoming increasingly acidic with each passing year.

How acid rain is formed

In high temperature combustion processes most of the nitric oxide originates from atmosphere and some Nitric oxide also released by burning of wood and as a result of microbial nitrification in the soil. Lightning is another source of Nitric oxide.

In day time Nitric oxide is oxidized by oxygen, ozone

$$NO + O_3 \longrightarrow NO_2 + O_2$$
$$H_2O \longrightarrow H^+ + OH^-$$
$$NO_2 + OH^- \longrightarrow HNO_3$$

Similarly, formulation of H_2SO_4 acid in the atmosphere can take place with a wide range of reduced as well as partially oxidised Sulphur compounds, H_2S, CS_2 etc. These compounds are released from oceans and soil under reducing conditions. The production of H_2SO_4 from SO_2 may take place homogeneously in the gas phase as :-

$$SO_2 + OH^- + X \longrightarrow HSO_3^- + X$$

Where $X = O_2$ or N_2 in atmosphere.

The HSO_3 so formed can undergo a number of reactions, some of which produce sulphuric acid.

$$HSO_3^- + O_2 + X \longrightarrow HOO^- + SO_3 + X$$
$$SO_3 + H_2O \longrightarrow H_2SO_4$$

The hydroperoxy radical HOO^- can also react to give HNO_3

$$NO + HOO^- \longrightarrow NO_2 + OH^-$$
$$NO_2 + OH^- + X \longrightarrow HNO_3 + X^-$$

Thus a number of reactions are taken place, forming different acids.

The chemical composition of rain is highly variable and depends on the geographic location and the influence of natural, anthropogenic chemical processes on the atmosphere of that region.

Effects of acid rain

Acid rain exerts both direct and indirect effects on the organisms and materials it comes in contact with. The dry deposition attacks building material, steel and other metals. When deposited in gaseous form it causes direct damage to plants and trees. Wet deposition has direct and indirect effects. It increases the acidity of lakes and rivers and effects aquatic as well as terrestrial ecosystems. Some of the effects may be described as :-

1. A significant reduction in fish population accompanied by decrease in the variety of species in food chains have been observed.
2. Adirondack ponds having high acidity levels, were among the first to lose fish population.
3. Different species reacts differently to acidified lakes. Adult fish can survive in more acidic water having high concentration aluminium than dry fish.
4. Many bacteria and blue green algae are killed due to acidification, disrupting the whole ecological balance.
5. In 1958 at Europe pH of rain water was 5.0 and in Netherland (1962) was 4.5. It damaged the leaves of plants and trees.

6. Forests of West Germany, Switzerland, Czechoslovakia, Swedish were severely effected by acid rain.
7. In North America and Europe, acid rain destroyed crops and forests, reducing agricultural productivity.
8. Acid rain has retarded the growth of pea, beams, radish, potato, spinach, carrots etc.
9. Modern researches show that acid rain leaches Potassium, Calcium, Magnesium etc essential elements from the top soil.
10. Acid lakes have low levels of phytoplankton.
11. The activity of bacteria and other microscopic animals is reduced in acidic water.
12. Broad leafed pond weeds do not grow in acid water.
13. Acid rain causes extensive damage to buildings and structural materials of marble, lime stone, mortar etc.

    Lime stone attacked as :-

    $$CaCO_3 + H_2SO_4 \longrightarrow CaSO_4 + H_2O + CO_2$$

    The attack on marble is termed as **Stone-leprosy.**
14. The Taj Mahal in agra is suffering from SO_2, H_2SO_4 and other fumes, pollutants released from Mathura refinery.
15. Acid rain corrodes houses, monuments, statues, bridges, fences, railways etc.
16. Acidification can play havoc with human nervous, respiratory and digestive systems by making the person an easy pray to neurological diseases.

OZONE LAYER DEPLETION

Stratosphere – Troposphere is the part of atmosphere where humans live and other life processes also occur. The stratosphere is the region of space between approximately 15-50 kms above the earth's surface. The gas molecules in the stratosphere act as absorbing centres, moderating the transmission of the solar radiation to the earth. The qualitative as well as quantitative effect of this is an important determining factor with respect to life processes.

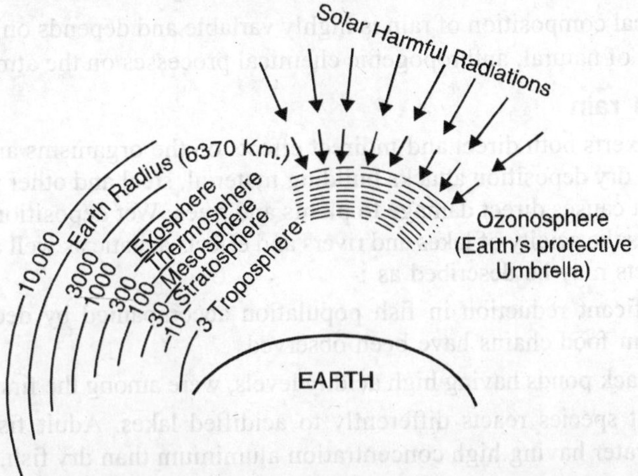

Fig. Atmospheric ozone-earth's protective umbrella.

CREATION OF OZONE LAYER

Ozone is naturally occurring gas found through out atmosphere, with a maximum mixing ratio at the altitude ranging from 15-30 km above the earth. This region is known as ozone layer. Ozone can be toxic to plants and animals but increased concentration has a profound beneficial effect. Both, atmosphere and earth surface are subjected to radiation from sun. These certain radiations are absorbed by atmospheric gases leading to ionization or dissociation of gases. In the lower mesosphere, atmospheric oxygen gets dissociated and subsequently combines with molecular oxygen forming ozone in stratosphere.

The presence of ozone layer in stratosphere is of vital significance for all biota, because the harmful solar radiations such as ultraviolet rays, which are lethal to life on earth are not allowed to enter the earth's atmosphere by ozone layer(or ozone umbrella). In the absence of this ozone layer all the UV rays of sun will reach the earth surface to increase the temperature so that *biological furnace* will turn into *blast furnace*. **Thus the ozone layer strongly absorbs or blocks the short wave ionising ultraviolet rays and so protect the life on earth.** In stratosphere, ozone is an effective filter capable of absorbing UV radiations with wavelength between 200nm and 315 nm.

FORMATION OF OZONE

In the lower mesosphere, the atmospheric oxygen absorbs UV radiation < 240 nm and photo-dissociates in to two oxygen atoms. These atoms subsequently combine with molecular oxygen of upper stratosphere producing ozone. Ozone is also capable of absorbing short wave length UV radiations releasing oxygen atoms.

$$O_2 + h\nu\ (\lambda < 240\ \text{nm}) \longrightarrow O + O$$

$$O + O_2 \longrightarrow O_3$$

Decomposition $$O_3 + h\nu\ (\lambda = 230 - 320\ \text{nm}) \longrightarrow O_2 + O$$

$$O + O_3 \longrightarrow O_2 + O_2$$

This mechanism does not necessarily upset the ozone equilibrium because ozone (loss) is compensated by creation of ozone. As a result ozone occurs in 10 ppm concentration in the form of layer in stratosphere. **The thickness of ozone layer is measured in Dobson units (DU), where 1 DU = 0.01 mm of the compressed gas 0°C and 760 mm mercury pressure.**

The average thickness of ozone layer in stratosphere has been estimated to be about 230 DU. It varies marginally with latitude.

MECHANISM OF OZONE DEPLETION

There are two processes –

1. Natural process
2. Anthropogenic process

1. Natural process – A dynamic equilibrium existing between the production and decomposition of ozone molecule constitutes one of the most important mechanism. The heat generated during the reaction causes a rise in temperature. Secondly, the photochemical process absorbs most of the harmful solar UV radiations.

Hence, the atmosphere is heated because of this absorption and the earth's biosphere is shielded from these lethal radiations. Without it, life on the earth would be completely destroyed. The ozone layer acts UV radiations from sun.

Ozone acts as powerful oxidising agent i.e. has ability to remove electrons from other molecules. Surface ozone is also produced by the action of UV radiations in sunlight involving both nitrogen oxides(NO_x) and volatile organic matter (VOM). Ozone also occur in air and water purification plants, oil wax, textiles and in synthetic industries.

2. Anthropogenic process – Some of the natural species moving in to stratosphere has been augmented in recent years by a number of human activities. Many of the processes, which are responsible for ozone layer depletion share a general mechanism of the type –

$$X + O_3 \longrightarrow XO + O_2$$

$$XO + O \longrightarrow X + O_2 \qquad \text{Where x = reacting species}$$

$$\text{Net reaction} \quad O + O_3 \longrightarrow O_2 + O_2$$

The most of the common species (above x) have been identified to be free radicals like HO_x, NO_x and ClO_x. Each of these species is capable of destroying the ozone layer. For example, near the stratosphere, the HO_x radicals are responsible for about 70% of the total mechanism of ozone destruction, including the oxygen only process. Around 30 KM atmosphere, the NO_x catalytic decomposition process dominates the ozone destruction.

EFFECTS OF OZONE DEPLETION

(1) With the ozone layer depletion, there is danger of the increase in the flux of ultraviolet radiation over earth's biosphere. They are harmful for man's life.

(2) UV radiations effects biological systems in two ways – one is confined to patches of skin while the other develops in the immune system as a whole.

(3) These kinds of skin cancer, **Basal** cell carcinoma, squamous cell carcinoma and melanoma caused by UV rays.

(4) UV radiations cause sun burns, leukemia and breast cancer.

(5) UV radiations absorbed by cornea and lens in the eye leading to **Photokeratitia** and cataracts.

(6) Ozone at ground level (of low concentration) exerts its toxic effects directly on the lungs.

(7) Ozone exposure has been shown to be associated with lung cancer, DNA breakage.

(8) Photochemical smog is the measure cause of ozone exposure causing urban air pollution posing a threat to human health.

(9) Many micro – phytoplankton's would die because of their exposure to UV solar radiation. The marked reduction in the productivity of phytoplankton's would in turn adversely effect zoo planktons.

(10) The loss of fish population would directly effect the inhabitants of coastal areas.

(11) Ozone is reported to be highly toxic to fish.

(12) UV radiations increases the mortality rate of larvae of zoo plankton's.

(13) Ozone flecking is observed with the plants of grape, citrus and tobacco. At 00.02 ppm it damage tomato, pea, pine and other plants.

(14) Plant proteins are also susceptible to UV injury.

(15) In plants O_3 enters through stomata. It causes visible damage to leaves thereby reducing their photosynthetic rate.

(16) Due to ozone reduction, intense UV radiation causes greater evaporation of surface water through the stomata of the leaves and decreases the soil moisture content.

(17) Ozone reacts with many fibres, such as cotton, nylon and polyester, dyes etc.

NUCLEAR ACCIDENTS AND HOLOCAUST

Japanese towns of Hiroshima and Nagasaki. The first atom bomb was exploded about 580 metres in the atmosphere over ill fated Hiroshima on August 6, 1945. the second atom bomb was detonated 507 metres high in air over Nagasaki. At least 100,000 people were reported killed, severely injured and missing in Hiroshima alone, where the bomb virtually demolished all structures and buildings in about 15 square km. area. In Nagasaki 49000 civilians were killed, injured and disappeared while an area of 6 to 7 km. was devastated.

The atom bomb exploded on Hiroshima used Uranium (U – 235) with a half life period of 8.5 x 10^8 years, while the Nagasaki bomb had plutonium (Pu – 239) as an explosive man-made radio-nuclide with half life of 24,000 years.

Exhaustive studies conducted in Hiroshima show that heavily exposed hibakusha, bomb effected people. Have a 30% greater chance than normal of dying from cancer.

This lethal explosion caused millions to die, while thousands became unable to breathe, unable to see and eat. Public suffered from pulmonary oedema, anoxia, brain damage and increased risk of sterility. However, several cases reported high incidence of still births and congenital deformities among pregnant women. Excess cases of leukemia began appearing in 1948, but by the early 1970's the rate leveled off nearly to that of the general population.

The first hydrogen bomb was exploded in 1954 on Bikini Island in the Pacific. The radioactive fallout from this explosion severely affected the crew of a Japanese fishing boat, the Lucky Dragon about 150 km. away from the site of explosion. Several persons were hospitalized, killed and disappeared, while in Bikini Island the explosion caused the entire toll vanished.

In 1957 and 1958, the USA, Soviet Union and Great Britain detonated nuclear weapons whose total yield was about 85 megatons. These weapons were equal to 4250 Hiroshima sized atom bombs. They caused several dangerous effects on man.

In 1961, Russia detonated a bomb of 57 megatons that could obliterate a city more than three hundred times the size of Hiroshima.

Case study

CHERNOBYL ACCIDENT

Chernobyl was the first officially acknowledged nuclear accident in USSR and first reported to the world. April 26, 1986 was a sad day for nuclear power generation when a major accident occurred at 1.23 A.M. in the nuclear reactor at Chernobyl, in the Ukraine area of the Soviet Union. It resulted in clouds of radioactive smoke over a large area in Scandinavian countries which are 2000 km. away in the Russian region itself. There was a devastating fire in the reactor which caused few casualties and severe damage to the nuclear plant. On finding the fire uncontrollable the soviet authorities sought the help of West Germany and other nuclear nations to tackle the situation. Presumably, the core of the nuclear reactor had melted.

The explosion at the **Chernobyl power plant** in Soviet Ukraine, USSR confirmed the worst nuclear disaster. Poor design of the reactor magnified with operator negligence caused the havoc. The operators ignored warnings from various sensors and even disconnected the emergency core cooling systems.

Neutrons went out of control and enormous steam built up in pipes. The explosion sent the graphite slabs of the reactor core through the roof, setting it a fire and spewing radioactive materials around the world.

Twenty percent of the plant's radioactive iodine escaped along with 15 to 20% of radioactive caesium, hazardous plutonium and mixture of several radioisotopes. Radiation level reached 100 times than the normal. Samples of radioactive debris collected in Sweden indicated that some percentage of radiation in the reactor core actually escaped due to which air became frosty and the environment critically polluted. The USSR authorities claimed later that the fire had been controlled and put out.

The vast cloud of radiations caused considerable anxiety in Western Europe. The Soviet Union reported in a statement on April 29, 1986 that two persons were killed and a few injured.

But the senior US official told that the death toll was probably much higher. On May 2nd, the Soviet authorities admitted that 18 persons affected by the leakage of Chronic radiation were in incident were credible. He made this remark after studying a data relayed by US spy satellite which over flew the Ukrainian area where the fire and radiation leak occurred.

A top Swedish nuclear official estimated that "thousands of people could have been killed" around the Chernobyl reactor, 100 km. north of the Ukrainian capital of Kiev in USSR, if all radiations had leaked into the atmosphere. He also reported that water supplies to the homes of 3.5 million persons of Kiev had probably contaminated with acute radionuclides.

Scientists have recently discovered that worms exposed to excess radiations have started copulating instead of reproducing asexually. The **Chernobyl nuclear disaster** has not only affected human being, but has even drastically changed the lives of worms. The finding is the clear evidence of how wildlife is affected by radioactive pollution. Although there is a wealth of knowledge on the impact of ionizing radiation on humans, the effects on wildlife have never been assessed. Many researchers are now focusing on how wildlife has been affected by the radioactive pollution that spewed from a reactor which exploded 17 years ago in Chernobyl. Researchers from Ukraine found that the worms from a lake near Chernobyl had received 20 times more radiation as compared to those from the lake situated far apart. They found some remarkable changes in the sexual habits of worms. According to the scientists, the worms have started sexually reproducing to protect themselves from radiation. Sexual reproduction allows natural selection of genes that offer better protection from radiation damage. A change in sexual behavior may have increased the resistance of the worms.

The **Chernobyl power plant,** the world's worst nuclear accident best symbolizes the potential dangers of atomic energy, has been officially closed on December 15, 2000 – 14 years after the plant exploded and sent a cloud of radioactive dust over Europe. More than 4000 Ukrainians who took part in the hasty clean up effort since the 1986 disaster have died and 70,000 were disabled by radiation, according to government figures. About 3.4 million of Ukraine's 50 million people, including some 1.26 million children, have been affected.

WASTELAND RECLAMATION

The area or land like salt affected, sandy, barren-hill-ridge snow covered or glacial which is economically unproductive suffer from environmental deterioration are called waste lands. These are for one reason or the other do not fulfill their life sustaining potential. Therefore, wasteland should be reclaimed and put to some productive use. About half of our country's geographical area is lying as wasteland. Maximum wasteland area in our country lie in Rajasthan, followed by Madhya Pradesh and Andhra Pradesh.

There are two modes of formation of wastelands :-

1. *Natural Process.* These include undulating uplands, snow covered lands, coastal saline areas, sandy areas etc.

2. *Anthropogenic(Man-made) activities.* These are deforestation, overgrazing, mining and erroneous agricultural practices.

Due to increasing population more land is required for agricultural and forestry. Opposite to it quality land is decreasing due to erosion, desertification, water logging, salinity, toxic effects of agrochemicals and industrial effluents. Now the only way to increase the land resources is by reclamation and developing degraded land. Govt. also planning and giving funding for conservation and regeneration of forest resources. Faulty irrigation specially canal, drainage practices, water logging are also main causes to form wasteland. Indira Gandhi Canal project in Rajasthan has changed water-starved wasteland to water-soaked wasteland.

Unscientific mining with no environmental protection have developed large tracts that lost productivity. In U.P, Bihar, M.P, Orissa, A.P, Punjab, a number of mining operations have started since long, which effects not only cultivated land but forest also. Ecological problems developed in coal mine areas in Ranchi, Bihar, Singrauli, M.P and other places due to high power and transmission lines, rails, road tracks and townships. Due to industrialization (Cement, Power, Auto etc) and increased multi national companies in metros and other states, the land for cultivation is shrinking.

For reclamation of wasteland in 1985 National Wasteland Development Board (NWLDB) was established. The basic functions of this board, were to check the soil erosion, deforestation and plantation. But later on in 1989-90, the aims of board were re-examined and modified. Now the main objects of board are :-

1. To prevent soil erosion, landslides and flood.
2. To improve the quality, physical structure of marginal soils.
3. To increase the availability of Biomass.
4. To conserve the biological resources of the land for sustainable use.

Wasteland reclamation is development of degraded, mined and unused land for productive purpose. The following are some of the reclamation practices:-

1. *Soil Sodicity* – Sodicity in soil is due to excess of sodium. This can be reduced by use of Gypsom in soil. The calcium of gypsum will replace sodium by exchangeable process.
2. *Use of green – manures and Biofertilizers* – Dhaincha (Sesbania aculeate), Sun hemp (Corolania juncea) or guar etc. should be used as green manures to reduce salinity of soil. Farm yard manure and biofertilizers like blue green algae are also used to reduce salinity.
3. *Irrigation practices* – High frequency, thin and frequent irrigations have been found more useful. Surface irrigation with precise land leveling and smoothening also help to reduce water logging and salinity.
4. *Drainage* – Water logged soils are reclaimed by removing excess of water by artificial drainage. When water stands in fields after heavy rains, the ditches (30 – 40 cm deep at distances of 30 – 50 metres) are provided to runoff the excess of water. This is known as *surface–drainage.* But in *sub-surface drainage* method, perforated PVC pipes or open jointed pipes graved 2-3 metre below the land surface are used. This is better method.
5. *Leading and land development* – this is done in salt affected soil. After land levelling, apply excess of water, which pushes down excess of salts. This loss of extra salts by downward moving water is called leaching.
6. *Selection of tolerant crops* – Tolerance of crops to salt is found to range from sensitive, semi–tolerant, tolerant to highly tolerant. Date-palm, sugar barley are highly tolerant crops, which are not affected by high salinity. Soyabeen, coconut mustard, wheat etc.are salt tolerant. Rice, maize, pulses, sunflower, sugarcane, vegetables etc. are semi tolerant. The different combinations of crops should be grown in saline soil.

There are two successful cases of reclamation of mined area, one is Neyveli Lignite Corporation Ltd in Tamilnadu and other is Stone Quarries of Sayaji Iron Works in Gujarat. The NWLD Board and environment and forest ministry planned a programme for wasteland reclamation. In first phase they selected the five districts-Almorah, Purulia, Bellari, Dungarpur and Sundergarh. During 8th five year programme 13.75 crores of rupees sectioned for 234 projects to NGO's. In 1999-2000, 8 crores of rupees were sanctioned for 3000 hactare wasteland. With the help of retired military personals, Wasteland Development Task Force, 1994 in Morena district of M.P. was started to develop the 12000 hactare Chambal revines.

Consumerism and Waste Products

Consumerism is the consumption of natural resources by the human beings. Exponential increase in consumption of natural resources are there now days as comparison to early period. This is due to increase in life style and population growth. Earlier much simpler life was but now in modern period needs and consumption of resources increased.

In different countries the consumerism of natural resources are not the same. On the basis of consumerism, population have two categories, one is people over population and consumption over population. This is nothing but population influence consumerism of natural resources and generation of wastes.

(A) People over-population – This occurs in less develop countries (LDC) where per capita consumption is less. Because as compared to resources population is high i.e. more people than available supply of food, water and other resources. India has 16% of the world population, hence per capita consumption is 3% waste generation has direct correlation with the rate of consumerism. In India wastes generated are about 3% very less than more developed countries(MDC).

(B) Consumption over Population – It occurs in more developed countries where population is less. Naturally the natural resources will be more as per population. Hence per capita consumption will be high. In other words more food, water and natural resources will be available, results very high wastes generation. The life style will be good as compared to less developed countries. USA, a more developed country has about 4.7% of world population and about 25% waste generation. About 400 million metric tonnes of industrial wastes (other than mining and mineral) and 180 million metric tonnes municipal wastes are in USA.

Thus tremendous amount of stuff thrown away. Much of the wastes can be reused or recycled in other useful products. But in practice easy to throw than environmentally responsible.

The relationship between population size, consumerism and environmental impact was explained by *Paul Ehrlich and John Hodlern (1972)* in following ways –

Environmental Impact = Population size × Per capita use of resources
× Waste generated per unit of resource used

THE ENVIRONMENT (PROTECTION) ACT, 1986

INTRODUCTION

Since the sixties concern over the state of environment has grown the world over. There has been substantive decline in environment quality due to increasing population, loss of vegetal cover and biological diversity, excessive concentrations of harmful chemicals in the ambient atmosphere and in food chains, growing risks of environmental accidents and threats to life support systems. The decisions which were taken at the United Nations conference on the human environment held in Stockholm in June 1972 were based on the world community's resolve to protect and enhance the environmental quality. While participating in the said conference Govt. of India strongly voiced the environmental concerns. Although several measures had been taken for environmental protection both before and after the conference, it was found necessary to enact a comprehensive law on the subject to implement the decision of the conference. Accordingly the ENVIRONMENT (PROTECTION) BILL was introduced in the parliament. This bill having been passed by both the houses of parliament received the assent of the President on 23rd May 1986. it came in to force on 19th Nov. 1986 as **The Environment (Protection) Act 1986 (29 of 1986)**.

THE ENVIRONMENT (PROTECTION) ACT, 1986 (29 OF 1986) [23RD MAY, 1986]

An Act to provide for the protection and improvement of environment and for matters connected therewith.

WHEREAS decisions were taken at the United Nations Conference on the Human Environment held at Stockholm in June, 1972, in which India participated, to take appropriate steps for the protection and environment of human environment;

AND WHEREAS it is considered necessary further to implement the decisions aforesaid in so far as they relate to the protection and improvement of environment and the prevention of hazards to human beings, other living creatures, plants and property;

Be it enacted by Parliament in the Thirty-seventh year of the Republic of India as follows:-

CHAPTER 1

PRELIMINARY

1. Short title, extent and commencement – (1) This Act may be called the Environment (Protection) Act, 1986.

(2) It extends to the whole of India.

(3) It shall come into on such date as the Central Governemnt may, by notification in the Official Gazette, appoint and different dates may be appointed for different provisions of this Act and for different areas.

GENERAL POWERS OF THE CENTRAL GOVERNMENT

2. Power of Central Government to take measures to protect and improve environment–

(1) Subject to the provisions of this Act, the Central Government shall have the power to take all such measures as it deems necessary or expedient for the purpose of protecting and improving the quality of the environment and preventing controlling and abating environmental pollution.

(2) In particular, and without prejudice to the generality of the provisions of the sub-section (1) Such measures may include measures with respect to all or any of the following matters, namely:-

(i) co-ordination of action by the State Government, officers and other authorities–

(a) under this Act, or the rules made there under; or

(b) under any other law for the time being in force which is relatable to the objects of this Act;

(ii) planning and execution of a nation-wide programme for the prevention, control and abatement of environmental pollution;

(iii) laying down standards for the quality of environment in its various aspects.

(iv) laying down standards for emission or discharge of environmental pollutants from various sources whatsoever:

Provided that different standards for emission or discharge may be laid down under this clause from different sources having regard to the quality or composition of the emission or discharge of environmental pollutants from such sources;

(v) restriction of areas in which any industries, operations or processes or class of industries, operations or processes shall not be carried out or shall be carried out subject to certain safeguards;

(vi) laying down procedures and safeguards for the prevention of accidents which may cause environmental pollution and remedial measures for such accidents;

(vii) laying down procedures and safeguards for the handling of hazardous substances;

(viii) examination of such manufacturing processes, materials and substances as are likely to cause environmental pollution;

(ix) carrying out and sponsoring investigations and research relating to problems of environmental pollution;

(x) inspection of any premises, plant, equipment, machinery, manufacturing or other processes, materials or substances and giving, by order, of such directions to such authorities, officers or persons as it may consider necessary to take steps for the prevention, control and abatement of environmental pollution;

(xi) establishment or recognition of environmental laboratories and institutes to carry out the functions entrusted to such environmental laboratories and institutes under this Act;

(xii) collection and dissemination of information in respect of matters relating to environmental pollution;

(xiii) preparation of manuals, codes or guides relating to the prevention, control and abatement of environmental pollution;

(xiv) such other matters as the Central Government deems necessary or expedient for the purpose of securing the effective implementation of the provisions of this Act.

3. Rules to regulate environmental pollution – (1) The Central Government may, by notification in the Official Gazette, make rules in respect of all or any of the matters referred to in section 3.

(2) In particular, and without prejudice to the generality of the foregoing power, such rules may provide for all or any of the following matters, namely:-

(a) the standards of quality of air, water or soil for various areas and purposes;

(b) the maximum allowable limits of concentration of various environmental pollutants (including noise) for different areas;

(c) the procedures and safeguards for the handling of hazardous substances;

(d) the prohibition and restrictions on the handling of hazardous substances in different areas;

(e) the prohibition and restrictions on the location of industries and the carrying on of processes and operations in different areas;

(f) the procedures and safeguards for the prevention of accidents which may cause environmental pollution and for providing for remedial measures for such accidents.

PREVENTION, CONTROL AND ABATEMENT OF ENVIRONMENTAL POLLUTION

1. Persons carrying on industry, operation, etc., not to allow emission or discharge of environmental pollutants in excess of the standards – No person carrying on any industry, operation or process shall discharge or emit or permit to be discharged or emitted any environmental pollutant in excess of such standards as may be prescribed.

2. Persons handling hazardous substances to comply with procedural safeguards – No person shall handle or cause to be handled any hazardous substance except in accordance with such procedure and after complying with such safeguards as may be prescribed.

3. Furnishing of information to authorities and agencies in certain cases – (1) Where the discharge of any environmental pollutant in excess of the prescribed standards occurs or is apprehended to occur due to any accident or other unforeseen act or event, the person responsible for such discharge and the person in charge of the place at which such discharge occurs or is apprehended to occur shall be found to prevent or mitigate the environmental pollution caused as a result of such discharge and shall also forthwith –

(a) intimate the fact of such occurrence or apprehension of such occurrence; and

(b) be bound, if called upon, to render all assistance, to such authorities or agencies as may be prescribed.

(2) On receipt of information with respect to the fact or apprehension of any occurrence of the nature referred to in sub-section (1), whether through intimation under that sub-section or otherwise, the authorities or agencies referred to in sub-section (1) shall, as early as practicable, cause such remedial measures to be taken as are necessary to prevent or mitigate the environmental pollution.

(3) The expenses, if any, incurred by any authority or agency with respect to the remedial measures referred to in sub-section (2), together with interest (at such reasonable rate as the Government may, by order, fix) from the date when a demand for the expenses is made until it is paid, may be recovered by such authority or agency from the person concerned as arrears of land revenue or of public demand.

4. Powers of entry and inspection – (1) Subject to the provisions of this section, any person empowered by the Central Government in this behalf shall have a right to enter, at all reasonable times with such assistance as he considers necessary, any place –

(a) for the purpose of performing any of the functions of the Central Government entrusted to him;

(b) for the purpose of determining whether and if so in what manner, any such functions are to be performed or whether any provisions of this Act or the rules made thereunder or any notice, order, direction or authorization served, made, given or granted under this Act is being or has been complied with;

(c) for the purpose of examining and testing any equipment, industrial plant, record, register, document or any other material object or for conducting a search of any building in which he has reason to believe that an offence under this Act or the rules made thereunder has been or is being or is about to be committed and for seizing any such equipment, industrial plant, record, register, document or other material object if he has reasons to believe that it may furnish evidence of the commission of an offence punishable under this Act or the rules made there under or that such seizure is necessary to prevent rules mitigate environmental pollution.

(2) Every person carrying on any industry, operation or process or handling any hazardous substance shall be bound to render all assistance to the person empowered by the Central Government under sub-section (1) for carrying out the functions under that sub-section and if he fails to do so without any reasonable cause or excuse, he shall be guilty of an offence under this Act.

(3) If any person wilfully delays or obstructs any person empowered by the Central Government under sub-section (1) in the performance of his functions, he shall be guilty of an offence under this Act.

(4) The provisions of the Code of Criminal Procedure, 1973 (2 of 1974), or, in relation to the State of Jammu and Kashmir, or any area in which that Code is not in force, the provisions of any corresponding law in force in that State or area shall, so far as may be, apply to any search or seize under this section as they apply to any search or seizure made under the authority of a warrant issued under section 94 of the said Code or, as the case may be, under the corresponding provision of the said law.

5. Power to take sample and procedure to be followed in connection therewith - (1) The Central Government or any officer empowered by it in this behalf, shall have power to take, for the purpose of analysis, samples of air, water, soil or other substance from any factory, premises or other place in such manner as may be prescribed.

(2) The result of any analysis of a sample taken under sub-section (1) shall not be admissible in evidence in any legal proceeding unless the provisions of sub-sections (3) and (4) are complied with.

(3) Subject to the provisions of sub-section (4), the person taking the sample under sub-section (1) shall –

(a) serve on the occupier or this agent or person in charge of the place, a notice, then and there, in such form as may be prescribed, of his intention to have it so analysed;

(b) in the presence of the occupier or his agent or person, collect a sample for analysis;

(c) cause the sample to be placed in a container or containers which shall be marked and sealed and shall also be signed both by the person taking the sample and the occupier or his agent or person;

(d) send without delay, the container or the containers to the laboratory established or recognized by the Central Government under section 12.

(4) When a sample is taken for analysis under sub-section (1) and the person taking the sample serves on the occupier or his agent or person, a notice under clause (a) of sub-section (3), then –

(a) in a case where the occupier, his agent or person wilfully absents himself, the person taking the sample shall collect the sample for analysis to be placed in a container or containers which shall be marked and sealed and shall also be signed by the person taking the sample, and

(b) in a case where the occupier or his agent or person present at the time of taking the sample refuses to sign the marked and sealed container or containers of sample as required under clause (c) of sub-section (3), the marked and sealed container or containers shall be signed by the person taking the samples, and the container or containers shall be sent without delay by the person taking the sample for analysis to the laboratory established or recognized under section 12 and such person shall inform the Government Analyst appointed or recognized under section 13 in writing, about the wilful absence of the occupier or his agent or person, or, as the case may be, his refusal to sign the container or containers.

MISCELLANEOUS

1. **Protection of action taken in good faith** – No suit, prosecution or other legal proceeding shall lie against the Government or any officer or other employee of the Government or any authority constituted under this Act or any member, officer or other employee of such authority in respect of anything which is done or intended to be done in good faith in pursuance of this Act or the rules made or orders or directions issued thereunder.
2. **Cognizance of offences** – No court shall take cognizance of any offence under this Act except a complaint made by –
   (a) the Central Government or any authority or officer authorized in this behalf by that government; or
   (b) any person who has given notice of not less than sixty days, in the manner prescribed, of the alleged offence and of his intention to make a complaint, to the Central Government or the authority or officer authorized as aforesaid.

3. Information, reports or returns – The Central Government may, in relation to its functions under this Act, from time to time, require any person, officer, State Government or other authority to furnish to it or any prescribed authority or officer any reports, returns, statistics, accounts and other information and such person, officer, State Government or other authority shall be bound to do so.

4. Members, officers and employees of the authority constituted under section 3 to be public servants – All the members of the authority, constituted, if any, under section 3 and all officers and other employees of such authority when acting or purporting to act in pursuance of this Act or the rules made or orders or directions issued thereunder shall be deemed to be public servants within the meaning of section 21 of the Indian Penal Code (45 of 1860).

5. Bar of jurisdiction – No civil court shall have jurisdiction to entertain any suit or proceeding in respect of anything done, action taken or order or direction issued by the Central Government or any other authority or officer in pursuance of any power conferred by or in relation to its or his functions under this Act.

6. Power to delegate – Without prejudice to the provisions of sub-section (3) of section 3, the Central Government may, by notification in the Official Gazette, delegate, subject to such conditions and limitations as may be specified in the notification, such of its powers and functions under this Act [except the power to constitute an authority under sub-section (3) of section 3 and to make rules under section 25] as it may deem necessary or expedient, to any officer, State Government or other authority.

7. Effect of other laws – (1) Subject to the provisions of sub-section (2), the provisions of this Act and the rules or orders made therein shall have effect notwithstanding anything inconsistent therewith contained in any enactment other than this Act.

(2) Where any act or omission constitutes an offence punishable under this Act and also under any other Act then the offender found guilty of such offence shall be liable to be punished under the other Act and not under this Act.

8. Power to make rules – (1) The Central Government may, by notification in the Official Gazette, make rules for carrying out the purposes of this Act.

(2) In particular, and without prejudice to the generality of the foregoing power, such rules may provide for all or any of the following matters, namely:-

(a) the standards in excess of which environmental pollutants shall not be discharged or emitted under section 7;

(b) the procedure in accordance with and the safeguards in compliance with which hazardous substances shall be handled or cause to be handled under section 8;

(c) the authorities or agencies to which intimation of the fact of occurrence or apprehension of occurrence of the discharge of any environmental pollutant in excess of the prescribed standards shall be given and to whom all assistance shall be bound to be rendered under sub-section (1) of section 9;

(d) the manner in which samples of air, water, soil or other substance for the purpose of analysis shall be taken under sub-section (1) of section 11;

(e) the form in which notice of intention to have a sample analysed shall be served under clause (a) of sub-section (3) of section 11;

(f) the functions of the environmental laboratories, the procedure for the submission to such laboratories of samples of air, water, soil and other substances for analysis or test; the form of laboratory report; the fees payable for such report and other matters to enable such laboratories to carry out their functions under sub-section (2) of section 12;

(g) the qualifications of Government Analyst appointed or recognized for the purpose of analysis of samples of air, water, soil or other substances under section 13;

(h) the manner in which notice of the offence and of the intention to make a complaint to the Central Government shall be given under clause (b) of section 19;

(i) the authority or officer to whom any reports, returns, statistics, accounts and other information shall be furnished under section 20;

(j) any other matter which is required to be, or may be, prescribed.

9. Rules made under this Act to be laid before Parliament – Every rule made under this Act shall be laid, as soon as may be after it is made, before each House of Parliament, while it is in session, for a total period of thirty days which may be comprised in one session or in two or more successive sessions, and if, before the expiry of the session immediately following the session or the successive sessions aforesaid, both Houses agree in making any modification in the rule or both Houses agree that the rule should not be made, the rule shall thereafter have effect only in such modified form or be of no effect, as the case may be; so, however, that any such modification or annulment shall be without prejudice to the validity of anything previously done under that rule.

THE AIR (PREVENTION AND CONTROL OF POLLUTION) ACT 1981

INTRODUCTION – With increasing industrialization and the tendency of the majority of industries to congregate in areas which are already heavily industrialised, the problem of air pollution had begun to be felt in the country. The various pollutants discharged from certain human activities connected with traffic, heating, use of domestic fuel, refuse, incinerations etc. also have detrimental effects on the health of the people as also on animal life, vegetation and property. In view of decisions taken at the June 1972 United Nations Conference hold at Stockholm, the Govt. decided to implement those decisions related to the preservation of the quality of air and control of air pollution. Accordingly the **Air (Prevention and Control of Pollution) Bill** was introduced in the parliament and passed by both the houses of parliament. It came into force on the **16th day of May 1981 as THE AIR (PREVENTION AND CONTROL OF POLLUTION) ACT 1981 (14 of 1981).** Its amendment act 1987 (47 of 1987) came into force.

THE AIR (PREVENTION AND CONTROL OF POLLUTION) ACT, 1981 (14 OF 1981) [29TH MARCH 1981]

An Act to provide for the prevention, control and abatement of air pollution, for the establishment, with a view to carrying out the aforesaid purposes, of Boards, for conferring on and assigning to such Boards, powers and functions relating thereto and for matters connected therewith.

WHEREAS decisions were taken at the United Nations Conference on the Human Environment held in Stockholm in June, 1972, in which India participated, to take appropriate steps for the preservation of the natural resources of the earth which, among other things, include the preservation of the quality of air and control of air pollution;

AND WHEREAS it is considered necessary to implement the decisions aforesaid in so far as they relate to the preservation of the quality of air and control of air pollution;

BE it enacted by Parliament in the Thirty-second Year of the Republic of India as follows:-

CHAPTER I
PRELIMINARY

1. Short title, extent and commencement – (1) This Act may be called the Air (Prevention and Control of Pollution) Act, 1981.

(2) It extends to the whole of India.

(3) It shall come into force on such date as the Central Government may, by notification in the Official Gazette, appoint.

CENTRAL AND STATE BOARDS FOR THE PREVENTION AND CONTROL OF AIR POLLUTION

Central Pollution Control Board – The Central Pollution Control Board constituted under section 3 of the Water (Prevention and Control of Pollution) Act, 1974 (6 of 1974), shall, without prejudice to the exercise and performance of its power and functions under the Act, exercise the powers and perform the functions of the Central Pollution Control Board for the prevention and control of air pollution under this Act.

State Pollution Control Boards – In any State in which the water (Prevention and Control of Pollution) Act, 1974 (6 of 1974), is in force and the State Government has constituted for that state a State Pollution Control Board under section 4 of that Act, such State Board shall be deemed to be the State Board for the Prevention and Control of Air Pollution constituted under section 5 of this Act, and accordingly that State Pollution Control Board shall, without prejudice to the exercise and performance of its powers and functions under that Act, exercise the powers and perform the functions of the State Board for the prevention and control of air pollution under this Act].

Constitution of State Boards – In any State in which the Water (Prevention and Control of Pollution) Act, 1974 (6 of 1974), is not in force, or that Act is in force but the State Government has not constituted a [State Pollution Control Board] under that Act, the State Government shall, with effect from such date as it may, by notification in the Official Gazette, appoint constitute a [State Pollution Control Board] under such name as may be specified in the notification, to exercise the powers conferred on, and perform the functions assigned to, that Board under this Act.

Central Board to exercise the powers and perform the functions of a State Board in the Union territories – No State Board shall be constituted for a Union territory and in relation to a Union territory, the Central Board shall exercise the powers and perform the functions of a State Board under this Act for that Union territory:

Provided that in relation to any Union territory the Central Board may delegate all or any of its powers and functions under this section to such person or body of persons as the Central Government may specify.

POWERS AND FUNCTIONS OF BOARDS

Functions of Central Board – (1) Subject to the provisions of this Act, and without prejudice to the performance of its functions under the Water (Prevention and Control of Pollution) Act, 1974 (6 of 1974), the main functions of the Central Board shall be to improve the quality of air and to prevent, control or abate air pollution in the country.

(2) In particular and without prejudice to the generality of the foregoing functions, the Central Board may –

(a) advise the Central Government on any matter concerning the improvement of the quality of air and the prevention, control or abatement of air pollution;

(b) plan and cause to be executed a nation-wide programme for the prevention, control or abatement of air pollution;

(c) co-ordinate the activities of the State Boards and resolve disputes among them;

(d) provide technical assistance and guidance to the State Boards, carry out and sponsor investigations and research relating to problems of air pollution and prevention, control or abatement of air pollution;

[dd) perform such of the functions of any State Board as may be specified in an order made under sub-section (2) of section 18;]

(e) plan and organize the training of persons engaged or to be engaged in programmes for the prevention, control or abatement of air pollution on such terms and conditions as the Central Board may specify;

(f) organize through mass media a comprehensive programme regarding the prevention, control or abatement of air pollution;

(g) collect, compile and publish technical and statistical data relating to air pollution and the measures devised for its effective prevention, control or abatement and prepare manuals, codes or guides relating to prevention, control or abatement of air pollution;

(h) lay down standards for the quality of air;

(i) collect and disseminate information in respect of matters relating to air pollution;

(j) perform such other functions as may be prescribed.

(3) The Central Board may establish or recognize a laboratory or laboratories to enable the Central Board to perform its functions under the section efficiently.

(4) The Central Board may –

(a) delegate any of its functions under this Act generally or specially to any of the committees appointed by it;

(b) do such other things and perform such other acts as it may think necessary for the proper discharge of its functions and generally for the purpose of carrying into effect the purposes of this Act.

Functions of State Boards – (1) Subject to the provisions of this Act, and without prejudice to the performance of its functions, if any, under the Water (Prevention and Control of Pollution) Act, 1974), the functions of a State Board shall be –

(a) to plan a comprehensive programme for the prevention, control or abatement of air pollution and to secure the execution thereof;

(b) to advise the State Government on any matter concerning the prevention, control or abatement of air pollution;

(c) to collect and disseminate information relating to air pollution;

(d) to collaborate with the Central Board in organising the training of persons engaged or to be engaged in programmes relating to prevention, control, or abatement of air pollution and to organise mass-education programme relating thereto;

(e) to inspect, at all reasonable times, any control equipment, industrial plant or manufacturing process and to give, by order, such directions to such persons as it may consider necessary to take steps for the prevention, control, or abatement of air pollution;

(f) to inspect air pollution control areas at such intervals as it may think necessary, assess the quality of air therein and take steps for the prevention, control or abatement of air pollution in such areas;

(g) to lay down, in consultation with the Central Board and having regard to the standards for the quality of air laid down by the Central Board, standards for emission of air pollutants into the atmosphere from industrial plants and automobiles or for the discharge of any air pollution into the atmosphere from any other source whatsoever not being a ship or an aircraft;

Provided that different standards for emission may be laid down under this clause for different industrial plants having regard to the quantity and composition of emission of air pollutants into the atmosphere from such industrial plants;

(h) to advise the State Government with respect to the suitability of any premises or location for carrying on any industry which is likely to cause air pollution;

(i) to perform such other functions as may be prescribed or as may, from time to time, be entrusted to it by the Central Board or the State Government;

(j) to do such other things and to perform such other acts as it may think necessary for the proper discharge of its functions and generally for the purpose of carrying into effect the purposes of this Act.

(2) A State Board may establish or recognize a laboratory or laboratories to enable the State Board to perform its functions under this section efficiently.

PREVENTION AND CONTROL OF AIR POLLUTION

(i) Power to declare air pollution control areas.
(ii) Power to give instructions for ensuring standards for emission from automobiles.
(iii) Restrictions on use of certain industrial plants.
(iv) Persons carrying an industry etc. not to allow emission of air pollutants in excess of the standard laid down by State Board.
(v) Power to take samples of air or emission and procedure to be followed.
(vi) Reports of analysis.
(vii) Appeals.

PENALTIES AND PROCEDURE

Failure to comply with the provisions of section 21 or section 22 or with the directions issued under section 31A - (1) Whoever fails to comply with the provisions of section 21 or section 22 or directions issued under section 31A, shall, in respect of each such failure, be punishable with imprisonment for a term which shall not be less than one year and six months but which may extend to six years and with fine, and in case the failure continues, with an additional fine which may extend to five thousand rupees for every day during which such failure continues after the conviction for the first such failure.

(2) If the failure referred to in sub-section (1) continues beyond a period of one year after the date of conviction, the offender shall be punishable with imprisonment for a term which shall not be less than two years [but which may extend to seven years and with fine.]

Penalties for certain acts – Whoever –

(a) destroys, pulls down, removes, injures or defaces any pillar, post or stake fixed in the ground or any notice or other matter put up, inscribed or placed, by or under the authority of the Board, or

(b) obstructs any person acing under the orders or directions of the Board from exercising his powers and performing his functions under this Act, or

(c) damages any works or property belonging to the Board, or

(d) fails to furnish to the Board or any officer or other employee of the Board any information required by the Board or such officer or other employee for the purpose of this Act, or

(e) fails to intimate the occurance of the emission of air pollutants into the atmosphere in excess of the standards laid down by the State Board or the apprehension of such occurrence, to the State Board and other prescribed authorities or agencies as required under sub-section (1) of section 23, or

(f) in giving any information which he is required to give under this Act, makes a statement which is false in any material particular, or

(g) for the purpose of obtaining any consent under section 21, makes a statement which is false in any material particular,

Shall be punishable with imprisonment for a term which may extend to three months or with fine which may extend to [ten thousand rupees] or with both.

Penalty for contravention of certain provisions of the Act – Whoever contravenes any of the provisions of this Act or any order or direction issued thereunder, for which no penalty has been elsewhere provided in this Act, shall be punishable with imprisonment for a term which may extend to three months or with fine which may extend to ten thousand rupees or with both, and in the case of continuing contravention, with an additional fine which may extend to five thousand rupees for every day during which such contravention continues after conviction for the first such contravention.]

Power to amend the Schedule – *[Rep. by the Air (Prevention and Control of Pollution) Amendment Act, 1987 (47 of 1987), sec. 22 (w.e.f. 1-4-1988)]*

Maintenance of register – (1) Every State Board shall maintain a register containing particulars of the persons to whom consent has been granted under section 21, the standards for emission laid down by it relation to each such consent and such other particulars as may be prescribed.

(2) The register maintained under sub-section (1) shall be open to inspection at all reasonable hours by any person interested in or affected by such standards for emission or by any other person authorized by such person in this behalf.

Effect of other laws – Save as otherwise provided by or under the Atomic Energy Act, 1962 (33 of 1962), in relation to radioactive air pollution the provisions of this Act shall have effect not withstanding anything inconsistent therewith contained in any enactment other than this Act.

Power of Central Government to make rules – (1) The Central Government may, in consultation with the Central Board, by notification in the Official Gazette, make rules in respect of the following matters, namely:-

(a) the intervals and the time and place at which meetings of the Central Board or any committee thereof shall be held and the procedure to be followed at such meetings, including the quorum necessary for the transaction of business threat, under sub-section (1) of section 10 and under sub-section (2) of section 11;

(b) the fees and allowances to be paid to the members of a committee of the Central Board, not being members of the Board, under sub-sction (3) of section 11;

(c) the manner in which and the purposes for which persons may be associated with the Central Board under sub-section (1) of section 12;

(d) the fees and allowances to be paid under sub-section (3) of section 12 to persons associated with the Central Board under sub-section (1) of section 12;

(e) the functions to be performed by the Central Board under clause (j) of sub-section (2) of section 16;

[(f) the form in which and the time within which the budget of the Central Board may be prepared and forwarded to the Central Government under section 34;

(ff) the form in which the annual report of the Central Board may be prepared under section 35;]

(g) the form in which the accounts of the Central Board may be maintained under sub-section (1) of section 36.

(2) Every rule made by the Central Government under this Act shall be laid, as soon as may after it is made, before each house of Parliament, while it is in session, for a total period of thirty days which may be comprised in one session or in two or more successive sessions, and if, before the expiry of the session immediately following the session or the successive sessions aforesaid,

both Houses agree in making any modification in the rule or both Houses agree that the rule should not be made, the rule shall thereafter have effect only in such modified form or be of no effect, as the case may be; so, however, that any such modification or annulment shall be without prejudice to the validity of anything previously done under that rule.

Power of State Government to make rules – (1) Subject to the provisions of sub-section (3), the State Government may, by notification in the Official Gazette, make rules to carry out the purposes of this Act in respect of matters not falling within the purview of section 53.

(2) In particular, and without prejudice to the generality of the foregoing power, such rules may provide for all or any of the following matters, namely:-

[(a) the qualifications, knowledge and experience of scientific, engineering or management aspects of pollution control required for appointment as member-secretary of a State Board constituted under the Act;]

[(aa)] the terms and conditions of service of the Chairman and other members (other than the member-secretary) of the State Board constituted under this Act under sub-section (7) of section 7;

(b) the intervals and the time and place at which meeting of the State Board or any committee thereof shall be held and the procedure to be followed at such meetings, including the quorum necessary for the transaction of business threat, under sub-section (1) of section 10 and under sub-section (2) of section11;

(c) the fees and allowances to be paid to the members of a committee of the State Board, not being members of the Board under sub-section (3) of section 11;

(d) the manner in which and the purpose for which persons may be associated with the State Board under sub-section (1) of section 12;

(e) the fess and allowances to be paid under sub-section (3) of section 12 to persons associated with the State Board under sub-section (1) of section 12;

(f) the terms and conditions of service the member-secretary of a State Board constituted under this Act under sub-section (1) of section 14;

(g) the powers and duties to be exercised and discharged by the member-secretary of a State Board under sub-section(2) of section 14;

(h) the conditions subject to which a State Board may appoint such officers and other employees as it considers necessary for the efficient performance of its functions under sub-section (3) of section 14;

(i) the conditions subject to which a State Board may appoint a consultant under sub-section (5) of section 14;

(j) the functions to be performed by the State Board under clause (i) of sub-section (1) of section 17;

(k) the manner in which any area or areas may be declared as air pollution control area or areas under sub-section (1) of section 19;

(l) the form of application for the consent of the State Board, the fees payable thereof, the period within which such application shall be made and the particulars it may contain, under sub-section (2) of section 21;

(m) the procedure to followed in respect of an inquiry under sub-section (3) of section 21;

(n) the authorities or agencies to whom information under sub-section (1) of section 23 shall be furnished;

(o) the manner in which samples of air or emission may be taken under sub-section (1) of section 26;

(p) the form of the notice referred to in sub-section (3) of section 26;

(q) the form of the report of the State Board analyst under sub-section (1) of section 27;

(r) the form of the report of the Government analyst under sub-section (3) of section 27;

(s) the functions of the State Air Laboratory, the procedure for the submission to the said Laboratory of samples of air or emission for analysis or tests, the form of Laboratory's report thereon, the fees payable in respect of such report and other matters as may be necessary or expedient to enable that Laboratory to carry out its functions, under sub-section (2) of section 28;

(t) the qualifications required for Government analysts under sub-section (1) of section 29;

(u) the qualifications required for State Board analysts under sub-section (2) of section 29;

(v) the form and the manner in which appeals may be preferred, the fees payable in respect of such appeals and the procedure to be followed by the Appellate Authority in disposing of the appeals under sub-section (3) of section 31;

[(w) the form in which and the time within which the budget of the State Board may be prepared and forwarded to the State Government under section 34;

(ww) the form in which the annual report of the State Board may be prepared under section 35;]

(x) the form in which the accounts of the State Board may be maintained under sub-section (1) of section 36;

[(xx) the manner in which notice of intention to make a complaint shall be given under section 43;]

(y) the particulars which the register maintained under section 51 may contain;

(z) any other matter which has to be, or may be, prescribed.

(3) After the first constitution of the State Board, no rule with respect to any of the matters referred to in sub-section (2) (other than those referred to in [in clause (aa)] thereof), shall be made, varied, amended or repealed without consulting that Board.

THE SCHEDULE

[Omitted by the Air (Prevention and Control of Pollution) Amendment Act, 1987 (47 of 1987) sec. 25 (w.e.f. 1-4-1988)]

THE WATER (PREVENTION AND CONTROL OF POLLUTION) ACT, 1974

INTRODUCTION – As a result of growth of industries and the increasing tendency to urbanization the problem of pollution of rivers and streams had assumed considerable importance. It had become essential to ensure that the domestic and industrial effluents are not allowed to be discharged in to the water courses without adequate treatment. To draw a draft enactment for the prevention of water pollution a committee was set up in 1962. Later on a draft bell was prepared and put up for consideration at the joint session of Central Council of local self Govt., ministers of Town and Country planning hold in 1965. After long discussion, some resolutions were passed by the legislatives of some states. To give effect to these resolutions, the Water (Prevention and Control of Pollution) bill was introduced in the parliament, and having been passed by both the houses, received the assent of the President on 23 March 1974. It came on the Statute book as **THE WATER (PREVENTION AND CONTROL OF POLLUTION ACT 1974 (6 OF 1974).** Later on, it was amended twice (44 of 1978) and (53 of 1988).

THE WATER (PREVENTION AND CONTROL OF POLLUTION) ACT, 1974

(6 OF 1974) [23RD MARCH, 1974]

An Act to provide for the prevention and control of water pollution and the maintaining or restoring of wholesomeness of water, for the establishment, with a view to carrying out the purposes aforesaid, of Boards for the prevention and control of water pollution, for conferring on and assigning to such Boards powers and functions relating thereto and for matters connected therewith.

WHEREAS it is expedient to provide for the prevention and control of water pollution and the maintaining or restoring of wholesomeness of water, for the establishment, with a view to carrying out the purposes aforesaid, of Boards for the prevention and control of water pollution and for conferring on and assigning to such Boards powers and functions relating thereto;

AND WHEREAS Parliament has no power to make laws for the States with respect to any of the matters aforesaid except as provided in Articles 249 and 250 of the Constitution;

AND WHEREAS in pursuance of clause (1) of Article 252 of the Constitution resolutions have been passed by all the Houses of the Legislatures of the States of Assam, Bihar, Gujarat, Haryana, Himachal Pradesh, Jammu and Kashmir, Karnataka, Kerala, Madhya Pradesh, Rajasthan, Tripura and West Bengal to the effect that the matters aforesaid should be regulated in those States by Parliament by law;

BE it enacted by Parliament in the Twenty-fifth Year of the Republic of India as follows:-

PRELIMINARY

1. Short title, application and commencement – (1) This Act may be called the Water (Prevention and Control of Pollution) Act, 1974.

(2) It applies in the first instance to the whole of the States of Assam, Bihar, Gujarat, Haryana, Himachal Pradesh, Jammu and Kashmir, Karnataka, Kerala, Madhya Pradesh, Rajasthan, Tripura and West Bengal and the Union territories; and it shall apply to such other State which adopts this Act by resolution passed in that behalf under clause (1) of Article 252 of the Constitution.

(3) It shall come into force, at once in the States of Assam, Bihar, Gujarat, Haryana, Himachal Pradesh, Jammu and Kashmir, Karnataka, Kerala, Madhya Pradesh, Rajasthan, Tripura and West Bengal and in the Union territories, and in any other State which adopts this Act under clause (1) of Article 252 of the Constitution of the date of such adoption and any reference in this Act to the commencement of this Act shall, in relation to any State or Union territory, means the date on which this Act comes into force in such State or Union territory.

POWERS AND FUNCTIONS OF BOARDS

Functions of Central Board – (1) Subject to the provisions of this Act, the main function of the Central Board shall be to promote cleanliness of streams and wells in different areas of the States.

In particular and without prejudice to the generality of the foregoing function, the Central Board may perform all or any of the following functions, namely:-

(a) advise the Central Government on any matter concerning the prevention and control of water pollution;

(b) co-ordinate the activities of the State Boards and resolve disputes among them;

(c) provide technical assistance and guidance to the State Boards, carry out and sponsor investigations and research relating to problems of water pollution and prevention, control or abatement of water pollution;

(d) plan and organize the training persons engaged or to be engaged in programmes for the prevention, control or abatement of water pollution on such terms and conditions as the Central Board may specify;

(e) organize through mass media a comprehensive programme regarding the prevention and control of water pollution;

[(ee) perform such of the functions of any State Board as may be specified in an order made under sub-section (2) of section 18;]

(f) collect, compile and publish technical and statistical data relating to water pollution and the measures devised for its effective prevention and control and prepare manuals, codes or guides relating to treatment and disposal of sewage and trade effluents and disseminate information connected therewith;

(g) lay down, modify or annual, in consultation with the State Government concerned, the standards may be laid down for the same stream or well or for different streams or wells, having regard to the quality of water, flow characteristics of the stream or well and the nature of the use of the water in such stream or well or streams or wells;

(h) plan and cause to be executed a nation-wide programme for the prevention, control or abatement of water pollution;

(i) perform such other functions as may be prescribed.

The Board may establish or recognize a laboratory or laboratories to enable the Board to perform its functions under this section efficiently, including the analysis of samples of water from any stream or well or of samples of any sewage or trade effluents.

Functions of State Board – (1) Subject to the provisions of this Act, the functions of a State Board shall be –

(a) to plan a comprehensive programme for the preventions, control or abatement of pollution of streams and wells in the State and to secure the execution thereof;

(b) to advise the State Government on any matter concerning the prevention, control or abatement of water pollution;

(c) to collect and disseminate information relating to water pollution and the prevention, control or abatement thereof;

(d) to encourage, conduct and participate investigations and research relating to problems of water pollution and prevention, control or abatement of water pollution;

(e) to collaborate with the Central Board in organizing the training of persons engaged or to be engaged in programmes relating to prevention, control or abatement of water pollution and to organize mass education programmes relating thereto;

(f) to inspect sewage or trade effluents, works and plants for the treatment of sewage and trade effluents and to review plans, specifications or other data relating to plants set up for the treatment of water, works for the purification thereof and the system for the disposal of sewage or trade effluents or in connection with the grant of any consent as required by this Act;

(g) to lay down, modify or annul effluent standards for the sewage and trade effluents and for the quality of receiving waters (not being water in an inter-State stream) resulting from the discharge of effluents and to classify waters of the State;

(h) to evolve economical and reliable methods of treatment of sewage and trade effluents, having regard to the peculiar conditions of soils, climate and water resources of different regions and more especially the prevailing flow characteristics of water in streams and wells which render it impossible to attain even the minimum degree of dilution;

(i) to evolve methods of utilization of sewage and suitable trade effluents in agriculture;

(j) to evolve efficient methods of disposal of sewage and trade effluents on land, as are necessary on account of the predominant conditions of scant stream flows that do not provide for major part of the year the minimum degree of dilution;

(k) to lay down standards of treatment of sewage and trade effluents to be discharged into any particular stream taking into account the minimum fair weather dilution available in that stream and the tolerance limits of pollution permissible in the water of the stream, after the discharge of such effluents;

(l) to make, vary or revoke any order –

(i) for the prevention, control or abatement of discharges of waste into streams or wells;

(ii) requiring any person concerned to construct new systems for the disposal of sewage and trade effluents or to modify, alter or extend any such existing system or to adopt such remedial measures as are necessary to prevent, control or abate water pollution;

(m) to lay down effluent standards to be complied with by persons while causing discharge of sewage or sullage or both and to lay down, modify or annul effluent standards for the sewage and trade effluents;

(n) to advise the State Government with respect to the location of any industry the carrying on of which is likely to pollute a stream or well;

(o) to perform such other functions as may be prescribed or as may, from time to time, be entrusted to it by the Central Board or the State Government.

The Board may establish or recognize a laboratory or laboratories to enable the Board to perform its functions under this section efficiently, including the analysis of samples of water from any stream or well or of samples of any sewage or trade effluents.

PREVENTION AND CONTROL OF WATER POLLUTION

These are as follows –

1 – (19)	Power of State Govt. to restrict the application of the act to certain areas.
2 – (20)	Power to obtain information.
3 – (21)	Power to take the samples of effluents and procedure to be followed in connection herewith.
4 – (22)	Reports of the results of analysis on samples taken under section 21.
5 – (23)	Power of entry and inspection.
6 – (24)	Prohibition on use of stream or well for disposal of polluting matter.
7 – (25)	Restrictions on new outlets and new discharges.
8 – (26)	Provision regarding existing discharge of sewage or trade effluent.
9 – (27)	Refusal or withdrawal of consent by State Board.
10 – (28)	Appeals.
11 – (29)	Revisions.
12 – (30)	Power of State Board to carry out certain works.
13 – (31)	Furnishing of information to State Board and other agencies in certain cases.
14 – (32)	Emergency measures in case of pollution of stream or well.
15 – (33)	Power of Board to make application to counts for restraining apprehended pollution of water in streams or wells.
16 – (33-A)	(Ins. By Act 53 of 1988 section 18) – Power to give directions.

PENALTIES AND PROCEDURE

1. Failure to comply with directions under sub-section (2) or sub-section (3) of section 20, or orders issued under clause (C) of sub-section (1) of section 32 or directions issued under sub-section (2) of section 33 or section 33A -

(1) Whoever fails to comply with the direction given under sub-section (2) or sub-section (3) of section 20 within such time as may be specified in the direction shall, on conviction, be punishable with imprisonment for a term which may extend to three months or with fine which may extend to ten thousand rupees or with both and in case the failure continues, with an additional fine which may extend to five thousand rupees for every day during which such failure continues after the conviction for the first such failure.

(2) Whoever fails to comply with any order issued under clause (C) of sub-section (1) of section 32 or any direction issued by a court under sub-section (2) of section 33 or any direction issued under section 33A shall, in respect of each such failure and on conviction, be punishable with imprisonment for a term which shall not be less than one year and six months but which may extend to six years and with fine, and in case the failure continues, with an additional fine which may extend to five thousand rupees for every day during which such failure continues after the conviction for the first such failure.

(3) If the failure reffered to in sub-section (2) continues beyond a period of one year after the date of conviction, the offender shall, on conviction, be punishable with imprisonment for a term which shall not be less than two years but which may extend to seven years and with fine.

2. Penalty for certain acts –

(1) Whoever –

(a) destroys, pulls down, removes, injures or defaces any pillar, post or stake fixed in the ground or any notice or other matter put up, inscribed or placed, by or under the authority of the Board, or

(b) obstructs any person acting under the orders or directions or the Board from exercising his powers and performing his functions under this Act, or

(c) damages any works or property belonging to the Board, or

(d) fails to furnish to any officer or other employee of the Board any information required by him for the purpose of this Act, or

(e) fails to intimate the occurrence of any accident or other unforeseen act or event under section 31 to the Board and other authorities or agencies as required by that section, or

(f) in giving any information which he is required to give under this act, knowingly or wilfully makes a statement which is false in any material particular, or

(g) for the purpose of obtaining any consent under section 25 or section 26, knowingly or wilfully makes a statement which is false in any material particular, shall be punishable with imprisonment for a term which may extend to three months or with fine which may extend to [ten thousand rupees] or with both.

(2) Where for the grant of a consent in pursuance of the provisions of section 25 or section 26 the use of meter or gauge or other measure or monitoring device is required and such device is used for the purposes of those provisions, any person who knowingly or wilfully alters or interferes with that device so as to prevent it from monitoring or measuring correctly shall be punishable with imprisonment for a term which may extend to three months or with fine which may extend to [ten thousand rupees] or with both.

3. Penalty for contravention of provisions of section 24 – Whoever contravenes the provisions of section 24 shall be punishable with imprisonment for a term which shall not be less than [one year and six months] but which may extend to six years and with fine.

4. Penalty for contravention of section 25 or section 26 – Whoever contravenes the provisions of section 25 or section 26 shall be punishable with imprisonment for a term which shall not be less than [one year and six months] but which may extend to six years and with fine.

5. Enhanced penalty after previous conviction – If any person who has been convicted of any offence under section 24 or section 25 or section 26 is again found guilty of an offence involving a contravention of the same provision, he shall, on the second and on every subsequent conviction, be punishable with imprisonment for a term which shall not be less than [two years] but which may extend to seven years and with fine;

Provided that for the purpose of this section no cognizance shall be taken of any conviction made more than two years before the commission of the offence which is being punished.

6. Penalty for contravention of certain provisions of the Act – Whoever contravenes any of the provisions of this Act or fails to comply with any order or direction given under this Act, for which no penalty has been elsewhere provided in this Act, shall be punishable with imprisonment which may extend to three months or with fine which may extend to ten thousand rupees or with both and in the case of a continuing contravention or failure, with an additional fine which may extend to five thousand rupees for every day during which such contravention or failure continues after conviction for the first such contravention or failure.]

7. Publication of names of offenders.
8. Offences by companies.
9. Offences by Govt. Departments.
10. Cognizance of offences.
11. Members, officers and servants of boards to be the public servants.

Power of Central Government to make rules – (1) The Central Government may, simultaneously with the constitution of the Central Board, make rules in respect of the matters specified in sub-section (2):

Provided that when the Central Board has been constituted, no such rule shall be made, varied, amended or repealed without consulting the Board.

(2) In particular, and without prejudice to the generality of the foregoing power, such rules may provide for all or any of the following matters, namely:-

(a) the terms and conditions of service of the members (other than the chairman and member-secretary) of the Central Board under sub-section (8) of section 5;

(b) the intervals and the time and place at which meetings of the Central Board or of any committee thereof constituted under this Act, shall be held and the procedure to be followed at such meetings, including the quorum necessary for the transaction of business under section 8, and under sub-section (2) of section 9;

(c) the fees and allowances to be paid to such members of a committee of the Central Board as are not members of the Board under sub-section(3) of section 9;

(d) the manner in which and the purposes for which persons may be associated with the Central Board under sub-section (1) of section 10 and the fees and allowances payable to such person;

(e) the terms and conditions of service of the chairman and the member-secretary of the Central Board under sub-section (9) of section 5 and under sub-section (1) of section 12;

(f) conditions subject to which a person may be appointed as a consulting engineer to the Central Board under sub – section (4) of section 12;

(g) the powers and duties to be exercised and performed by the chairman and the member-secretary of the Central Board;

(h) the form of the report of the Central Board analyst under sub-section (1) of section 22;

(i) the form of the report of the Government analyst under sub-section (3) of section 22;

[(j) the form in which and the time within which the budget of the Central Board may be prepared and forwarded to the Central Government under section 38;

(k) the form in which the annual report of the Central Board may be prepared under section 39;]

(l) the form in which the accounts of the Central Board may be maintained under section 40;

[(m) the manner in which notice of intention to make a complaint shall be given to the Central Board or officer authorized by it under section 49;]

(n) any other matter relating to the Central Board, including the powers and functions of that Board in relation to union territories;

(o) any other matter which has to be, or may be, prescribed.

(3) Every rule by the Central Government under this Act shall be laid, as soon as may be after it is made, before each House of Parliament while it is in session for a total period of thirty days which may be comprised in one session or in two or more successive sessions, and if, [before the expiry of the session immediately following the session or the successive sessions or the successive sessions aforesaid], both Houses agree in making any modification in the rule or both Houses agree that the rule should not be made, the rule shall thereafter have effect only in such modified form or be of no effect, as the case may be; so, however, that any such modification or annulment shall be without prejudice to the validity of anything preciously done under that rule.

Power of State Government to make rules. (1) The State Government may, simultaneously with the constitution of the State Board, make rules to carry out the purposes of this Act, in respect of matters not falling within the purview of section 63:

Provided that when the State Board has been constituted, no such rule shall be made, varied, amended repealed without consulting that Board.

(2) In particular, and without prejudice to the generality of the foregoing power, such rules may provide for all or any of the following matters, namely:-

(a) the terms and conditions of service of the members(other than the chairman and the member-secretary) of the State Board under sub-section (8) of section 5;

(b) the time and place of meetings of the State Board or of any committee of that Board constituted under this Act and the procedure to be followed at such meeting, including the quorum necessary for the transaction of business under section 8 and under sub-section (2) of section 9;

(c) the fees and allowances to be paid to such members of a committee of the State Board as are not members of the Board under sub-section (3) of section 9;

(d) the manner in which and the purposes for which persons may be associated with the State Board under sub-section (1) of section 10 [and the fees and allowances payable to such persons];

(e) the terms and conditions of service of the chairman and the member-secretary of the State Board under sub-section (9) of section 5 and under sub-section (1) of section 12;

(f) the conditions subject to which a person may be appointed as a consulting engineer to the State Board under sub-section (4) of section 12;

(g) the powers and duties to be exercised and discharged by the chairman and the member-secretary of the state Board;

(h) the form of the notice referred to in section 21;

(i) the form of the report of the State Board analyst under sub-section (1) of section 22;

(j) the form of the report of the Government analyst under sub-section (3) of section 22;

(k) the form of application for the consent of the State Board under sub-section (2) of section 25, and the particulars it may contain;

(l) the manner in which inquiry under sub-section (3) of section 25 may be made in respect of an application for obtaining consent of the State Board and the matters to be taken into account in granting or refusing such consent;

(m) the form and manner in which appeals may be filed, the fees payable in respect of such appeals and the procedure to be followed by the appellate authority in disposing of the appeals under sub-section (3) of section 28;

[(n) the form in which and the time within which the budget of the State Board may be prepared and forwarded to the State Government under section 38;

(o) the form in which the annual report of the State Board may be prepared under section 39.

(p) the form in which the accounts of the State Board may be maintained under sub-section (1) of section 40.

(q) the manner in which notice of intention to make a complaint shall be given to the State Board or officer authorized by it under section 49.

(r) any other matter which has to be or may be prescribed.

THE WILD LIFE (PROTECTION) ACT, 1972 (NO. 53 OF 1972)

An Act to provide for the protection of wild animals, birds and plants, and for matters connected therewith or ancillary or incidental thereto with a view to ensuring the ecological and environmental security of the Country.

Preliminary

Short title, extent, and commencement

(1) This Act may be called the Wild Life (Protection) Act, 1972.

[(2) It extends to the whole of India, except the State of Jammu and Kashmir.]

(3) It shall come into force in a State or Union Territory to which it extends, on such date as the Central Government may, by notification, appoint, and different dates may be appointed for different provisions of this Act or for different States or Union Territories.

Standing Committee of the National Board –

(1) The National Board may, in its discretion, constitute a Standing Committee for the purpose of exercising such powers and performing such duties as may be delegated to the Committee by the National Board.

(2) The Standing Committee shall consist of the Vice-Chairperson, the Member-Secretary, and not more than ten members to be nominated by the Vice-Chairperson from amongst the members of the National Board.

(3) The National Board may constitute committees, sub-committees or study groups, as may be necessary, from time to time in proper discharge of the functions assigned to it.

Functions of the National Board –

(1) It shall be the duty of the National Board to promote the conservation and development of wild life and forests by such measures as it thinks fit.

(2) Without prejudice to the generality of the foregoing provision, the measures referred to therein may provide for –

(a) framing policies and advising the Central Government and the State Governments on the ways and means of promoting wild life conservation and effectively controlling poaching and illegal trade of wild life and its products;

(b) making recommendations on the setting up of and management of national parks, sanctuaries and other protected areas and on matters relating to restriction of activities in those areas;

(c) carrying out or causing to be carried out impact assessment of various projects and activities on wild life or its habitat;

(d) reviewing from time to time, the progress in the field of wild life conservation in the country and suggesting measures for improvement thereto; and

(e) preparing and publishing a status report at least once in two years on wild life in the country.

Duties of the State Board for Wild Life – It shall be the duty of State Board for Wild Life to advise the State Government,

(a) in the selection and management of areas to be declared as protected areas;

(b) in formulation of the policy of protection and conservation of Wild Life and specified plants;

(c) in any matter relating to any schedule;

[(cc) in relation to the measures to be taken for harmonizing the needs of the tribals and other dwellers of the forest with the protection and conservation of wild life; and*]

(d) in any matter that may be referred to it by the State Government.

Hunting of Wild Animals

Prohibition of Hunting – No person shall hunt any wild animal specified in Schedule I,II,III and IV except as provided under section 11 and section 12.]

Hunting of wild animals to be permitted in certain cases –

(1) Notwithstanding anything contained in any other law for the time being in force and subject to the provisions of Chapter IV -

(a) the Chief Wild Life Warden may; if he is satisfied that any wild animal specified in Sch. I has become dangerous to human life or is so disabled or diseased as to be beyond recovery, by order in writing and stating the reasons therefore, permit any person to hunt such animal or cause animal to be hunted;

Provided that nothing in this sub-section shall exonerate any person who, when such defence becomes necessary, was committing any act in contravention of any provision of this Act or any rule or order made thereunder.

Provided that no wild animal shall be ordered to kill unless the Chief Wild Life Warden is satisfied that such animal cannot be captured, tranquilized or translocated.

Provided further that no such captured animal shall be kept in captivity unless the Chief Wild Life Warden is satisfied that such animal cannot be rehabilitated in the wild and the reasons for the same are recorded in writing.

Explanation - For the purposes of clause (a), the process of capture or translocation, as the case may be, of such animal shall be made in such a manner as to cause minimum trauma to the said animal.

(b) the Chief Wild Life Warden or the authorized officer may, if he is satisfied that any wild animal specified in Sch II, Sch. III or Sch IV has become dangerous to human life or to property (including standing crops on any land) or is so disabled or diseased as to be beyond recovery, by order in writing and stating the reasons therefore, permit any person to hunt such animal or group of animals in a specified area or cause such animal or group of animals in that specified area to be hunted.

(2) The killing or wounding in good faith of any wild animal in defence of oneself or of any other person shall not be an offence.

(3) Any wild animal killed or wounded in defence of any person shall be Govt. property.

Protection of Specified Plants

Prohibition of picking, uprooting, etc., of specified plants – Save as otherwise other wise provided in this chapter, no person shall –

(a) wilfully pick, uproot, damage, destroy, acquire or collect any specified plant from any forest land and area specified, by notification, by the Central Government,

(b) possess, sell, offer for sale, or transfer by way of gift or otherwise, or transport any specified plant, whether alive or dead, or part or derivative thereof:

Provided that nothing in this section shall prevent a member of a scheduled tribe, subject to the provisions of Chapter IV, from picking, collecting or possessing in the district he resides any specified plant or part or derivative thereof for his *bona fide* personal use.

Grant of permit for special purposes – The Chief Wild Life Warden may with the previous permission of the State Government, grant to any person a permit to pick, uproot, acquire or collect from a forest land or the area specified under section 17A or transport, subject to such conditions as may be specified therein, any specified plant for the purpose of –

(a) education;

(b) scientific research;

(c) collection, preservation and display in a herbarium of any scientific institutions; or

(d) propagation by a person or an institution approved by the Central Government in this regard.

Cultivation of specified plants without licence prohibited –

(1) No person shall cultivate a specified plant except under, and in accordance with a licence granted by the Chief Wild Life Warden or any other officer authorized by the State Government in this behalf;

Provided that nothing in this section shall prevent a person, who immediately before the commencement of the Wild Life (Protection) Amendment Act, 1991, was cultivating specified plant, from carrying on such cultivation for a period of six months from such commencement, or where he has made an application within that period for the grant of a licence to him, until the licence is granted to him, or he is informed in writing, that a licence cannot be granted to him.

(2) Every licence granted under this section shall specify the area in which and the conditions if any, subject to which the licensee shall cultivate a specified plant.

Dealıng in specified plants without licence prohibited –

(1) No peson shall, except under and in accordance with a license granted by the Chief Wild Life Warden or any other officer authorized by the State Government in this behalf, commence or carry on business or occupation as a dealer in a specified plant or part or derivative thereof:

Provided that nothing in this section shall prevent a person, who, immediately before the commencement of the Wild Life (Protection) Amendment Act, 1991, was carrying on such business or occupation for a period of sixty days from such commencement, or where he has made an application within that period for the grant of a license to him, until the licence is granted to him or he is informed in writing that a licence cannot be granted to him.

(2) Every licence granted under this section shall specify the premises in which and the conditions if any, subject to which the licenses shall carry on his business.

Possession, etc., of plants by licensee – No licensee under this chapter shall –

(a) Keep in his control or possession -

(i) any specified plant, or part or derivative therof in respect of which a declaration under the provisions of section 17E has to be made, but has not been made;

(ii) any specified plant, or part or derivative thereof which has not been lawfully acquired under the provisions of this Act or any rule, or order made thereunder;

(b) (i) pick, uproot, collect or acquire any specified plant, or

(ii) acquire, receive, keep in his control, custody or possession, or sell, offer for sale or transport, any specified plant or part or derivative thereof, except in accordance with the conditions subject to which the licence has been granted and such rules as may be made under this Act.

Purchases, etc., of specified plants – No person shall purchase, receive or acquire any specified plant or part or derivative thereof otherwise than from a licensed dealer: Plants are Govt. property.

Declaration of Sanctuary and Protection to Sanctuaries –

(1) When the State Government declares its intention under sub-section (1) of section 18 to constitute any area, not comprised within any reserved forest or territorial waters under that sub–section, as a sanctuary, the provisions of sections 27 to 33A (both inclusive) shall come into effect forthwith.

(2) Till such time as the rights of affected persons are finally settled under sections 19 to 24 (both inclusive), the State Government shall make alternative arrangements required for making available fuel, fodder and other forest produce to the persons affected, in terms of their rights as per the Government records.

Appointment of Collectors – The State Government shall appoint an officer to act as Collector under the Act, within ninety days of coming in to the force of the Wild Life (Protection) Act 1972 or 30 days of the issue of notification under section 18.

Trade or Commerce in Wild Animals, Animal Articles and Trophies

Wild Animals, etc. to be Government property –

(1) Every

(a) wild animal, other than vermin, which is hunted under Sec. 11 or Sec 29 or sub-section (6) of sec 35 or kept or [bred in captivity or hunted] in contravention of any provisions of this Act or any rule or order made thereunder, or found dead, or killed by mistake;

(b) animal article, trophy or uncured trophy or meat derived from any wild animal referred to in Cl.(a) in respect of which any offence against act or any rule or order made there under has been committed;

(c) ivory imported into India and an article made from such ivory in respect of which any offence against this Act or any rule or order made thereunder has been committed.

(d) vehicle, vessel, weapon, trap or tool that has been used for committing an offence and has been seized under the provision of this Act.

shall be the property of the State Government and, where such animal is hunted in a sanctuary or National Park declared by the Central Government, such animal or any article, trophy, uncured trophy or meat[derived from such animal or any vehicle, vessel. weapon, trap, or tool used in such hunting, shall be the property of Central Government.

(2) Any person who obtains, by any means, the possession of Government property, shall, within forty-eight hours of obtaining such possession, report it to the nearest police station or authorized officer and shall, if so required, hand over such property to the officer in – charge of such police station or such authorized officer, as the case may be.

(3) No person shall, without the previous permission in writing of the Chief Wild Life Warden or the authorized officer.-

(a) acquire or keep in his possession, custody, or control, or

(b) transfer to any person, whether by way of gift, sale or otherwise, or

(c) destroy or damage such Government property.

Purchase of animal, etc., by licensee – No licensee under this Chapter **shall** –

(a) Keep in his control, custody, or possession –

(i) any animal, animal article, trophy or uncured trophy in respect of which a declaration under the provisions of sub section (2) of Sec. 44 has to be made but has not been made;

(ii) any animal or animal article, trophy, uncured trophy or meat which has not been lawfully acquired under the provisions of this Act or any rule of order made thereunder.

(b) (i) capture any wild animal, or

(ii) acquire, receive, keep in his control, custody, or possession, or sell, offer for sale, or transport, any captive animal specified in Sch.I or Part II of Sch. II or any animal article, trophy or uncured trophy, or meat derived therefrom, or serve such meat, or put under a process or taxidermy or make animal article containing part or whole of such animal, except in accordance with such rules as may be made under this Act:

Provided that where the acquisition, or possession, or control, or custody of such animal or animal article, trophy or uncured trophy entails the transfer or transport from one State to another, no such transfer or transport shall be effected except with the previous permission in writing of the Director or any other officer authorized by him in this behalf.

Provided further that no such permission under the foregoing provision shall be granted unless the director of the officer authorized by him is satisfied that the animal or article aforesaid has been lawfully acquired.

Restriction on transportation of Wild Life – No person shall accept any wild animal (other than vermin) or any animal article, or any specified plant or part or derivative thereof, for transportation except after exercising due care to ascertain that permission from the Chief Wild

Life Warden or any other officer authorized by the State Government in this behalf has been obtained for such transportation.

Purchase of captive animal, etc. by a person other than a licensee – No person shall purchase, receive or acquire any captive animal, wild animal other than vermin, or any animal article, trophy, uncured trophy, or meat derived therefrom otherwise than from a dealer or from a person authorized to sell or otherwise transfer the same under this Act.

Power of entry, search, arrest and detention –

(1)Notwithstanding, anything contained in any other law for the time being in force, the Director or any other officer authorized by him in this behalf or the Chief Wild Life Warden or the authorized officer or any forest officer or any police officer not below the rank of a sub-inspector may, if he has reasonable grounds for believing that any person has committed an offence against this Act.-

(a) require any such person to produce for inspection any captive animal, wild animal, animal article, meat, trophy, uncured trophy, specified plant or part or derivative thereof in his control, custody or possession, or any licence, permit or any other document granted to him or required to be kept by him under the provisions of this Act;

(b) stop any vehicle or vessel in order to conduct search or inquiry or enter upon and search any premises, land, vehicle, or vessel in the occupation of such person, and open and search any baggage or other things in his possession;

(c) seize any captive animal, wild animal, animal article, meat, trophy or uncured trophy, or any specified plant or pert or derivative there of in respect of which an offence against this Acts appears to have been committed, in the possession of any person together with any trap, tool, vehicle, vessel, or weapon used for committing any such offence and unless he is satisfied that such person will appear and answer any charge which may be preferred against him arrest him without warrant and detain him.

Provided that where a fisherman, residing within ten kilometers of a sanctuary or National Park, inadvertently enters on a boat not used for commercial fishing, in the territorial waters in that sanctuary or National Park, a fishing tackle or net no such boat shall not be seized.]

Certain conditions to apply while granting bail – When any person accused of the commission of any offence relating to Schedule I or part II of Schedule II or offences relating to hunting inside the boundaries of National Park or wild life sanctuary or altering the boundaries of such parks and sanctuaries, is arrested under the provisions of the Act, then not withstanding anything contained in the Code of Criminal Procedure, 1973 (2 of 1974) no such person who had been previously convicted of an offence under this Act shall, be released on bail unless –

(a) the Public Prosecutor has been given an opportunity of opposing the release on bail; and

(b) where the Public Prosecutor opposes the application, the court is satisfied that there are reasonable grounds for believing that he is not guilty of such offence and that he is not likely to commit any offence while on bail.

Power of Central Government to make rules

(1) the Central Govcernment may, by notification, make rules for all or any of the following matters, namely:

(a) conditions and other matters subject to which a licensee may keep any specified plant in his custody or possession under section 17F;

(ai)* the term of office of members other than those who are members *ex. Officio;* the manner of filling vacancies, the procedure to be followed by the National Board under sub-section (2) and allowances of these members under sub-section (3) of section 5A;

(b) the salaries and allowances and other conditions of appointment of chairperson, members and member-secretary under sub-section (5) of Section 38B;

(c) the terms and conditions of service of the officer and other employees of the Central Zoo Authority under sub-section (7) of section 38B;

(d) the form in which the annual statement of accounts of Central Zoo Authority shall be prepared under sub-section (4) of section 38E;

(e) the form in which and the time at which the annual report of Central Zoo Authority shall be prepared under section 38F;

(f) the form in which and the fee required to be paid with application for recognition of a zoo under sub-section (2) of section 38H;

(g) the standards, norms and other matters to be considered for granting recognition under sub-section (4) of section 38H;

(h) the form in which declaration shall be made under sub-section (2) of section 44;

(i) the matters to be prescribed under clause (b) of sub-section (4) of section 44;

(j) the terms and conditions which shall govern transaction referred to in clause (b) of section 48;

(k) the manner in which notice may be given by a person under clause (c) of section 55;

(l) the matters specified in sub-section (2) of section 64 in so far as they relate to sanctuaries and National Parks declared by the Central Government.]

Power of State Government to make rules –

(1) The state Government may, by notification, make rules for carrying out the provisions of this Act in respect of matters which do not fall within the purview of sec. 63.

(2) In particular and without prejudice to the generality of the foregoing power, such rules may provide for all or any of the following matters, namely:

(a) the term of office of members other than those who are members, ex. *Officio,* the manner of filling vacancies and the procedure to be followed by the Board under sub-section (2) of section 6;

(b) allowances referred to in sub-section (3) of section 6;

(c) the forms to be used for any application, certificate, claim, declaration, license, permit, registration, return, or other document, made, granted, or submitted under the provisions of this Act and the fees, if, any therefore;

(d) the conditions subject to which any licence or permit may be granted under this Act;

(dd) the conditions subject to which the officers will be authorized to file cases in the court;

(e) the particulars of the record of wild animal (captured or killed) to be kept and submitted by the licensee;

(ee) the manner in which measures for immunization of livestock shall be taken;

(f) regulation of the possession, transfer, and the scale of captive animals, meat, animal articles, trophies, and uncured trophies;

(g) regulation of taxidermy;

(ga) the manner and conditions subject to which the Administrator shall receive and manage the property under sub-section (2) of section 58G;

(gb) the terms and conditions of service of the Chairman and other members under sub-section (3) of section 58N;

(gc) the fund from which and the manner in which payment of reward under section 6OB shall be made;,

(h) any other matter which has to be, or may be, prescribed under this Act.

FOREST CONSERVATION ACT 1980

An act to provide for the conservation of forests and for matters connected therewith or ancillary or incidental thereto.

STATEMENT OF OBJECTS AND PURPOSE OF LEGISLATION

Large scale deforestation causes ecological imbalance and leads to environmental degradation. In post independence era deforestation had been taking place on a large scale in the country and it had caused widespread concern in the Govt. and public. With a view to check further deforestation, the President of India promulgated the Forest (conservation) Ordinance on 25th October 1980. The ordinance made the prior approval of the Central Govt. necessary for denotification of RF/PF/ unclassed forests and for use of forest land for non-forestry purposes. The ordinance also provided for the constitution of a "Forest Appraisal Committee" – to advice the Central Government with regard to grant of such approval. The bill seeks to replace the aforesaid ordinance.

The present act replaced the said ordinance and contains the similar provisions. The Forest Conservation Act extends to the whole of India except the state of Jammu and Kashmir and came into force on 25th October 1980.

The act has been amended in 1988 to incorporate penal provisions for the violation of the section – 2 of the F.C.A.

IMPORTANT POINTS OF THIS ACT ARE –

(1) Restriction on the dereservation of forests or use of forest land for non- forest purposes.

(2) Except with prior permission of the Central Government deforestation is impermissible under provisions of F.C.A.

(3) Notification of forest land as "reserve forest" not necessary.

(4) Meaning of expression "forest" and "forest land"

(5) Invoking of restrictions on transfer of forest land to persons organization, committee, trust, companies etc.

(6) Mining in forest land.

(7) Principal of natural justice and fundamental rights in forest conservation act cases.

(8) Breaking up of forest land after promulgation of F.C.A.

(9) Dereservation of Forest.

(10) Breaking of forest land when it has been notified under section – 9 of coal bearing areas (Acquisition & Development) Act 1957.

(11) Projects involving both forest land and non-forest land.

(12) Status of forest land after it has been diverted under section – 2 of F.C.A.

(13) Legal status of water bodies formed as a result of the submergence of forest land.

(14) Ownership of the forest crop raised on diverted forest land.

(15) No restriction on Miohtar rights under forest conservation act.

(16) Prior approval of Central Government for harvesting of plantations in Government land.

(17) Applicability of F.C.A. in respect of building construction on private forest land.

(18) Remove of minerals after the expiry of lease or "forest clearance" from the lease area.

(19) Assignment of forest land to corporations JFM committees, trusts authority or quasi Government bodies.

(20) Applicability of F.C.A. on forest villages.

(21) Applicability of F.C.A. on the areas utilized for settlement of refugees of erstwhile east Pakistan and Tibet.

(22) Correction of revenue and forest records to avoid violation of F.C.A.

Power to make rules – The Central Government may by notification in the official Gazette make rules for carrying out the provisions of this act.

Repeat and savings –

(i) The forest (conservation) ordinance 1980 is herby replaced.

(ii) Notwithstanding such repeat any thing done or any action taken under the provisions of the said ordinance shall be deemed to have been done or taken under the corresponding provisions of this act.

FOREST (CONSERVATION) RULES 1981

In exercise of the powers conferred by sub-section (1) of section 4 of the F.C.A. 1980 (ACT LXIX of 1980) the Central Government hereby make the following rules namely the **Forest Conservation Rules 1981.**

They shall extend to the whole of India except the state of Jammu & Kashmir.

GUIDELINES ISSUED BY THE GOVT. OF INDIA, MINISTRY OF ENVIRONMENT AND FOREST UNDER PROVISIONS OF FOREST CONSERVATION ACT AND FOREST CONSERVATION RULES

1. Application of Forest (Conservation) Act, 1980

1.1 Definitions

(i) The term 'Forest land' mentioned in Section-2 of the Act refers to reserved forest, protected forests or any area recorded as forest in the government records. Lands which are notified under section-4 of the Indian Forest Act would also come within the purview of the Act. (Supreme Court's judgement in NTPC's case). All proposals for diversions of such areas to any non-forest purpose, even if the area is privately owned, would require the prior approval of the Central Government.

(ii) The term **"tree"** for the purpose of this Act will have the same meaning as defined in Section-2 of the Indian Forest Act, 1927 or any other Forest Act which may be in force in the forest area under question.

1.2. Clarifications

(i) The cases in which specific orders for dereservation or diversion of forest areas in connection with any project were issued by the State Government prior to 25.10.1980, need not be referred to the Central Government. However, in cases where only administrative approval for the project was issued without specific orders regarding dereservation and/ or diversion of forest lands, a prior approval of the Central Government would be necessary.

(ii) Harvesting of fodder, grasses, legumes etc, which grow naturally in forest areas, without removal of the tree growth, will not require prior approval of the Central Government. However, lease of such areas to any organization or individual would necessarily require approval under the Act.

1.3. Investigation and Survey -

(i) Investigations and surveys carried out in connection with development projects such as transmission lines, hydro-electric projects, seismic surveys, exploration for oil drilling etc. will not attract the provisions of the Act as long as these surveys do not involve any clearing of forest or cutting of trees, and operations are restricted to clearing of bushes and lopping of tree branches for purpose of sighting.

(ii) If, however, investigations and surveys involve clearing of forest area or felling of trees, prior permission of the Central Government is mandatory.

(iii) Notwithstanding the above survey, investigation and exploration shall not be carried out in wildlife sanctuaries, national parks and sample plots demarcated by the forest Department without obtaining the prior approval of the Central Government, whether felling of trees is involved or not.

(iv) The work of actual construction would however, fully attract the provisions of the Act and prior clearance of the Central Government must be obtained even if such work does not require felling of trees.

(v) It is clarified that the permission to survey, exploration or prospection would not *ipso facto* imply any commitment on the part of the Central Government for diversion of forest land.

Comments

Permission for survey and investigations in the Wildlife sanctuaries and national parks – The Ministry of Environment & Forests, GOI vide No. 11 -5/99-FC dated 16.11.1999 has clarified on the above mentioned subject, that all proposals requiring permission for survey and investigations in sanctuaries and national parks involving forest area up to 20 ha. is to be processed by the concerned Regional Office. However, before processing examination and only thereafter the proposals should be processed.

1.4. Explanation regarding Non-Forestry Purposes -

(i) Cultivation of tea, coffee, spices, rubber and palm is a non-forestry activity, attracting the provisions of the Act.

(ii) Cultivation of fruit – bearing trees or oil – bearing plants or medicinal plants would also require prior approval of the Central Government except when,

(a) The species to be planted are indigenous to the area in question, and

(b) Such planting activity is part of an overall afforestation programme for the forest area in question.

1.5. Tusser Cultivation –

(i) Tusser cultivation in forest areas by the tribals as a means of their livelihood without undertaking monocultural Asan or Arjun plantations shall be treated as a forestry activity. Therefore, no prior approval of the Central Government under the Act is necessary.

(ii) Tusser cultivation in forest areas for which specific plantation of Asan or Arjun trees are undertaken for providing host trees to the silk cocoons shall be treated as forestry activity not requiring prior approval of the Central Government provided such plantation activity does not involve any felling of existing trees; provided further that while undertaken such plantations at least three species are planted, of which no single species shall cover more than 50% of the planted area.

(iii) Plantation of mulberry for silkworm rearing is a non-forestry activity, attracting the provisions of the Act. Because silkworm feeds the mulberry and give silk.

1.6. Mining -

(i) Mining including underground mining is a non-forestry activity. Therefore, prior approval of the Central Governement is essential before a mining lease is granted in respect of any forest area. The Act would apply not only to the surface area which is used in the mining but also to the entire underground mining area beneath the forest. A renewal of an existing mining lease in a forest area also requires the prior approval of the Central Government. Continuation or resumption of mining operation on the expiry of a mining lease without prior approval would amount to contravention of the Act.

(ii) The advice of the Ministry of Law, Government of India in regard to the Hon'ble Supreme Court order in Civil appeal No. 2349 of 1984 dated 07.05.1985 is provided at Addendum-3.

(iii) Boulders, bajri, stone, etc. in the riverbeds located within forest areas would constitute a part of the forest land and their removal would require prior approval of the Central Government.

Comments

Applicability of Forest Conservation Act, 1980 to the forest area to be broken up after 25.10.80 under a continued mining lease – In the aforementioned context the MOEF, GOI vide No.5-5/86-FC-Pt. 1 dated 21.12.1993 has clarified that-

(i) In respect of mining lease granted/renewed before 25.10.1980 prior approval of the Central Government under the Forest Conservation Act, 1980 will not be required for continuing mining activities in areas already broken up before 25.10.1980 during continuance of the lease period.

(ii) In respect of mining leases granted/renewed before 28.10.1980 if area is to be broken up afresh after commencement of the Act i.e. 25.10.1980, prior approval of the Central Government will be required under the Forest Conservation Act, even during the continuance of lease period.

(iii) The prior approval of the Central Government under Section 2 of the F.C.A. would be required when a mining lease granted before commencement of the Act, is renewed after its coming into force. It applies even to such cases where mining is to be continued in already broken up areas.

1.7. Clarification on the sub-clause 2(iii) of the Act -

(i) The sub-clause shall not be attracted when any forest land or any portion thereof is assigned to any authority, corporation, agency or any other organization wholly owned, managed or controlled by the concerned State / Union Territory Government and /or the Central Government owned, managed or controlled authority /corporation / agency, which has been assigned such forest land shall not reassign it or any part thereof to any other organization or individual.

(ii) Any scheme or project which involves assignment of any forest land by way of lease or similar arrangement for any purpose whatsoever, including afforestation, to any private person or to any authority/agency/organization not wholly owned, managed or controlled by the Government (such as private or joint section ventures) shall attract provisions of this sub-clause.

1.8. Clarifications on the sub-clause (iv) of the Act -

(i) Sub-clause 2(iv) of the Act prohibits clearing of naturally grown trees in forest land for the purpose of using it for reforestation. The provisions of this sub-clause will be attracted if the forest area in question bears naturally grown trees and are required to be clear-felled, irrespective of their size, for harnessing existing crop and / or raising plantation through artificial regeneration techniques, which may include coppicing, pollarding or any other mode or vegetative propagation.

(ii) All proposals involving clearing of naturally grown trees in any forest area, including for the purpose of deforestation, shall be sent by the concerned State / UT Government in the form of Management Plans/Working Plans to the Regional Chief Conservator of Forest of the concerned Regional Office of the Ministry of Environment and Forests.

(iii) All proposals in respect of sanction of Working Plans / Management Plans shall be finally disposed of by the Regional Office, under Section-2 of the Act. While examining the proposal, the Regional Office would ensure that the final decision is in conformity with the National Forest Policy, Working Plan Guidelines and other relevant rules and guidelines issued by the Central Government form time to time. The Regional Office will however, invariably seek prior clearance of the Ministry whenever the proposal involves clear-felling of forest area having density above 0.4 irrespective of the area involved. Also, prior clearance would be required when the proposal is for clear felling of an area of size more than 20 ha. in the plains and 10 ha. in the hilly region, irrespective of density.

[However in National parks and sanctuaries where fellings are carried for improvement of wildlife and its habitate only, forests would be managed according to a scientifically prepared management plan approved by the Chief Wildlife Warden. But in cases where large scale felling / removal of timber and non timber produce is required in a national park / sanctuary, which need disposal through sales, approval of the Central Government would be necessary].

1.9. Clarifications of the Section-3B of the Act -

(i) Each case of the violation of the Act shall be reported by the concerned State / Union Territory Government to the Central Government.

(ii) The report of violation shall be described in a self-contained note and supported by requisite documents, including particularly the names and designations of the officials / persons who are *prima-facie* responsible for the contravention of the Act.

(iii) In case it is not possible to fix the responsibility for commission / omission of any action leading to the violation of the Act, a full explanation with relevant supporting documents shall be appended to the report.

(iv) Any person and / or authority nominated by the Central Government may be required to discharge any of the duties, including prosecution under the Act in any Court as may be deemed appropriate for this purpose. In such an eventuality, the Government of the concerned State / Union Territory shall make available all such records or documents as may be called upon by the investigation officer.

1.10. Diversion of Forest Land for Regularisation of Enhancements –

Detailed guidelines issued in this regard vide Circular No. 13.1/90 – F.P. (1) dated 18.09.90 of Ministry of Environment and Forests shall be strictly followed. These are included in Addendum – 4 of this Chapter.

2. Submission of Proposals –

Rules -4 of the Forest Conservation Rules, 1981 prescribes the procedure for submission of proposals for seeking prior approval of the Central Government. Proposals in the prescribed format and complete in all respects will be considered.

3. Compensating Afforestation is one of the more important conditions stipulated by the Central Government while approving proposals for dereservation or diversion of forest land for non forest use.

4. Any proposal for diversion of forest land is submitted, it should be accompanied by a resolution of the "Aam Sabha" of gram panchayat local body of the area endorsing the proposal that the project is in the interest of people living in and around the proposed forest land.

Issues involved in enforcement of environmental legislation

During the last hundred of years several laws have been enacted, which deals with environmental, eco-conservation and protection. Sixth plan documents emphasized the need to ensure conservation of environmental resources for sustainable development. Department of Environment was initiated in 1893. All this indicate that Government making conscious efforts to repair the damage and take a view to save the environment for future. But inspite of these, we are not able to achieve the target. It seems that there are some drawbacks in environmental legislations and problems in their effective implementation. In a number of countries the concern for disappearing wild life, deteriorating conditions of environment and over exploitation of resources led to the establishment of a body of laws. To these laws were added rules restricting practices which adversely affect human environment. It is fact that national action alone can not effectively serve the purpose. In third world countries, the national laws were environmentally blind. Environmental legislation in these countries was developed after the international legal framework had come into being.

Pity and compassion for animals can be seen in the common law of many countries which prohibit cruelty against animals. Animals are being used to pull over burdened carts or carriages without any regard to the capability or health of the animals. Such practices should be prohibited under the articles on Prevention of Cruelty against animals. There are provisions in law to restrict practices causing nuisance, and suffocating smoke in the atmosphere. But these provisions are rarely used for the purpose of protecting animals or restricting other environmentally damaging practices.

Till date the law of many countries did not possess provisions which could be effectively used to protect wild life, natural resources or the environment. This was simple because of the fact that the negative effects of the process of growth and development of a modern society had never been felt so deeply. The legal frame work, at national levels can be analysed in following categories –

1. Legislation restricting hunting, killing or over exploitation of a species or a group of species.

In addition to law made for wild elephants in 1873 AD and 1879 AD, the following important acts were framed –

(1) Madras Wild Elephant Preservation Act 1873.

(2) All India Elephant Preservation Act 1879.

(3) Wild Birds and Animals Preservation Act 1912.

(4) Bengal Rhinoceros Preservation Act 1932.

(5) Assam Rhinoceros Preservation Act 1954.

(6) Indian Wild Life (Protection) Act 1972.

These laws vested the state with the powers to act against those responsible for capture, injury or death of wild animals. Most of these laws were like a "Penel code" which specifies penalties for offences committed against wild life. *But the exploitation of wild animals and plants continued illegally or at times even under the state permit. Many Governments locked the will or adequate machinery to implement these laws. The lure of spectacular profits, vested interests and corruption often cause in the way of effective enforcement of these laws.*

2. Legislation for the protection of wild life preservation of natural habitats and environment in general.

In previous mentioned laws destruction of habitats, pollution of environment or factors which caused much more damage to wild life were ignored. Earlier some laws against ecologically damaging human activities were enacted. But it was the inconvenience caused to the general public which was responsible for their implementation. Some of the acts thus promulgated were –

(1) Bengal Smoke Nuisance Act 1905.
(2) Indian Ports Act 1907.
(3) The Motor Vehicle Act 1908.
(4) Maharashtra Prevention of Water Pollution Act 1953.
(5) Orissa River Pollution Act 1954.
(6) Gujarat smoke Nuisance Act 1963.

Most of these laws were ineffective due to Government inefficiency and apathy while human Society's obsession for economic growth and industrialization enfeabled public protests whatsoever. *In the light of population growth ecological issues are often ignored. Inspite of these Acts, ecologically human activities continued – after all there has to be a place for man also in the biosphere.*

3. Legislation recognizing rights of people to a healthy environment and provision of compensation to people affected thereof.

There is provision to fix responsibilities by Government for agencies responsible for ecologically damaging practices in laws but have no role for directly or indirectly affected people by damaged environment. Little help could be obtained from the judiciary as the law did not specifically recognize the right of the people to clean air for breathing, clean water for drinking or a healthy environment to live in. After amendment in Article 51-A, people could now go to the court to seek legal help to guard their rights concerning issues of the environment. The Indian judiciary on the other hand, also started assisting people in their effort to maintain a healthy environment and wild life by admitting litigations against state's continued neglect of ecological rights of its citizens. The principle that the "Polluter must pay" is on its way to become a reality now.

When matters concerning human rights are frontally posed, it becomes the duty of judiciary to face & resolve them. A bill called the National Environment Tribunal Bill 1992 was introduced in Parliament in August 1992 to decide the cases related with wild life and environment in addition to providing compensation to the people for death injury or damage to property or to the environment by constituting a high power body. Excessive centralization of powers and authority lindens efficient implementation. Various problems like the financial implications and lack of measurement data, appropriate instrumentation and technology creep in the path of enforcement of environmental legislation.

PUBLIC AWARENESS

Environmental pollution, environmental degradation, environmental deterioration, environmental crisis etc. are few words which becoming day by day a subject of concern in every walk of life. This is all due to industrialization, rapid population growth, urbanization, changing life style etc. The formulation of various acts and legislations to control pollution and conserve or protect environment, underlines the will and concern of the Government. But incomplete knowledge, informations and ignorance about many aspects of environment has led to misconceptions. Therefore, it is necessary to make people aware about the laws and legislations and to save environment. There is no single subject by which we can have complete knowledge of environmental aspects. Simultaneously, it can not be done by single man, agency or institution. It is of the people, by the people and for the people. Thus public awareness means, making the people conscious about the physical, social and aesthetic aspects of environment.

Methods –

To protect and conserve the environment is the basic duty of all sections of people because environment belongs to all and every individual matters. Instead of searching the solution, it is necessary to find the permanent solution of environmental and ecological problems. It can be done by following means –

(1) Through mass–media – There are various means of mass communication to educate, entertain & give informations, instructions etc. to people. Radio was the first. Next is T.V. in terms of its reach to the masses. In addition to these newspapers, magazines etc. are also there. Since media has an important role in masses hence awareness among people can be propagate by articles, important serial programmes, instructions, stories etc. using these media. Documentary and films on environmental awareness may be displayed in cinema houses. Posters can also be displayed in streets, on roads and at common places for environmental awareness.

(2) Through Education – Students are the back bone of a country. If environmental education is started from grass root level i.e. from childhood stage, it will give good results. It will be done through formal and informal environmental education. A welcome step to introduce environmental studies paper at collage level by Government by the directive of Supreme Court is good in the direction of awareness.

(3) Through rallies, orientation and training programmes - To promote environmental awareness, environmental rallies with posters, handbills, programmes may be organized on certain occasions like 5th June as world environmental day and 1st week of October as wild life week. Some training programmes, orientation, workshops, seminars, meetings based on environment awareness may be organized for decision makers, planners, leaders also, so that they can also spread the message to protect & conserve the environment.

(4) Through voluntary organizations and NGOs – Due to having link between people and Government some voluntary organizations and non Government organizations can play an important role in the direction of environmental awareness in people by organizing educational, religious, plantation, musical, Nukkad Natak, competitions (essay, drawing, oral) etc. programmes. These organizations can also advise the government to implement effective programmes for environmental awareness. They can organize certain public monuments for conservation of environment. Chipko Movement, Narmada Bachao Andolan, etc. are such public monuments were organized by Dasholi Gram Swaraj Mandal in Gopeshwar and Kalpavriksha respectively. Some voluntary organisations working in this field are –

(1) Bombay Natural History Society (BNHS).

(2) Wild Life Preservation Society of India. (WPSI)

(3) Worldwide Fund for Nature – India (WWF - India).

(4) Centre for Science and Environment (CSE)

In addition to these some scientific and technical societies like Indian Science Congress, National Academy of Sciences, Institution of Chemists, The Indian Chemical Society, Biological Society of India, Pollution Research, CDRI, NEERI etc. are also doing work in the field of public awareness by organizing seminars, workshops, publishing articles, research papers etc.

NGOs usually consists of volunteers or group of individuals genuinely interested in the cause of conservation of wild life and protection of environment. They generate their own resources, contributory funds, grants etc. They play an important role in developing public awareness about environmental matters. These organizations can be grouped in to three categories –

(1) Specialized agencies with restricted membership consisting primarily of scientific and professional members. For example – International Council of Scientific Unions (ICSU), International Union for Conservation of Nature and Natural Resources (IUCN)

(2) Institutes and centres devoted to the task of collection of information, research and consultation. eg. International Institute for Environment and Development etc.

(3) Activist organizations with open membership for the cause of raising support and consensus for wild life protection, are doing for conservation of resources and improvement in the quality of environment. For example Sierra Club National Audubon Society, Bombay National History Societ etc.

QUESTIONS

Short answer type questions :–

1. Explain sustainable development.
2. What are the measures to attain sustainable development ?
3. Why it is essential to conserve water ?
4. Explain, how acid rain is formed.
5. In what cases, hunting of wild animals are permitted ?
6. Discuss the role of NGO's to protect environment.
7. Write about urban problems related to energy.
8. In acid rain, from where acid come ? Explain.
9. What do you understand by public awarenese. Explain.
10. What is acid rain ?

Long answer type questions :–

1. Discuss rain water harvesting. What are advantages from this technique ?
2. What do you mean by resettlement and rehabilitation of people ?
3. Describe environmental ethics.
4. Write an essay on "Global warming".
5. Differentiate between nuclear accidents and Holocaust.
6. Discuss the environmental (protection) Act 1986. How it helps to protect the environment?
7. Describe in detail the air (Prevention and Control of Pollution) Act 1981.
8. Discuss silient features of (a) wild life (protection) Act 1972 and (b) Forest (conservation) Act 1980.
9. Discuss the main issues involved in enforcement of environmental legislation.
10. Write short notes on :–
    (*i*) Water shed management
    (*ii*) Indira sagar project
    (*iii*) Green house effect
    (*iv*) Ozone layer depletion.
    (*v*) Chernobyl accident
11. Explain forest conservation Act.
12. Write an essay on water conservation.
13. Write an essay on Environment Protection Act.
14. Describe the role of public awareness in environmental protection.

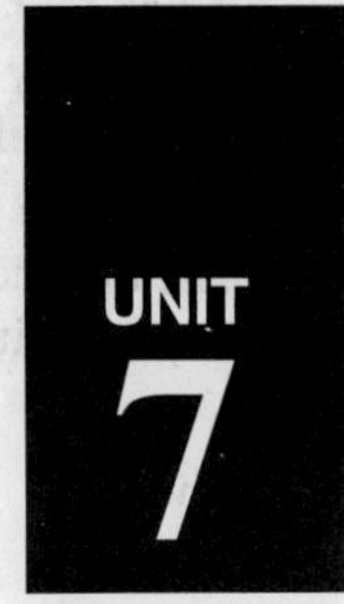

Human Population and the Environment

INTRODUCTION

A population may be defined as a group of organisms of the same species occupying a given area at the same time. It is subdivided in to demes or local populations, which are groups of interbreeding organisms or the smallest collective unit of a population. A population may consist of either *unitary* or *modular organisms*. Insects, fish amphibians, birds, mammals are examples of unitary organisms where each individual is produced from a zygote and the form and development of individuals is highly predictable. But in modular organisms, the zygote develops in to a unit of construction or module, which produces further modules to form a branching structure, therefore form and development of individuals are unpredictable. Sponges, Corals and plants are examples of modular organisms.

POPULATION GROWTH

The most important freatures of population is the growth i.e. the capacity of increase in individual members. By measuring the size or density of a given population from time to time, we can get rate of increase and can also predict future changes in its size. It can be defined in following ways—

(*a*) Logistic growth : When a population is allowed to grow in a limited space (environment) it shows logistic growth. If we plot a graph between number of bacteria or cells against time, we find a typical S shaped sigmoid curve called population growth curve.

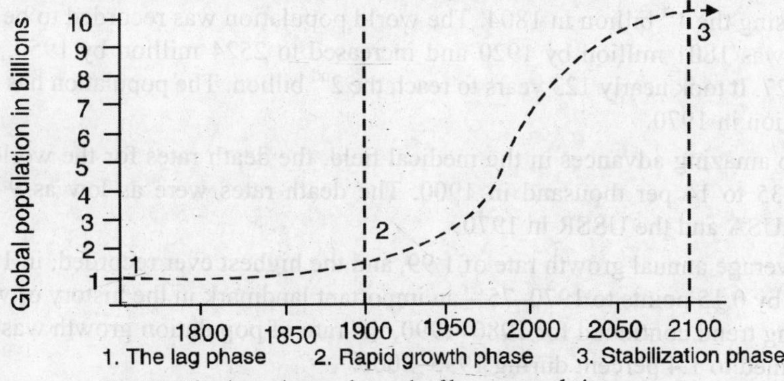

Fig. 7.1. The three phases of growth of human population

It has four phases i.e. Ist phase shows slow rate called lag period, second is accelerating stage followed by a phase of extremely rapid population. The last phase is accelerating multiplication followed by equilibrium phase where is essentially no net change in population called saturation level or carrying capacity. It is represented by letter K.

The logistic equation shows density dependent growth i.e. growth of a simple population in a limited space with limited resources. It may be written as—

$$dN/dt = \gamma N(1 - N/K)$$

where dN/dt = rate of growth of population

γ = intrinsic rate of increase (per individual of population)

N = Population size (No. of organisms in population at time t)

K = Carrying capacity of population

$(1 - N/K)$ = density – dependent factor.

(*b*) Exponential Growth : When a population growth curve quickly begins to rise very steeply, the population shows exponential growth. It is J shaped (Fig.7.1). A population growing exponentially increases accordingly to the equation

$$N_t = N_0 e^{rt}$$

where N_t = The number of individuals in the population after t units of time

N_0 = initial population size (t = 0)

r = exponential growth rate

e = the base of the natural logarithm (about 2.72)

(*c*) Geometric Growth : Geometric growth may be defined as the population growth in which the rate of increase is proportional to the number of individuals in the population at the beginning of the breeding session. When young ones are added to the population only at specific times of the year during well defined reproductive periods, the population is said to have geometric growth. The equation for this is

$$N_t = N_0 \lambda$$

where λ = the geometric growth rate.

POPULATION - GROWTH

Demographers have back projected it as (in olden days no reliable data was available) 275 million in 1000 AD and increased to 486 million in 1650 AD. During 1650 to 1750 the world population increased to 791 million. There was an addition of 187 million population in another 50 year period, i.e., from 1750–1800 AD, reaching to 978 million. From 1800–1850 it increased to 1260 million crossing the 1st billion in 1804. The world population was recorded to be 1650 million by 1900 AD. It was 1801 million by 1920 and increased to 2524 million by 1950, crossing the 2nd billion in 1927. It took nearly 123 years to reach the 2nd billion. The population has further increased to 3630 million in 1970.

Due to amazing advances in the medical field, the death rates for the world as a whole fell down from 35 to 14 per thousand in 1900. The death rates were as low as 9 per thousand in Europe, the USA and the USSR in 1970.

The average annual growth rate of 1.99, and the highest ever recorded, in 1960–65 declined to 1.84, i.e., by 0.15 points to 1970–75—an important landmark in the history of world population. The declining trend continued for 1980–1990, the rate of population growth was 1.7 per cent and further declined to 1.4 percent during 1995–2000.

The first billion (100 crores) was reached by 1804 AD (probably 29 lakhs year to reach the first billion on the earth).

The second billion was added only in next 123 years i.e., in 1927.

The third billion was reached in 1960, i.e., 33 years after 1927.

The fourth billion was added in 1974, in just 14 years.

The fifth billion was added in 1987, i.e., in 13 years.

The sixth billion was added in 1999 in just 12 years and it will be increased by 80 million people a year to reach 8th billion by 2025.

Doubling time : The concept of doubling time implies the time needed for the population to get doubled. It shows that more the growth rate, less will be the doubling time. (Table 7.1)

Table 7.1

Pop (billion) from 0.8 (1750)	*Years* to 1.6 (1900)	*Doubling* 150 time (years)
1.25 (1850)	2.5 (1950)	100
2.5 (1950)	5.0 (1987)	37
3.0 (1960)	6.6 (1993)	33
Growth rate %	**Population doubling time**	
Slow growth	< 0.5	139 years
Moderate growth	0.5–1	70–138 years
Rapid growth	1–1.5	47–69 years
Very rapid growth	1.5–2	35–46 years
Explosive growth	2–2.5	28–35 years
	3–3.5	20–23 years

On July 11th 2000, the world population was 6.5 billion. According to United Nations estimates the world population will reach 7 billion mark in 2011, 8 billion in 2025 and 9 billion in 2042.

If the population explosion occurs in such a way the world will be more crowded, more polluted, warmer, and more unequal between north and south and more tension ridden. Now itself half billion persons of the world are starving or semi-starving. The world is facing grave problems like less (or unequal distribution of) food grains per person, less living space, greater unemployment and underemployment, higher rates of inflation, lower physical quality of life (PQLI) of the poorest of the poor and increasing lawlessness and corruption every where. Added to this are the problems of deforestation, green house gases, low level ozone, chlorofluorocarbons, nitrous oxide, and methane and carbon dioxide.

Estimated world population

3,000 BC—2 crores	Date not available
1 AD—25 crores	Date not available
1650 AD—50 crores	Date not available
1800 AD—100 crores	Date not available

VARIATION AMONG NATIONS (DEVELOPED AND DEVELOPING COUNTRIES)

The population of Europe was 163 million in 1750 and increased to 276 million in 1850. Th mortality conditions in Europe began to improve as a result of socioeconomic development, due advances in medical technology and reforms in the field of environmental sanitation and public health. The population increased to 547 million in 1950 and to 728 million in 1999 and is projected to decline to 701 million in 2025.

The share of North America was 2 million in 1750 but by 1900 reached to 82 million and by 1999 to 303 million. This is mainly by immigration. The population of Latin America and Caribbean islands was 16 million in 1750, 38 million in 1850 and 166 million in 1950, further increasing to 512 million in 1999.

The population of Asia was 502 million in 1750 AD increased to 809 in 1850 AD and further increased to 1402 million in 1950. The population of Asia increased its share of world population between 1950–1999 from 56 percent to 61 percent. The two largest Asian countries (Asian elephants) China and India along added 28.6 million to the world population. The population of Africa was 106 million in 1750; fastest growth of population is seen in Africa.

The growth rates are not uniform in the world. In many countries (Europe) the growth rate is less than 0.5 percent per year. In developing countries the growth rates are excessive. For example, India, the second largest will soon overcome China, the present largest, as China is restricting population by rigorous population control and advocating one-child family norm to bring down the population. But impressive progress was seen in East and South East Asia and in Latin America.

The fertility rates of South Korea and Singapore are below the replacement level (NRR—1) and that of Thailand is 2.1, Sri lanka 2.3, Indonesia 2.8. In India the fertility rate decline to 3.4, whereas in sub-Saharan Africa, The TFR was more than 6.0. In many developing countries, the key factors for fertility decline are changes in govt. attitude towards population growth, increased availability of contraception and the extension of services offered through family planning programme as well as marked changes in marriage patterns.

The population explosion is the major obstacle to the economic and social development of most of the under developed countries.

Table 7.2. AD 2000 Crude birth rate and death rate (and Growth rate)

Country	BR	DR	GR%
India	24	9	1.5
China	16	7	0.9
Pakistan	36	8	2.8
Bangladesh	28	10	1.8
Sri Lanka	18	6	1.2
USA	14	8	0.6
Russia	10	14	-0.4
Germany	9	11	-0.2
UK	12	11	0.1
France	12	9	0.3
Afghanistan	52	21	3.1
Tanzania	41	15	2.6
Sierra Leone	46	25	2.1

POPULATION GROWTH IN INDIA

India's population was estimated as 120 million in 1800, 194 million in 1860 and 255 million in 1871. The modern population census was started in the year 1881; since then the population count has been taken regularly every 10 years. The population of India in 1891 was 267 million which declined to 238.4 million in 1901, due to severe famine, weak monsoons (failures) and epidemic of plague.

In 1911, the population of our country was 252 million, which fell to 251.3 million in 1921. After the world war, the entire country was swept by the distinct waves of the world wide pandemic of influenza. Due to epidemic there was a slight decline in the population. The year 1921 is called *the year of great divide* because it distinguishes the earlier period of declined population growth from the period of moderately increasing growth. From here the absolute number of population added in each decade increased. It is the turning point which marks the beginning of a regular growth in India's population and also the beginning of a rapid and massive population growth in India.

As a result of improved agricultural techniques, advancement in medical and health technology to control epidemics and diseases, improved sanitary and health services, improved transport, communication and infrastructural facilities, the population of India progressively increased from 251.3 million in 1921 to 361 million in 1951, a net addition of 110 million people. From 1951, population started growing at a phenomenal rate. The population of India was more than doubled during 1951 and 1991. The net addition to the population was 485 million during this 40 years period, reaching to 846 million in 1991. India's population touched 1 billion mark on 11th of May 2000 and according to 2001 census the population of India was 1027 million (i.e., 102.7 crore).

The percentage decadal growth of the country as a whole declined from 23.86 during 1981–91 to 21.34 during 1991–2001, a decline of about 2.52 percent in the decadal growth rates.

Currently, India ranks second in size of the population, next to China. It is estimated that by 2025 AD India's population will touch 1414.3 million mark passing China's population and become the most populous country in the world.

India has only 2.4 percent of the world's land area, but it has to support 16.7 percent of the world's population with an annual growth rate of 1.93 percent (2001). Today for every 2 seconds, 1 baby is born in India and per year 17.6 million babies are born. The absolute addition to the population during the decade 1991–2001 was more than the estimated population of Brazil, the fifth most populous country in the world.

By state wise, Uttar Pradesh is the most populous state in the country with more than 166 million people (which is more than the population of Pakistan, the sixth most populous country in the world). Maharashtra (96.7 million), Bihar (82.8 million) are the second and third largest states in respect of population.

Almost half of the country's population lives in five states, namely, UP, Maharashtra, Bihar, West Bengal and Andhra Pradesh. Interestingly, in the last decade Andhra Pradesh achieved the sharpest decline in the annual growth rate of population (from 2.2 percent during the eighties to 1.3 percent during the nineties), mainly by increasing the level of contraception. However, some states like Bihar continue to register an increase in the annual rate of growth of population (from 2.1 percent in the eighties to 2.5 percent in the nineties).

POPULATION EXPLOSION

As we have seen growth rate of a population is expressed as the number of individuals by which the population increases divided by the amount of time that elapses i.e.

$$\text{Growth rate } (r) = \frac{\text{No. of birth } (b) - \text{Number of deaths } (d)}{\text{average population in time interval}}$$

There are many cases where b is substantially large than d for a period of time, following which conditions change so that d becomes much larger than b. This sort of variations are exponential called "**Population explosion**" during favourable conditions, followed by a **"Crash"** when conditions change. For example, diatom populations in Lake Michigan USA undergo such exponential increases at different time of years.

In 20th century population growth increased too much. This is also called population explosion. Economist Malthus said, resources increases 1, 2, 3, 4, ... while population increases 2, 4, 6, 8,... respectively. In India population growth is much more than twice. Fertility period is of 30 yrs (from 16 to 46 yrs of age). World population is also increasing day by day is 150 per minute 220,000 in one day. Growth rate is 2.2%, with this population will go 7 billion by 2010. World population increase by 9 crore 20 lac per year i.e. one Maxico every year. Do we have the resources and provisions for feeding, housing, educating and employing all those people being added every year. On 11th May 2000 we become one billion i.e. one person out of every 6 persons in this world.

Our resources like land, water, fuels, minerals, forests grasslands etc. are limited and due to population explosion these resources are getting exhausted. Social, economic, religious all type of reasons are responsible for the high rate birth in our country. The important reasons are lower marriage age, lack of education, joint family system, importance of male child, religous misbelieves, decline in death rate, increased protection of life from natural risks, increase life span, better means of transport and other facilities.

Due to overpopulation some serious problems are like food supply, space (accomodation), unemployment, education, human health, energy crisis etc. There is a fierce debate on population explosion to reduce fertility rates through world wide birth control programmes. This can be achieved by proper education, mass media, educational institutions, raising the marriage age from 18 to 22, providing the facilities like contraceptives, intra uterine devices, birth control bills, sterilization etc. Family planning programme which is Govt. sponsored programme is also one of the effective means to reduce fertility. It was started in India in 1951.

NATIONAL FAMILY WELFARE PROGRAMME

Previously this programme was known as National Family Planning Programme. In the year 1977 the name was changed to *National Family Welfare Programme.* Family planning programme was launched in India in 1952. India was the first country to do so.

Beginning of the programme was modest, i.e., establishment of few FP clinics, distribution of FP educational material, training of health functionaries and research. During the third 5-year-plan (1961-66) family planning was declared as *centre of planned development.* Then the emphasis was shifted from *clinic approach to extensive education approach* (i.e., motivating people about *small family norm*). A separate Department of Family Planning was created in 1966 in the Ministry of Health. In 1972, the MTP Act was passed. In April 1976, National Population Policy was framed.

During the emergency period (1976), forcible sterilisation campaign led to the defeat of Congress in 1977 elections. In June 1977, new Janata Government formulated *a new population policy* and made family planning as voluntary and renamed it as *Family Welfare Programme.*

The acceptance of primary health care approach as the key to the achievement of health for all by 2000 AD led to the formulation of National Family Welfare Programme in 1982.

Importance of Family Welfare Programme

1. The family welfare programme occupies an important position in the nation's socio-economic development.

2. Indian population which was 34 crores in 1947 has crossed 100 crore mark by 2000 AD. India has only 2.4% of world's land area but it supports about 15.5% of world's population.

3. India's population is increasing by 1.8 crores every year. To check this galloping growth, the country has laid down long-term demographic goal of achieving an NRR of one by the year 2000 AD.

4. Acceptance of the family welfare services is made voluntary.

5. The programme was 100% centrally sponsored scheme. FP programme was integrated with the MCH services.

ORGANISATIONAL SET UP

1. Central level

At central level Central Cabinet Subcommittee is present. It is headed by Prime Minister. Next level is Population Advisory Council. This is headed by Union Minister of Health and Family Welfare. Members are representatives of various professional bodies and some technical persons. Next level is Central Family Welfare Council, which is headed by union minister and ministers of health and family welfare of all states. It coordinates the work of the programme.

National Institute of Health and Family Welfare, situated in Delhi, is the apex institure. It undertakes research and training in family welfare. Directorate General of Health Services was the central programme officer for Family Planning. He advises Government of India on various aspects of family welfare.

2. State level

Ministry of Health and Family Welfare is the apex organization at the state level. This is headed by the minister of health and family welfare of the respective state. At the state level the family welfare work is organised by State Family Welfare Bureau. The State Family Welfare Bureau has three wings:

(a) Administrative wing (headed by state family welfare officer and associated by some officers)

(b) Education and information wing (headed by mass media information officer)

(c) Field operation and evaluation wing (headed by statistical officer).

3. District level

At district level the work of family welfare is organized by *District Family Welfare Bureau.* This has three wings like the state level. At some districts *Regional Family Welfare Training Centres* are present. These will undertake training of medical officers and para-medical staff.

4. Peripheral level

In rural areas the family welfare work is looked after by *rural family welfare centres* attached to PHC while in urban areas *urban family welfare centres* will took after this work.

5. Village level

At village level the MPHA(F) and MPHA(M) are mainly responsible for the programme. They will take the assistance of CHG, TBA and anganwadi workers.

Goals of National Population Policy

1. NRR 1 (which implies two-child norm)
2. Birth rate 21/100 population
3. Death rate 9 per 1000 population
4. Raising couple protection rate to 60%
5. Reduction of family size to 2.3
6. Decrease the IMR to 60 per 1000 live births.

Programme Strategies

1. Integrated approach
2. Cafetaria approach
3. Welfare approach

4. At risk approach

1. Integrated approach

It is the integration of all the activities like *raising the age of marriage, increasing female literacy, empowerment of women, raising the overall economy of the country etc.* To achieve these objectives, various anti-povery programmes have been started.

2. Cafetaria approach

The programme is made voluntary and all pressurizations are removed. It stresses more on motivation and education of the people about the benefits of family welfare programme.

3. Welfare approach

In 1977, the word *planning* was removed and changed to *family welfare programme.* This indicates the government's commitment to family welfare rather than family planning.

4. At risk approach

Like any other programme *NFWP stresses to ensure maximum support to those who are in need.*

Now family welfare has been given priority in the health development. The Government of India has evolved a comprehensive population policy.

ENVIRONMENT AND HUMAN HEALTH

Environment is the main determinant of health status of a community. Poor housing is a contributor to low physical and mental efficiency. The relation of poor housing and prevalence of disease is easily recognisable. Certainly if we aimed at obtaining optimum conditions for physical and mental well being, in addition to preventing disease, we must include improvement of housing conditions in the programme. Since poor housing is related to poverty, public opinion has tended to consider it an unavoidable evil.

Environment is defined as, "all the external factors (living or non living, material or non-material) present around man." So it is the entire medium in which the population lives and interacts. The environment may be divided into four components:

1. Physical environment
2. Biological environment
3. Social environment
4. Cultural environment

1. Physical environment

Physical environment is defined as, "All those non living things and physical forces present around man". The important components of physical environment are water, air, housing, temperature, lighting, noise and vibration, radiation, refuse and waste such as human excreta etc.

2. Biological environment

Biological environment is defined as, "All those living things (plants, animals, rodents, insects, microbes and other human) present around us".

3. Social environment

Social environment is defined as, "Social interactions between the individual such as their socioeconomic status, religion and the way of living, standard of living and availability and utilisation of health care facilities."

4. Cultural environment

It is the culture in which the individual lives. It includes their knowledge, attitude, beliefs, practices, customs, behaviour etc.

Environmental sanitation is defined by WHO as, "the control of all those factors in man's physical environment which exercise or may exercise a deleterious effect on his physical development, health and survival". Dictionary meaning of sanitation is : "The science of safeguarding health". The world sanitation covers the whole field of controlling environment with a view to prevent disease. It is a known fact that in the countries where environmental sanitation is good (e.g., USA, European countries), there the communicable disease problem is less, whereas in countries where environmental sanitation is poor (like India, African countries etc.), the communicable disease prevalance is very high.

PHYSICAL ENVIRONMENT

The following factors

1. Housing and health
2. Water
3. Air pollution
4. Waste disposal
5. Excreta disposal
6. Ventilation and light.
7. Noise
8. Radiation.

Human Rights Movement

(Early Political, Religious, and Philosophical Sources)

The concept of human rights has existed under several names in European thought for many centuries, at least since the time of King John of England. After the king violated a number of ancient laws and customs by which England had been governed, his subjects forced him to sign the *Magna Carta*, or Great Charter, which enumerates a number of what later came to be thought of as human rights. Among them were the right of the church to be free from governmental interference, the rights of all free citizens to own and inherit property and be free excessive taxes. It established the right of widows who owned property to choose not to remarry, and established principles of due process and equality before the law. It also contained provisions forbidding bribery and official misconduct.

The political and religious traditions in other parts of the world also proclaimed what have come to be called human rights, called on rulers to rule justly and compassionately, and delineating limits on their power over the lives, property, and activities of their citizens.

In the eighteenth and nineteenth centuries in Europe several philosophers proposed the concept of "natural rights," rights belonging to a person by nature and because he was a human being, not by virtue of his citizenship in a particular country or membership in a particular religious or ethnic group. This concept was vigorously debated and rejected by some philosophers as baseless. Others saw it as a formulation of the underlying principle on which all ideas of citizens' rights and political and religious liberty were based.

In the late 1700s two revolutions occurred which drew heavily on this concept. In 1776 most of the British colonies in North America proclaimed their independence from the British Empire in a document which still stirs feelings, and debate, the U.S. Declaration of Independence.

We hold these truths to be self-evident; that all men are created equal, that they are endowed by their creator with certain unalienable rights, that among these are life, liberty and the pursuit of happiness.

In 1789 the people of France overthrew their monarchy and established the first French Republic. Out of the revolution came the "Declaration of the Rights of Man."

The term natural rights eventually fell into disfavor, but the concept of universal rights took root. Philosophers such as Thomas Paine, John Stuart Mill, and Henry David Thoreau expanded the concept. Thoreau is the first philosopher I know of to use the term, "human rights", and does so in his treatise, *Civil Disobedience.* This work has been extremely influential on individuals as different as Leo Tolstoy, Mahatma Gandhi, and Martin Luther King. Gandhi and King, in particular, developed their ideas on non-violent resistance to unethical government actions from this work.

Other early proponents of human rights were English philosopher John Stuart Mill, in his *Essay on Liberty*, and American political theorist Thomas Paine in his essay, *The Rights of Man*.

The middle and late 190th century saw a number of issues take center stage, many of them issues we in the late 20th century would consider human rights issues. They included slavery, serfdom, brutal working conditions, starvation wages, child labor, and, in the Americas, the "Indian Problem", as it was known at the time. In the United States, a bloody war over slavery came close to destroying a country founded only eighty years earlier on the premise that, "all men are created equal." Russia freed its serfs the year that war began. Neither the emancipated American slaves nor the free Russian serfs saw any real degree of freedom or basic rights for many more decades, however.

For the last part of the nineteenth and first half of the twentieth century, though, human rights acitvism remained largely tied to political and religious groups and beliefs. Revolutionaries pointed at the atrocities of governments as proof that their ideology was necessary to bring about change and end the government's abuses. Many people, disgusted with the actions of governments in power, first got involved with revolutionary groups because of this. The governments then pointed at bombings, strike-related violence, and growth in violent crime and social disorder as reasons why a stern approach toward dissent was necessary.

Neither group had any credibility with the other and most had little or not credibility with uninvolved citizens, because their concerns were generally political, not humanitarian. Politically partisan protests often just encouraged more oppression, and uninvolved citizens who got caught in the crossfire usually cursed both sides and made no effort to listen to the reasons given by either.

Nevertheless many specific civil rights and human rights movements managed to affect profound social changes during this time. Labor unions brought about laws granting workers the right to strike, establishing minimum work conditions, forbidding or regulating child labour, establishing a Forty-hour work in a week in the United States and many European countries, etc. The women's rights movement succeeded in gaining for many women the right to vote. National liberation movements in many countries succeeded in driving out colonial powers. One of the most influential was Mahatma Gandhi's movement to free his native India from British rule. Movements by long-oppressed racial and religious minorities succeeded in many parts of the world, among them the U.S. Civil Rights movement was one of them.

In 1961 a group of lawyers, journalists, writers, and others, offended and frustrated by the sentencing of two Portugese college students to twenty years in prison for having raised their glasses in a toast to "freedom" in a bar, formed Appeal for Amnesty, 1961. The appeal was announced on May 28 in the London Observer's Sunday Supplement. The appeal told the stories of six "prisoners of conscience" from different countries and of different political and religious backgrounds, all jailed for peacefully expressing their political or religious beliefs, and called on governments everywhere to free such prisoners. It set forth a simple plan of action, calling for strictly impartial, non-partisan appeals to be made on behalf of these prisoners and any who, like them, had been imprisoned for peacefully expressed beliefs.

The response to this appeal was larger than anyone had expected. The one-year appeal grew, was extended beyond the year, and Amnestry International and the modern human rights movement were both born.

The modern human rights movement didn't invent any new principles. It was different from what preceeded it primarily in its explicit rejection of political ideology and partisanship, and its demand that governments everywhere, regardless of ideology, adhere to certain basic principles of human rights in their treatment of their citizens.

This appealed to a large group of people, many of whom were politically inactive, not interested in joining a political movement, not ideologically motivated, and didn't care about creating "the perfect society" or perfect government. They were simply outraged that any government dared abuse, imprison, torture, and often kill human beings whose only crime was in believing differently from their government and saying so in public. They (naively, according to many detractors) took to writing letters to governments and publicizing the plights of these people in hopes of persuading or embarrassing abusive governments into better behavior.

Like the early years of many movements, the early years of the modern human rights movement were rocky. "Appeal for Amnesty, 1961" had only the most rudimentary organization. The modern organization named Amnesty International gained the structure it has mostly by learing from mistakes. Early staff members operated with no oversight, and money was wasted. This led to establishing strict financial accountability. Early staff members and volunteers got involved in partisan politics while working on human rights violations in their own countries. This led to the principle that AI members were not, as a matter of practice, asked or permitted to work on cases in their country. Early campaigns failed because Amnesty was misinformed about certain prisoners. This led to the establishment of a formidable research section and the process of "adoption" of prisoners of conscience only after a thorough investigation phase.

The biggest lesson Amnesty learned, and for many the distinguishing feature of the organization, however, was to stick to what it knew and not go outside its mandate. A distinguished human rights researcher I know once said to me that, "Amnesty is an organization that does only one or two things, but does them extremely well." Amnesty International does not take positions on many issues which many people view as human rights concerns (such as abortion) and does not endorse or criticize any form of government. While it will work to ensure a fair trial for all political prisoners, it does not adopt as prisoners of conscience anyone who has used or advocated violence for any reason. It rarely provides startical data on human rights abuses, and never compares the human rights records of one country with another. It sticks to work on behalf of individual prisoners, and work to abolish specific practices, such as torture and the death penalty.

A lot of people found this too restrictive. Many pro-democracy advocates were extremely upset when the organization dropped Nelson Mandela (at the time a black South African anti-apartheid activist in jail on trumped-up murder charges) from its list of adopted prisoners, because of his endorsing a violent struggle against apartheid. Others were upset that Amnesty would not criticize any form of government, even one which (like Soviet-style Communism, or Franco-style fascism) appeared inherently abusive and incompatible with respect for basic human rights. Many activists simply felt that human rights could be better served by a broader field of action.

Over the years combinations of these concerns and others led to formation of other human rights groups. Among them were groups which later merged to form Human Rights Watch, the first of them being Helsinki Watch in 1978. Regional human rights watchdog groups often operated under extremely difficult conditions, especially those in the Soviet Block. Helsinki Watch, which later merged with other groups to form Human Rights Watch, started as a few Russian activists who formed to monitor the Soviet Union's compliance with the human rights provisions in the

Helsinki accords. Many of its members were arrested shortly after it was formed and had little chance to be active.

Other regional groups formed after military takeovers in Chile in 1973, in East Timor in 1975, in Argentina in 1976, and after the Chinese Democracy Wall Movement in 1979.

Although there were differences in philosophy, focus, and tactics between the groups, for the most part they remained on speaking terms, and a number of human rights activists belonged to more than one.

Recognition for the human rights movement, and Amnesty International in particular, grew during the 1970s. Amnesty gained permanent observer status as an NGO at the United Nations. Its reports became mandatory reading in legislatures, state departments and foreign ministries around the world. Its press releases received respectful attention, even when its recommendations were ignored by the governments involved. In 1977 it was awarded the Nobel Peace Prize for its work.

Unfortunately, the Nobel Peace Prize didn't impress the governments Amnesty most wanted to get through to. That year the Argentine military dictatorship reportedly claimed that Amnesty was a front organization for the Soviet KGB. This supposedly occurred the same week that the Soviet government claimed Amnesty was run by the U.S. CIA, to the amusement of human rights activists and, presumably, embarassment of certain people in Argentina and the Soviet Union.

VALUE EDUCATION

INTRODUCTION—Man acts to satisfy his needs or wants. Any thing which satisfies a human need becomes thereby a thing of **Value.** It is the element of desirability and satisfaction that is common to all values, material or non material. In psychology the term value is generally employed to designate a dominant interest, motive or broad evaluative attitude. Value has been defined variously by different educationists, but on the whole it is interpreted to be either a set of feeling or an action. Human behaviour is governed by his values. These are socially approved desires or goals, conceptions or standards by which things are approved or disapproved. Value is a dynamic term used in different aspects. Indian philosophy has used it in sense of state free from pleasure and pain, psychologists in the sense of "psychic energy", sociologists in the sense of "use of time, energy and money" for certain ends. The last theory is named as "Integral theory".

The progress and development of a nation depends upon the quality of the values cherished by its citizens. One of the serious criticism against our educational system is that it lacks value orientation. Our 1986 National Policy on Education and its modifications have strongly advocated value education.

IMPORTANT VALUES

Important values may be described as follows—

(*i*) **Religious Value :** It is defined in terms of faith in God. The outward acts of behaviour expressive of this value are going on pilgrimage, is linking in simple life, having faith in religious leaders, worshipping God and speaking the truth. Students (Higher studies) prefer least the religious value.

(*ii*) **Social Value :** It is defined in terms of cherity, kindness, love and sympathy for the people, efforts to serve God through the service of mankind, sacrificing personnel comforts and gain to relieve the needy and affected of their misery.

(*iii*) **Democratic Value :** This value is characterized by respect for individuality, absence of discrimination among persons on the basis of sex, language, religion, caste, colour, race and family status, ensuring equal social, political and religious rights to all and respect for all democratic institutions.

(*iv*) **Aesthetic Value :** It is characterized by appreciation of beauty, from proportion and harmony, love for fine arts, drawing painting, music, dance, sculpture, poetry and architecture, love for literature, decoration and the surroundings. It is also the least preferred values in schools.

(*v*) **Economic Value :** This value stand for desire for money and material gains. A man with high economic value is guided by consideration of money and material gain in the choice of his job.

(*vi*) **Knowledge Value :** This value stand for love of knowledge or theoretical principles of an activity and love of discovery of truth. A man with this value considers a knowledge of theoretical principles underlying a work essential for success in it. He values hard work in studies.

(*vii*) **Hedonistic Value :** It is the conception of desirability of loving pleasure and avoiding pain. For a hedonist the present is more important than the future. He indulges in pleasure of senses and avoids pain.

(*viii*) **Power Value :** It is defined as the conception of desirability of ruling over others and also of leading others. A man with this value prefers a job where he gets opportunity to exercise authority over the others.

(*ix*) **Family Prestige Value :** It is defined as the conception of desirability of such items of behaviour roles, functions and relationship as would become one's family status. It implies respect for roles which traditionally charateristic of different castes of Indian society.

(*x*) **Health Value :** It is the consideration for keeping the body in a fit state for carrying out one's normal duties and functions. It also implies the consideration for self preservation.

AIMS AND VALUES

People all over the world are going through "Value Crisis" due to science and technology. Future India is also most likely to suffer the 'Future shock' of value crisis. Because of this space age value crisis, the need for the preservation and inculcation of moral values of human life is keenly felt with a growing sense of urgency all over the world.

It is known, that aims of life are correlated with the aims of education. But, what aims are to be cherished? The aims of education is the value theory. Aims are value commitments. Aims which arise out of values are to be justified philosophically not based on dogmatic belief. It is suggested that modern civilization can not survive unless it is inspired by ethical and spiritual values. Among these values the most important, which they stressed was love, greed, hatred, violence, exploitation of others are the order of present society.

Different educational commissions and committees appointed in India in the context of either education in general or value education in particular, have strongly recommended various direct and indirect measures, means and methods for the inculcation of human values through education.

HIERARCHY OF VALUES

Modern India is passing through a transitional period. The old values are crumbling down without being substitued by the new ones. Tradition and modernity co-exist causing a good deal of duality in the minds of old and young alike. This confusion is most likely to be filtered down to the coming generations mainly through socialization practices and child-rearing techniques of the parents and instructors of the young learners. It is apprehended, that during this crisis period, the capcity for social interaction may be affected adversely. This is high time, to consider the priorities to be taken into account while passing on the values to the next generation.

Teaching is a value oriented activity. Teaching of value is unavoidable. All the activities in which teachers engage and governed by certain values. When we ask the students to read certain book, when the seating arrangement is done, when the topics are chosen, when we invite certain speakers to the school, when we hold the examinations, we are governed by the values we cherish. Every individual has got certain hierarchy of values in his mind. It is the duty of the teachers to enable the children develop proper priorities of values suited to the needs of Indian democracy. Freedom, equality, justice, brotherhood and secularism are to be regarded as top most values by the younger generation.

THE CONTENT AND METHODOLOGY

After forming proper hierarchy of values, we look to the content. As far as subject matter is concerned, any subject is loaded with question and issues about values. To ignore Sanskrit, Indian culture, Indian Democracy, Indian Languages, for example, is to ignore much that is rich and vital.

Family factors and educational climate, the kind of experiences the child has had in early years and several other ingredients of the child's growth and development go a long way towards developing the values that he finally possesses. Parents and teachers need to attempt to all such aspects of the child's life.

Now comes the question of methodology. Even though many teachers have not been trained to discuss values systematically, several strategies and techniques are available for analysing and teaching values. Let us discuss the main components of approaches to the realization of values by teachers and students in the institutions.

The main components of the method of value-orientations are the following three :

1. By having formal or informal dialogue.
2. By giving some written exercises, *i.e.* values sheets.
3. By holding individual or group-discussions.

The first approach is the clarifying approach. According to Louis E. Raths and his associates (as mentioned in Value and Teaching, pp. 58-64) value should satisfy the following criteria :

1. It has been freely chosen.
2. It is chosen from one or more available alternatives.
3. It has been chosen after due reflection.
4. It is prized and cherished.
5. It has been affirmed to others.
6. It has been incorporated into actual behaviour.
7. It is repeated in one's life.

During formal and informal dialogue the values are known and the students come to know that their judgements are correct or not.

Then comes the technique of writing. At this stage value sheets are given. A value sheet consists of a provocative statement and a series of questions, e.g., why learning is said to be the best of all wealths?

Then comes the stage of holding individual or group discussions. Some problems may be given to the class, e.g., Freedom or worship may be the problem. The students may be asked to discuss this problem. Discussion on situations and solutions may also be held. Discussion on India's religions may be interesting to students.

The above three strategies are to be adopted by means of several devices, e.g., films, pictures, songs, poetry, documents, dramatization, parable, fiction, letters, diaries, interviews, socio-drama, cartoons, posters, field trips, quest speaking, or debate. The use of these devices will make the value orientation through all the three approaches more interesting.

CO-CURRICULAR ACTIVITIES

The schools should plan their co-curricular activities also from the point of view of value-education. Through these activities many values can be inculcated profitably. Co-curricular activities form an integral part of the modern school curriculum. Various games and sports not only provide the best recreation but they have profound impact on the temperament and outlook of the players. They help in the sublimation of the personality of players. Training in physical education develops the high sense of obedience which leads a person to render his services sincerely. Besides games and sports, there are several out-door activities like N.C.C., P.E.C., N.S.S., scouting and guiding, mountaineering, trekking which train individuals in team work, self discipline, courage, bravery, obedience, integrity and friendship. Students may learn how to serve others and how to respect labour through these activities.

Apart from the above activities literacy and cultural activities may be organised in the schools for inculcating certain values. Poem recitation, debate, symposium, essay competition, work parliament, science club, story competition, melo-drama, mono-acting, socio-dance, music, fancy dress show, painting competition, group dance etc. will offer opportunities to children to learn qualities of leadership, self-discipline, co-operation etc. Morning Assembly has been found to be very good tool for clarification and communication of certain values. If it is conducted properly, it may prove to be very purposive and fruitful for fostering social moral and spiritual values. The sanctity and dignity of the morning assembly should be maintained both by teachers and students and all must participate in it.

Students' sensitivity to the feelings of others are to be increased. They should be provided with the opportunities in the school campus to talk about feelings, to identify with the feelings of other people and to react emotionally themselves. In short, teachers should encourage and help students to participate in experiences that allow them to feel different kinds of emotions, to come into contact with many different people, to do different things and then to share their perception of these experiences.

HIV/AIDS

Key words & Definitions :

AIDS = Acquired Immuno Deficiency Syndrome

Acquired = which is not present since birth but acquired after birth.

Immunodeficiency = Deficiency of immune functional cells; deficiency to perform the immunological function.

Syndrome = A group of diseases and signs and symptoms of illness.

HIV = Human Immunodeficiency Virus

HIV positive = The presence of antibodies against HIV in human body is termed as HIV positivity and the person is called HIV positive (Seropositive). It takes 6-12 weeks after infection for antibodies to rise to detectable levels. There is thus a window period during which the infected person may transmit the infection despite being seropositive.

The latest killer disease that has created nightmares for the medical experts is the Acquired Immuno Deficiency Syndrome (AIDS). Its terrifying spread has earned it the title of the "Pandemic" or an epidemic, which is out of control. This disease has wreaked the social & economic devastation. In the absence of medical defence against AIDS, public education is the only weapon in the fight to limit the spread the infection. Only by influencing personal behaviour & life style can we hope to maintain the ravages of AIDS throughout.

Virus :

HIV belongs to the family Retroviridae and subfamily Lentivirinae. Two different types of HIV have been recognized and are called as HIV-1 and HIV-2. Both differ in geographical distribution, biological & molecular characteristics and extent of transmissibility. These viruses store their genetic information as RNA. HIV-1 has three groups; HIV-1 major groups (HIV-1 M), outlier (HIV-1 O) and HIV-1 new (HIV-1 N) group. HIV-1 major groups can be further classified into different subtypes or clades designated A-K.

HIV is 120 nm icosahedral, enveloped, RNA virus. HIV comprises of an outer envelope consisting of lipid bilayer with uniformly arranged 72 spikes or knobs of gp 120 and gp41. Inside the protein core surrounding two copeis of RNA. Core also contains viral enzymes, Reverse transcriptase, integrase and protease, all essential for viral replication and maturation.

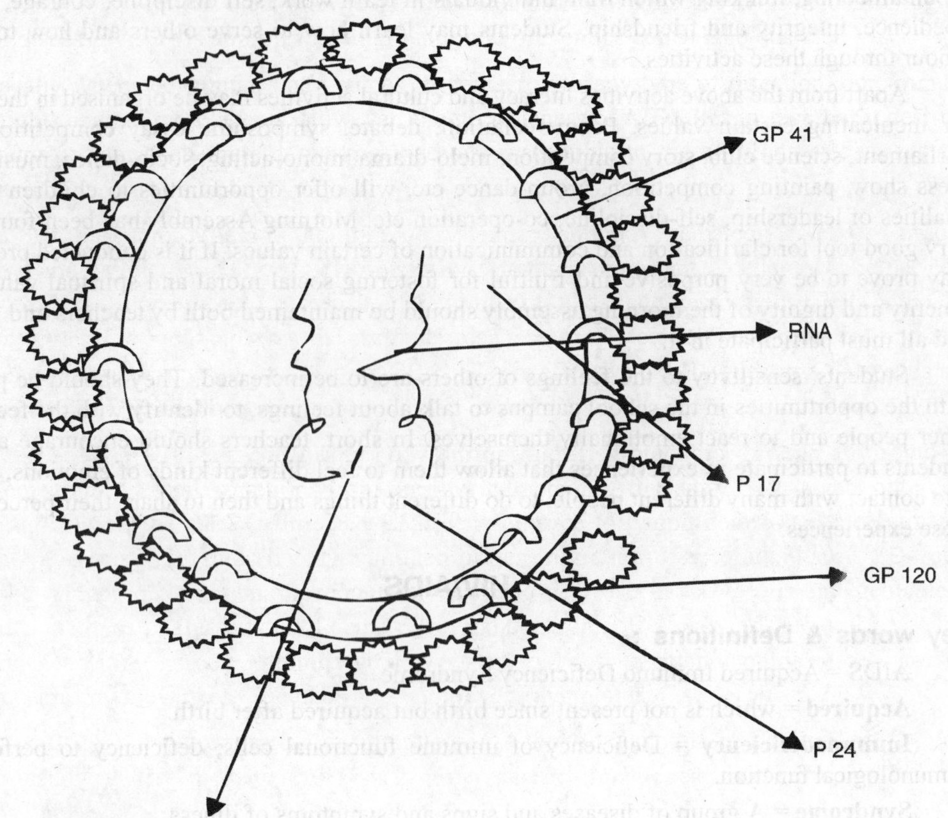

Major antigens of HIV-1 :		**Antigens :**
a.	Envelope antigen:	
	Spike antigen	gp 120
	Transmembrane Pedicle antigen	gp41
b.	Matrix antigen	p17
c.	Core antigen	p24

HIV replication:

Gp120 binds to host cell receptor

Reverse transcription

Proviral DNA synthesis

Integral with host cell DNA

Viral proteins synthesis

Virus assembly and budding

Maturation of core proteins

Transmission :

HIV infection can be transmitted by following modes and the efficiency of HIV transmission is:

Transmission Route	*Percent Efficiency*	*Percentage of total cases*	
		World over	*India*
Blood Transfusion	90-95	5	7
Perinatal	20-40	10	10
Sex	0.1-1	75	80
Injecting drug abusers	0.5-1.0	10	7.3

Sex: Homosexual and heterosexual contact with HIV infected partner

Parenteral: Infected Blood, Blood products; infected needle & syringes, IVDU

Perinatal: From infected mother to her child (Before 10-30%, during and after delivery 40-60%) and 30% through breast milk. Most common is the heterosexual contact.

Transmission cycle of HIV :

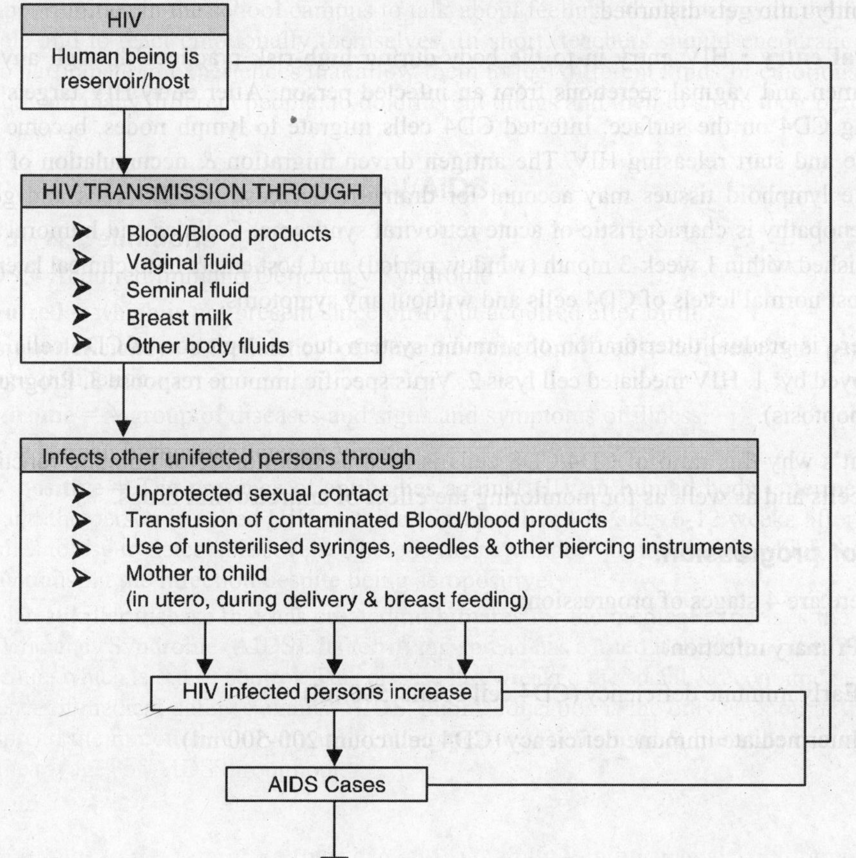

HIV IS NOT TRANSMITTED BY:

Shaking hands, hugging, dry kissing, sneezing, coughing, mosquito bite, toilet sharing, sharing of telephones, offices, playing, traveling together, sharing cups, living in same room, donating blood aseptically.

NATURAL HISTORY OF HIV INFECTION/AIDS:

The natural history of any disease refers to the stages through which a disease passes in the absence of any intervention. Clear knowledge of natural history of a disease helps in identifying the stages of diseasse vis a vis appropriate intervention to prevent or control the disease.

Usually whenever human body comes in contact with any type of infection, our immune system fights against these agents. However, the horribleness of AIDS lies in fact that it destroys the immune system itself thereby making the body susceptible to the all kinds of diseases and infections.

Further, the major pillars of the immune system are antibody mediated (B Cell or Humoral immunity) and cellular (T cell or CMI) immunity. Of which important T cells are T helper cells (Cells with CD4 epitope/receptors), which cause immunity, and T suppressor cells (Cells with CD8 epitope/receptors), which suppress the immune response of TH cells. These cells are normally present in equilibrium & with a ratio where CD4 cells are effective but HIV destroys CD4 cells and subsequently ratio gets disturbed.

Viral entry : HIV entry in to the body during high-risk practices through any route via blood, semen and vaginal secretions from an infected person. After entry HIV targets onto cells expressing CD4 on the surface. Infected CD4 cells migrate to lymph nodes, become activated, proliferate and start releasing HIV. The antigen driven migration & accumulation of CD4 cells within the lymphoid tissues may account for dramatic decrease in CD4 cells and generalised lymphadenopathy is characteristic of acute retroviral syndrome. Cellular and humoral respondes are established within 1 week-3 month (window period) and host enters in to clinical latency period with almost normal levels of CD4 cells and without any symptoms.

There is gradual deterioration of immune system due to depletion of CD4 cells. CD4 cells are destroyed by: 1. HIV mediated cell lysis 2. Virus specific immune response 3. Programmed cell death (Apoptosis).

That's why this ratio of CD4/CD8 cells is used as monitoring of immune functionality of infected cells and as well. as for monitoring the effect of antiretroviral drugs.

Stages of progression:

There are 4 stages of progression:

1. Primary infection
2. Early immune deficiency (CD4 cell count>500/ml)
3. Intermediate immune deficiency (CD4 cell count 200-500/ml)

4. Advance immune deficiency (CD4 cell count<200/ml)

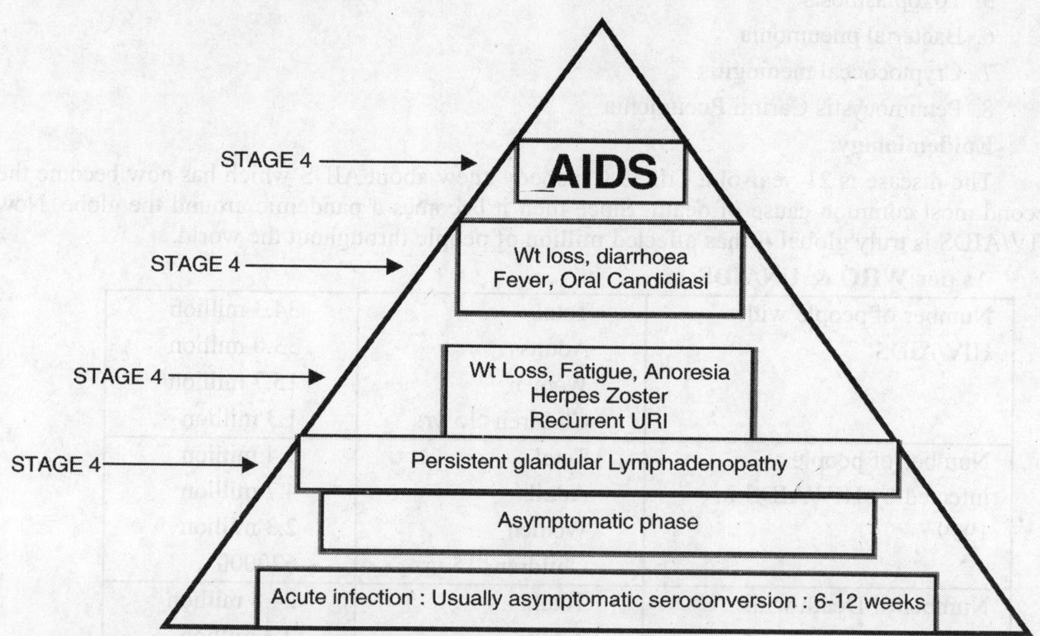

Laboratory investigations for HIV infection:

HIV infection can be detected in the laboratory either by detection of :

Antibodies to HIV	:	ELISA, rapid/simple, western blot (confirmatory test)
Antigen detection	:	p24 antigen
Direct detection of Virus	:	Microscopy
Detection of viral RNA	:	PCR

Isolation/culture of virus

The indirect predictors of HIV infection and disease

CD4 cell count

Beta 2 microglobulin

These are used as monitors of immunity status of patients and for monitoring the progression of the disease.

HIV testing strategy :

Unlinked anonymous testing

Voluntary confidential testing

Mandatory testing

Opportunistic Infections: Are those infections, which are non, infectious to healthy immune competent human being but on immunosuppression or immunodeficiency status leads to them become infectious. Common Opportunistic infections are:

1. Tuberculosis
2. Candidiasis
3. Crytosporidiosis

4. Herpes Zoster
5. Toxoplasmosis
6. Bacterial pneumonia
7. Cryptococcal meningitis
8. Penumocystis Carinii Pneumonia

Epidemiology:

The disease is 21-year-old. Till 1981 nobody knew about AIDS which has now become the second most common cause of death. Since then it becomes a pandemic around the globe. Now HIV/AIDS is truly global. It has affected million of people throughout the world.

As per WHO & UNAIDS:

Number of people with HIV/AIDS	Total Adults Women Children<15 yrs	34.3 million 33.0 million 15.7 million 1.3 million
Number of people infected by HIV/AIDS in 1999	Total Adults Women Children<15 yrs	5.4 million 4.7 million 2.3 million 620000
Number of Death with HIV/AIDS in 1999	Total Adults Women Children<15 yrs	24.8 million 2.3 million 1.2 million 500000

In India:

In India the first case was detected in 1986 in Tamil Nadu.

By the end of year 2000 it was estimated that 3.6 million Indians infected with HIV/AIDS.

More than 50% of all new infections take place among young adults below 25 years.

Every year 22837 newborn children are infected.

About 11,434 die due to HIV/AIDS.

Total 120,000 children are living with AIDS.

120,000 AIDS orphans living in country.

Conclusion:

The need of the hour is for dissemination of basic information in simple and intelligible terms about the causes of AIDS and the avenues of infection, the essential precautions to be taken, the facilities for treatment & all other related matters. The public must be alerted without generating hysteria about the real menace of AIDS epidemic & the measures required to combat it.

WOMEN AND CHILD WELFARE

History

The health of women and children began to receive separate attention early in this century, in recognition of their greater vulnerability, particularly to socioeconomic and environmental forces, and the interdependence of the child's health and that of the mother. In 1909 the first White House Conference on Child Health recommended the formation of the Children's Bureau, which proceeded to investigate the causes of infant mortality (more than 100 per 1,000 live births). The first direct support of health services for mothers and children came with the Shepard-Towner Act of 1921, which resulted in complete birth registration and the establishment of maternal and child health divisions in state and local health departments. Title V of the Social Security Act of 1935 extended services to handicapped children and further established the principle of public responsibility for the health of mothers and children.

In the 1960s and early 1970s additional programs were initiated by Congress. These included Medicaid, Early Periodic Screening, Diagnosis and Treatment (EPSDT), Neighborhood Health Centers, Maternity and Infant Care, Family Planning, Children and Youth Projects, Head Start, Title educational assistance, the Right to Education of the Handicapped (PL 94-142), and nutrition programs (WIC and School Lunch). While these laws expanded services at the state and local levels, the resultant programs were administered by a variety of different branches of government, diffusin_ responsibility and often leading to poor coordination and the undermining of maternal and child health (MCH) divisions.

MCH services were seriously weakened by the budget cuts of the Reagan years. Rather than cutting specific services, the programs were lumped into block grants that gave stage governments greater freedom to apportion the reduced funds as they saw fit. This has begun the trend of providing greater local autonomy in the establishment and administration of health programs for women and children. This trend may hold promise for the creation of innovative programs that more precisely address the community's needs, making greater use of local resources.

Managed care has emerged from the multiple efforts to control health care costs, and this new structure for the delivery of medical care has had a significant impact on maternal and child health services. Medicaid recipients are rapidly being shifted into managed care plans that restrict access to all but primary care services, causing health departments to lose their base of clients for immunization, family planning, and other health care service programs. Meanwhile the managed care plans that have enrolled these high-risk populations are not held accountable in the same manner as was the health department for service delivery. Cost control efforts led to postpartum hospital stays being reduced to 1 day until Congress intervened with a rule that allowed at least a 48-hour stay at the discretion of the mother and her physician.

Progress in MCH has been driven by the dual forces of research and advocacy. Multidisciplinary studies over the last 20 years have shed important light on the health problems of women and children and provided numerous examples of effective means to ameliorate those problems. Articulate and committed individuals and organizations have played a critical role in fostering public commitment to improve the lives of mothers and children. However, that support has been significantly reduced through the 1980s. As our knowledge of what to do has continued to grow, the political will to use that knowledge has shrunk.

Health Indicators

Various health indicators are used to assess the health status of mothers and children. The continuous monitoring of these indicators is in essential part of evaluating our progress in improving the health status of women and children.

Maternal mortality rates have reached such a low level that they are of little value now. Maternal health is better reflected in fertility rates and birth rates as well as in pregnancy-related morbidity rates. Pregnancy outcomes have become a more important measure of maternal health and quality of maternity services provided. Miscarriage, therapeutic abortion, stillbirth, and especially low-birth weight rates can be used to assess the success of the pregnancy. Prenatal care, place of delivery, attendant at delivery, type of delivery, complications, length of stay, and cost all measure the availability and quality of maternal health services.

The infant mortality rate remains an important though crude measure of MCH. Linking infant birth and death records has provided a far more precise way of assessing factors associated with pregnancy outcome, particularly when the causes of death are grouped by pregnancy-related conditions such as prematurity, rather than the organ system taxonomy of the ICD-9 codes. The new birth certificate form that was adopted in 1989 includes a wider array of information on both the mother and the child, offering opportunities for future exploration of the relationships between more extensive sociodemographic and medical information and various pregnancy outcomes. Childhood morbidity is less easily measured. Birth defects registries, neonatal intensive care use, discharge diagnoses, and national health surveys provide some estimates of morbidity. Immunization rates, school-based health data, and the data from such programs as EPSDT are also helpful indicators of child health, although they are not systematically collected at either the state or natinal level.

Larger social and demographic changes are also important indicators of the status of mothers and children. Over the last 20 years there have been dramatic increases in the percentage of mothers in the workforce, the percentage of marriages that end in divorce, the numbers of homeless mothers and children, and the percentage of children living in poverty. These social problems contribute directly or indirectly to most of the health problems of women and children.

Service Delivery

The goals of MCH services are (*a*) to encourage desired pregnancy, achieving the best possible outcome for the baby and the mother; (*b*) to promote healthy relationships within the family to nurture the growing child; (*c*) to optimize the normal developmental processes to allow the child to achieve his or her fullest potential; (*d*) to prevent child health problems and reduce the risks of adult health problems; and (*e*) to provide early intervention in the health problems of women and children so as to minimize morbidity and mortality in a cost-effective manner. To achieve these goals, attention must be paid to several basic principles of MCH. These principles are a product of the nature of women and children and the problems from which they suffer and therefore are a bit different from the general principles that underlie all health service delivery.

TWO CLIENTS

Maternal services are unique in that they simultaneously provide care for two, equally important, clients, the mother and the foetus. Balancing the needs of both to achieve the best possible outcome requires a thorough understanding of the complex interdependence of the maternal/foetal unit and the implications of events and treatments for both mother and baby.

Family-Centered Services

Because of the extreme importance of the family in the nurturing of the pregnant woman and the young child, it is essential that MCH services be delivered with attention to the family circumstances of the clients. The family influences growth, development, health-related behaviors, and lifestyle habits. Family resources influence the use and availabilty of health services and the ability to provide the care needed, particularly in chronic disease. The child is not merely the passive recipient of the influences of the family, but, rather, plays an increasingly interactive role in the family, shaping in part the environment in which he or she lives.

Developmental Perspective

The foetus and child are being continuously shaped by the normal developmental processes that result in a reasonably predictable series of changes from conception through adolescence. Progress over this course is a sensitive measure of both health and disease. These developmental forces can be potent allies in the management of chronic health problems. However, continued disruption of normal development can have progressively magnifying adverse effects on the fetus or child. Because of the importance of development, the dimension of time and the continuity of care over time become critical elements in the provision of MCH services. Prompt identification of problems and early intervention, therefore, hold the greatest promise for achieving the best outcome.

Health Promotion and Disease Prevention

There is great potential in childhood for health promotion and disease prevention to benefit both the current child and the future adult. However, careful attention must be paid to the immediate implications of interventions that are aimed at preventing problems in the distant future, making sure that the desired long-term benefits are not counter-balanced by short-term hazards.

Timely, Cost-Effective Treatment

The early identification and proper treatment of common health problems is a critical dimension of reducing morbidity and mortality in women and children. Simple early treatments can often prevent very expensive and serious problems, such as adolescent pregnancy, a premature birth, or a handicapping condition.

Integration of Principles

Perhaps the greatest challenge to delivery of MCH services is trying to inegrate all the principles articulated above into the care of each client. It is difficult for the provider to attend simultaneously to the treatment and prevention needs while considering the family and development issues in the care of both the mother and the child. However, the greatest success is achieved when all these concepts are addressed together.

Trends and Innovations in Services

The United States lags way behind all other developed countries in the provision of most services to mothers and children. Western Europe and Japan offer extensive maternity benefits, prenatal care, day care, and well-child care to all women and children. In infant mortality rates, the United States ranks 20th in the world, a fact that many attribute at least in part to the inadequate provision of maternal and infant services. The last decade has seen little progress in this arena, but some interesting innovations in the delivery of MCH services have recently emerged and warrant brief description. Many of these have not yet come into general practice but hold promise for the future, once they have been more carefully evaluated. Social trends of the last decade have also had an effect on MCH services and must be considered in the process of recommending improvements for the future.

Family Planning and Abortion

Optimum health for both mother and child has long been known to be related to maternal age, spacing of children, and the balance between family resources and family size. The ready availability of birth control and the option for abortion have provided means of achieving family planning. Norplant was hoped to be an easy means of contraception that required no action for 5 years after its insertion, but it has turned out to be unpopular due to its side effects. RU-487 will soon be available as an early pregnency abortion pill. Having such a non-surgical abortion option will make it very difficult to ever restrict abortion fully.

Preconceptional Health Promotion

Many of the critical phases of fetal development have already occurred before a woman is even aware that she is pregnant. Optimum fetal health, therefore, requires attention to maternal health and health-related behaviors even before conception. Efforts to counsel women before conception to avoid alcohol, drugs, tobacco and other fetal hazards are currently being tested to determine the impact on pregnancy outcome. The efficacy of preconceptional dietary folic acid in the prevention of neutral tube defects will likely soon result in routine folic acid fortification of bread.

Prenatal Care

Improving access to and quality of prenatal services continues to be a challenge with no obvious solution. Expansion of Medicaid to include women up to 185 percent of the federal poverty level in comprehensive services has made prenatal care more accessible. Some states have developed innovative programs to improve quality and access for the poor, and many local community-based projects have also been created with these goals in mind.

Human Immunodeficiency Virus

The vertical transmission of human immunodeficiency virus (HIV) from infected women to their babies will infect approximately one-quarter of these babies. However, treatment with azidothymidine (AZT) during pregnancy can reduce this transmission to as low as 8 percent. This creates the dilemma of whether pregnant women should be required to be HIV tested and, if positive, treated for the sake of their baby or whether they should merely be given the information and the option.

Immunization

The recently approved acellular pertussis vaccine offers hope that some of the complications of the whole cell vaccine might be avoided. The use of conjugated vaccine against *Haemophilus influenzae* in infants as young as 2 months of age has significantly reduced the incidence of meningitis in infant. The hepatitis B vaccine is now widely used in the prevention of both hepatitis and the consequent hepatic cancer—both major problems, particularly in develping countries. The eradication of polio from the Western Hemisphere has led to a recent switch from oral polio vaccine to a combination of oral and inactivated vaccines to reduce the incidence of paralytic reactions to the oral vaccine.

Day Care

As maternal employment continues to rise, we are falling further behind in providing adequate, affordable day care for young children. Most states provide little regulation of day care facilities, particularly home-based centers with few children. This aggravates concerns about spread of infection, injury, and child abuse in day care facilities, as well as potentially being a poor substitute for parents.

Sexual Abuse

Over the last decade we have been forced to recognize that child sexual abuse occurs far more frequently than we would like to believe. As this problem has come out of the closet, innovative programs have been developed to teach children how to avoid sexual exploitation and to identify and treat more effectively those who are victims. Many school systems have adopted curricula in prevention of sex abuse for children as young as those in kindergarten, and the open discussion of this problem has made it easier to inquire about such incidents as a part of routine health care. Many states are also experimenting with creative ways of humanely dealing with the child witness in court without unduly infringing on the constitutional rights of the accused. However, we are probably still studying only the tip of this iceberg as we begin to explore the consequences of the more subtle forms of abuse and try to gain a better understanding of the abusers and the factors that lead to abuse in some families.

Community-Based Social Support

Pregnant women and young children thrive best when they are surrounded by friends and relatives who provide companionship and assistance. As unwed motherhood becomes more common and the extended family further disintegrates, more young families face isolation and inadequate social supports. To remedy this, several programs have utilized home visitors to befriend and work closely with pregnant women and young families, offering the assistance and social support that are so often inadequate. Some of these programs have been able to demonstrate benefits in health and well-being associated with participation in the programme.

Future Directions

The 1960s and 1970s produced a number of centrally funded programs to help mothers and children. Through the 1980s the support for these programs eroded, and control over spending priorities for the diminished funding was shifted to the state capitals through the block grants. In the same period new initiatives for the elderly received increasing support. Despite renewed outcries about the plight of America's children, Congress was able to pass sweeping welfare reform, which is predicted to have significant negative impact on women and children.

As managed care appears to be cutting costs, there is a rising pressure for shifting MCH services from public sector provider systems into managed care plans where demand can be controlled. In those states where many of these services are provided by health department, the work and the funding of these departments will be dramatically reshaped by this trend. Both the consumers and the managed care providers will need to make adjustments if this new system is to simultaneously save money and still meet the health needs of this population. The tightening of other welfare benefits for the poor will additionally compromise the basic well-being of poor women and their children. Over the next decade it will be increasingly important to monitor the well-being of women and children to ensure that the systems established to improve their health are indeed succeeding. A recently released report of the Institute of Medicine of the National Academy of Sciences addresses ways in which communities might improve health through performance monitoring.

Role of Information Technology in Environment and Human Health

Just as chemical or metallurgical or electrical technologies enable the processing of raw materials in to usable goods, to satisfy man's and societies needs, so does information technology (IT) help the storage, processing, transmission and exploitation of information to satisfy a person's, company's, society's or Govt's needs for information. Information covers voice as in telephony, text as in fax, images as in video and data as between computers.

Information is knowledge and knowledge is power. Knowledge plus experience is wisdom and it is the wise use of information that gives advantage to those who have information.

Information technology as commonly picturised by computers in extending man's mind or brain or intellectual power. Information technology devices like microprocessors and becoming mass appliances from pace makers for the heart, hearing aids and efficiency enhances in automobile engines and devices to steer space vehicles on the moon. Like banking, trading, learning and teaching, librarying and other sectors of human activity, Information technology has tremendous use in the field of environment and human health. The print media has also highlighted and participated in some of the important environmental issues and helped in making the people aware. Apart from the development of softwares for environment & healthy studies facilities like internet, world wide web geographical information system (GIS), information through satellites are also developed. These all are also helpful in environment and health studies.

Media has the capability to influence people's opinion. It has highlighted and participated in some of the important environmental issues and has helped in making the people aware. The Ministry of Environment and Forests Govt. of India has taken up the task of compiling a DATABASE on various biotic communities. Database includes wildlife database, conservation database, forest cover database etc. Since database is the collection of inter-related data on various subjects, also available for chronic diseases like HIV/AIDS etc. They are in computerized form. The Ministry of Environment & Forests, Govt. of India has created Environmental Information System (ENVIS). ENVIS has many centres all over the country for generating a network of database in different areas like pollution, environmental management, wildlife etc.

www.mhhe.com/environmental science and multimedia Digital Content Manager (DCM) in the form of CD-ROM are the on line learning centre, which provides the most current and relevant information on environmental science.[2]

With the help of computers and internet, not only we can have knowledge of the patients at a glance within no time, but also we can get information about the diseases their medicines and alternative medicines. (Alopathic, Ayurvedic and Homoeopathic). CT-scanning (Computed Tomographic Scanning, CAT (Computed Axial Tomography) are such examples of information technology in human health. It also help in computer-aided monitoring, health testing, implantation of artificial body parts and in pharmaceutical designing of new drugs. IT is expanding rapidly with increasing applications in the field of environment and human health. For awareness, songs, drama, bhajans, advertisements in TV, radio, cultural programmes should be exhibited in urbans & rural sectors both.

Water Quality Standards

The Water Quality Standards as set by Union Health Ministry and followed by APHED are:

(i) Physical, (ii) Chemical, (iii) Bacteriological, (iv) Virological

PHYSICAL STANDARDS

No.	Characteristics	Acceptable*	Cause for Rejection*
(i)	Turbidity (units on J. T. U. Scale)	2.5	10
(ii)	Colour (units on platinum-cobalt scale)	5.0	25
(iii)	Taste and odour	Unobjectionable	Unobjectionable

CHEMICAL STANDARDS

No.	Characteristics	Acceptable*	Cause for Rejection*
(i)	pH	7.0-8.5	6.5-9.2
(ii)	Total dissolved solids (mg/*l*)	500	1500
(iii)	Total hardness (as $CaCO_3$) (mg/*l*)	200	600
(iv)	Chlorides (as Cl) (mg/*l*)	200	1000
(v)	Sulphates (as SO_4) (mg/*l*)	200	400
(vi)	Fluorides (as F) (mg/*l*)	0.1	1.5
(vii)	Nitrates (as NO_3) (mg/*l*)	45	45
(viii)	Calcium (as Ca) (mg/*l*)	75	200
(ix)	Magnesium (as Mg) (mg/*l*)	> 30 (If there are 250 mg/*l* of sulphates, Mg content can be increased to a maximum of 125 mg/*l* with the reduction of sulphates at the rate of 1 unit per every 2.5 units of sulphates)	150
(x)	Iron (as Fe) (mg/*l*)	0.1	1.0
(xi)	Maganese (as Mn) mg/*l*)	0.05	0.5
(xii)	Copper (as Cu) (mg/*l*)	0.05	1.5

(xiii)	Zinc (as Zn) (mg/*l*)	5.0	15.0
(xiv)	Phenolic compounds (as phenol) (mg/*l*)	0.001	0.002
(xv)	Anionic detergents (as MBAS) (mg/*l*)	0.2	1.0
(xvi)	Mineral oil (mg/*l*)	0.01	0.3
(xvii)	Arsenic (as As) (mg/*l*)	0.05	0.05
(xviii)	Cadmium (as Cd)(mg/*l*)	0.01	0.01
(xix)	Chromium (as hexavalent Cr) (mg/*l*)	0.05	0.05
(xx)	Cynides (as CN) (mg/*l*)	0.05	0.05
(xxi)	Lead (as Pb) (mg/*l*)	0.1	0.1
(xxii)	Selenium (as Se) (mg/*l*)	0.01	0.01
(xxiii)	Mercury (total as Hg) (mg/*l*)	0.001	0.001
(xxiv)	Polynuclear aromatic hydrocarbons (PAH) (pg/*l*)	0.2	0.2
(xxv)	Gross alpha activity (pCi/*l*)	3	3
(xxvi)	Gross beta activity (pCi/*l*)	30	30

Note 1 : The figures indicated under the column 'acceptable' are the limits up to which the water is generally acceptable to the consumers.

Note 2 : It is possible that some mine and spring waters may exceed these radio activity limits and in such cases it is necessary to analyze the individual radionuclides in order to assess the acceptability or otherwise for public consumption.

*Figures in excess of those mentioned under 'acceptable' render the water not acceptable, but still may be tolerated in the absence of alternative and better source up to the limits indicated under column 'cause for rejection' above which the supply will have to be rejected.

DISSOLVED OXYGEN IN WATER (DO)

Since Oxygen is necessary for all living/non-living organisms, but dissolved oxygen in water is vital to fish and other aquatic life. Oxygen is transferred from atmosphere to surface water as well as produced by aquatic plants, algae and phytoplankton as a bye product of photosynthesis. After dissolving in water oxygen diffuses or distributed throughout the water body. Distribution depends on movement of water, currents and thermal upwelling. Oxygen in water measured as dissolved oxygen (DO). It is measured as parts per million (ppm), which is the number of oxygen (O_2) molecules per million total molecules in a sample.

It is also defined as the number of moles of molecular oxygen (O_2) dissolved in a litre of water at a temperature. It is expressed as mg O_2 / *l*. Dissolved oxygen can range from 0-18 mg O_2 / *l*. Most natural water systems require 5-6 mg O_2 / *l*. The oxygen is used by plants and animals for respiration and by the aerobic bacteria which consume oxygen during the process of decomposition. A high percentage of dissolved oxygen is conducive to aquatic flora & fauna. A low percentage indicates a negative impact on a body of water which results in a abundance of worms and fly larvae.

Factors Affecting Dissolved Oxygen

The following are the main factors which affects the dissolved oxygen (DO) in the water :

1. Water temperature.
2. Flow.
3. Aquatic plant population.
4. Atmospheric pressure.
5. Human Activities.
6. Water discharge.
7. Organic waste.
8. Runoff from streets.

Winkler's Method for DO Determination

The most precise and reliable titrimetric procedure for DO analysis is the Winkler's method (1888). It is the same as iodometric technique. It is based on the fact that dissolved oxygen oxidises Potassium Iodide (KI) to Iodine, which is titrated against standard sodium thiosulphate (Hypo) solution using starch as an Indicator. Since dissolved oxygen is in the molecular state, her.ce can not oxidise as such. For that Manganese hydroxide is used as an oxgen carrier to bring out the reaction. Manganese hydroxide is obtained by the action of KOH with Manganese sulphate.

$$MnSO_4 + 2KOH \rightarrow Mn(OH)_2 + K_2SO_4$$
$$2Mn(OH)_2 + O_2 \rightarrow 2MnO(OH)_2$$
$$MnO(OH)_2 + H_2SO_4 \rightarrow MnSO_4 + 2H_2O + [O]$$
$$2KI + H_2SO_4 + [O] \rightarrow K_2SO_4 + H_2O + I_2$$
$$2Na_2S_2O_3 + I_2 \rightarrow Na_2S_4O_6 + 2NaI.$$

For oxidising and reducing agents present in water, two modification are given:

(i) Alsterberg's Modification : If oxidising agents like nitrate and ferric ions are present in water, they will oxiside I to I_2 and will give positive error. To over come with this problem sodium azide is used in alkaline solution to decompose them.

$$2NaN_3 + H_2SO_4 \rightarrow Na_2SO_4 + 2HN_3$$
$$HNO_3 + HN_3 \rightarrow N_2O + N_2 + H_2O$$

(ii) Rideat-Stewart Modification : If reducing agents like Fe^{+2}, SO_3^{-2}, S^{-2} etc. are present in sample. They will reduce I_2 to I^- and will produce negative error.

To overcome with this problem, $KMnO_4$ is used for pretreatment. Excess of $KMnO_4$ can be removed by reaction with Potassium oxalate.

CHEMICAL OXYGEN DEMAND (COD)

Chemical Oxygen Demand is a useful measure of water quality. It is defined as the amount of oxygen consumed under specified conditions in the oxidation of organic & oxidisable inorganic matter. COD expressed in milligrams per litre (*mg/l* or ppm).

COD of waste water is the number of *mg* of oxygen required to oxidise the impurities present in 1000 ml of waste water using strong oxidising agent like acidified $K_2Cr_2O_7$. COD represents the total amount of oxygen required to oxidise all oxidisable impurities in a given sample. Thus COD value for a sample is always higher than BOD value. Since time required for COD test is less, therefore it is always advantageous. In environmental chemistry COD test is indirect measure of organic compounds present in water.

Limitations of COD

COD test does not differentiate between bio-inert and biodegradable materials. It also not indicate the rate at which the biologically oxidisable material stabilize.

COD represents the total amount of oxygen required to oxidise all oxidisable impurities in a sample of sewage wastes COD is always greater than BOD since in COD measurement both biodegradable and non-biodegradable load are completely oxidised. The difference in COD and BOD is equivalent to the quantity of biologically resistant organic matter.

DETERMINATION OF COD

Principle

A known volume of the wastewater sample is refluxed with a known excess of $K_2Cr_2O_7$ solution in H_2SO_4 medium containing $HgSO_4$ (catalyst) and Ag_2SO_4 [which retains halides] for about $1\frac{1}{2}$ hr for the oxidation to be complete. A part of the $K_2Cr_2O_7$ is used up for the oxidation of impurities. The remaining $K_2Cr_2O_7$ is determined by titration with standard FAS (Ferrous Ammonium Sulphate) solution using ferroin as indicator. The endpoint is the change of colour from blue green to reddish brown.

A blanks is performed by titrating known volume of the acidified $K_2Cr_2O_7$ with the same FAS using the same indicator.

$$\text{COD of water sample} = \frac{(A - B) \times M\ 8000\ \text{ml}}{\text{Volume of sample}}\ \text{mg}/l$$

where A = Blank titre value of $K_2Cr_2O_7$ vs FAS and
B = Volume of FAS consumed for unreacted $K_2Cr_2O_7$ of the solution.
M = Molarity of FAS solution.

Procedure

25 ml of waste water is pipette out into a round bottomed flask. 10 ml of $K_2Cr_2O_7$ is pipette out into the same flask along with one test tube full of 1 : 1 H_2SO_4 containing $HgSO_4$ and Ag_2SO_4. The flask is filled with a reflux water condenser and the mixture is refluxed for 2 hours. The contents are cooled and transferred to a conical flask. 5 drops of ferroin indicator is added to it and titrated against FAS taken in the burette till the colour changes from blue green to red brown. Same volume of $K_2Cr_2O_7$ is pipette out, mixed with sulphuric acid and ferroin and titrated against same FAS to get blank titre value.

Calculation

$$\text{Chemical oxygen demand of water} = \frac{(A - B) \times M \times 8000}{V}$$

where A = FAS (ml) used for blank
B = FAS (ml) used for sample
M = Molarity of FAS
V = Volume of sample (ml)

BIOLOGICAL OXYGEN DEMAND (BOD)

Biological Oxygen Demand (BOD) is a measure of water quality. It is an important property. It is defined as a measure of oxygen needed (in mg/litre or ppm) by bacteria and other micro-organisms oxidise the organic matter present in water sample over a period. It may also be defined as the quantity of dissolved oxygen required by aerobic bacteria for the to oxidation of organic matter under aerobic conditions. The BOD of drinking water is less than one while sewages have more then several hundreds.

BOD is high, the dissolved oxygen becomes low. The greater the BOD, greater the pollution. Thus BOD is an indication of extent of pollution.

Micro-organisms such as bacteria and fungi are responsible for decomposing organic waste *i.e.,* dead plants, leaves, grass, manure, sewage or food waste. In this process much of the available dissolved oxygen is consumed by aerobic bacteria, robbing other aquatic organisms of the oxygen they need to live. The temperature of the water can also contribute to high BOD levels. Similarly Nitrates and Phosphates in a body of the water can contribute to high BOD levels.

Limitations of BOD

Effluents of industries like paper, pulp rayon & chemicals, have low value of BOD, although they contain enough organic matter. Thus BOD values should not be used as equivalent to organic load. In these cases COD reveals the real pollution potential.

BOD by Winkler's Method

Principle

Winkler's method is based on the fact that in alkaline medium, DO oxidises Mn^{2+} to Mn^{4+}, which in acidic medium oxidises I^- to free iodine. The amount of iodine released which can be titrated with a standard solution of sodium thiosulphate, is thus equivalent to the DO originally present.

The following reactions takes place:

Manganous sulphate reacts with the potassium hydroxide-potassium iodide (alkaline-iodide) to produce a white flocculent precipitate of manganous hydroxide:

$$MnSO_4 + 2KOH \rightarrow Mn(OH)_2 \text{ (white)} + K_2SO_4$$

DO in the water reacts with $Mn(OH)_2$ immediately to form a brownish manganic oxide flocculate (ppt)

$$2Mn(OH)_2 + O_2 \rightarrow 2MnO(OH)_2 \text{ (brown)}$$

If the precipitate is white there is no DO.

Manganic oxide reacts with added H_2SO_4 to give manganic sulphate $[Mn(SO_4)_2]$ as the product of this reaction:

$$2MnO(OH)_2 + H_2SO_4 \rightarrow 2Mn(SO_4)_2 + 6H_2O$$

The $Mn(SO_4)_2$ immediately reacts with the potassium iodide (KI) (added initially as part of alkali-iodide), liberating the iodine exactly equivalent to the number of moles of oxygen present in the sample. The release of iodine (12) imparts a brown coloration of iodine.

$$2Mn(SO_4)_2 + 4KI \rightarrow 2MnSO_4 + 2K_2SO_4 + 2I_2$$

The liberated iodine is titrated with thiosulphate and the reaction is :

$$2S_2O_3^{2-} + I_2 \rightarrow S_4O_6^{2-} + 2I^-$$

From the above stoichiometric equations, we can find that:

$$1 \text{ mole of } O_2 = 2 \text{ moles of } MnO(OH)_2 = 2 \text{ moles of } I_2 = 4S_2O_3^{2}$$

Therefore, after determining the number of moles of iodine produced we can determine the number of moles of oxygen molecules present in the water sample. Dissolved oxygen concentrations are generally expressed in mg O_2/L.

From the titre values BOD is calculated.

Procedure

A known volume of the sewage water is diluted to a known volume with fresh water. Equal quantities of the diluted water are taken in two BOD bottles. In the first bottle BOD is determined immediately as follows. To a known volume of diluted water 5 ml of $MnSO_4$ solution and 5 ml of alkaline KI are added and shaken for about 15 minutes till MnO_2H_2O gets precipitated. Then the precipitate is dissolved in 2 ml of 1 : 1 H_2SO_4 and the liberated I_2 is titrated against standard sodium thiosulphate. The titre value is called blank value and it indicates the total dissolved oxygen available at the start of the experiment.

The water in the second bottle is incubated under aerobic conditions for 5 days and BOD is determined as described above. The titre value indicates dissolved oxygen present after 5 days.

Calculation

When dilution water is not seeded:

$$BOD_5, \text{mg/l} = \frac{D_1 - D_2}{P}$$

where D_1 = DO of diluted sample immediately after preparation, mg/l,

D_2 = DO of dilute sample after 5 d incubation at 20°C, mg/l, and
P = decimal volumetric of sample used.

Air Quality Standards

The nation-wide programe was initiated in 1984. As on March 31, 1995 network comprised 290 stations covering over 90 towns/cities districts 24 States and 4 Union Territories. The National Ambient Air Quality Monitoring (NAAQM) network is operated through the respective States Pollution Boards, the National Environmental Engineering Research Institute Nagpur and also through the CPCB.

The pollutants monitored are Sulphur dioxide (SO_2), Nitrogen dioxide, Suspended Particulate Matter (SPM) besides the meteorological parame wind speed & direction, temperature and humidity. In addition to conventional parameters, NEERI monitors special parameters, like) (NH_3), Hydrogen Sulphide (H_2S), Respirable Suspended Particulate (RSPM) and Polyaromatic Hydrocarbons (PAH).

Based on Annual Mean Concentration (microgram per cubic meter of air) of SO_2, NO_2 and SPM and the Notified Ambient Air Quality Stand. Ambient Air Quality Status is described in terms of Low (L), Moderate (H) and Critical (C) for Industrial (I), Residential and mixed use (R) Cities/Towns in different States/UTs.

Annual Average : Annual Arithmetic Mean of minimum 104 measurements in a year taken twice a week 24-hourly at uniform interval.

24 Hours Average : 24-hourly/ 8-hourly values should be met 98% of the time in a year. However 2% of the time, it may exceeded but not two consecutive days.

1. The level of air quality necessary with an adequate margin of safety, to protect the public health, vegetation and property.
2. Whenever and wherever two consecutive values exceeds the limit specified above for the respective category, it shall be considered adequate reason to institute regular/continuous monitoring and further investigations.

National Ambient Air Quality Standards

Pollutants	Time weighed average	Concentration in Ambient air			Method of measurement
		Industrial Areas	Residential, Rural & other Areas	Sensitive Areas	
SulphurDioxide (SO_2)	Annual Average*	80 μg/m³	60 μg/m³	15 μg/m³	Improved West and Geake Method Ultraviolet Fluorescence
	24 hours"	120 μg/m³	80μg/m³	30 μg/m³	
Oxides of Nitrogen as (NO_2)	Annual Average*	80 μg/m³	60μg/m³	15 μg/m³	Jacob & Hochheiser Modified (Na Arsenite) Method
	24 hours	120 μg/m³	80 μg/m³	30 μg/m³	Gas Phase Chemiluminescence
Suspended Particulate Matter (SPM)	Annual Average*	360 μg/m³	140 μg/m³	70 ug/m³	High Volume Sampling, (Average flowrate not less than 1.1 m³/minute)
	24 hours**	500 μg/m³	200 μg/m³	100 μg/m³	
Respirable Particulate Matter (RPM) (size less than 10 microns)	Annual Average *	120 μg/m³	60 μg/m³	50 μg/m³	Respirable particulate matter sampler
	24 hours**	150 μg/m³	100 μg/m³	75 μg/m³	

Lead (Pb)	Annual Average*	1.0 μg/m^3	0.75 μg/m^3	0.50 μg/m^3	ASS Method after sampling using EPM 2000 or equivalent
	24 hours	1.5 μg/m^3	1.00 μg/m^3	0.75 μg/m^3	Filter paper
Ammonia	Annual Average*	0.1 mg/m^3	0.1 mg/ m^3	0.1 mg/m^3	
	24 hours"	0.4 mg/m^3	0.4 mg/m^3	0.4 mg/m^3	
Carbon Monoxide (CO)	8 hours"	5.0 mg/m^3	2.0 mg/m^3	1.0 mg/m^3	Non Dispersive Infra Red (NDIR)
	1 hour	10.0mg/m^3	4.0 mg/m^3	2.0 mg/m^3	Spectroscopy

NOTE :

National Ambient Air Quality Standard: The levels of air quality with an adequate margin of safety, to protect the public health, vegetation and property.

Whenever and wherever two consecutive values exceeds the limit specified above for the respective category, it would be considered adequate reason to institute regular/continuous monitoring and further investigations.

The standards for H_2S and CS_2 have been notified seperately vide GSR No.7, dated December 22, 1996 under Rayon Industry - for details please see SI. No. 65 of this document.

S.O. 384(E), Air (Prevention & Control Of Pollution) Act, 1981, dated April 11 1994 [EPA Notification: GSR 176 (E), April 02, 1996]

NATURAL GEYSERS

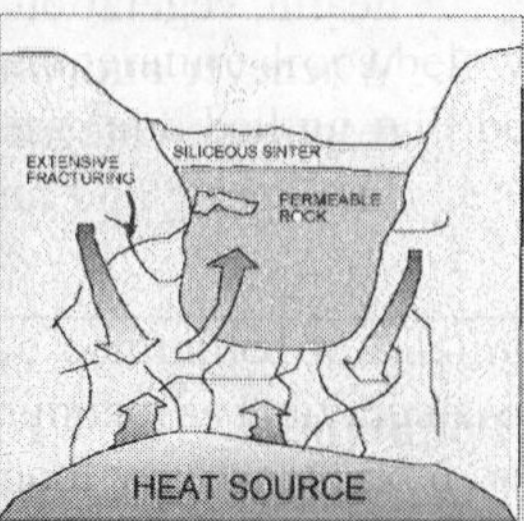

Natural geysers are hot springs that intermittently throw up a jet of hot water, stream, etc. The earth's crust has cracks and vents. Water occupies these and gets heated due to geothermal energy.

Due to heating, water turns into steam and pressure increases. In this case, if the opening of the vent/ crack is narrow, it erupts like a volcano, otherwise it flows like a stream. The 'Old Faithful' geyser at Yellowstone National Park in the US is one of the most famous.

Geysers

Geysers are **natural fountains** that throw up **jets** of hot water and steam at regular intervals through a vent in the surface. In some areas, rainwater seeps through cracks in the rocks and drains into a crevice or a large cave-like chamber so deep that it reaches hot rocks. The heat of these rocks comes from the molten rocks below. Eventually, the intense heat **boils the water,** which then turns into steam. This increases the pressure inside the crevice as bubbles of steam build up. Finally, the pressure is strong enough to shoot the water and steam upwards and out through a vent, high into the air. When the jet has died down, the crevice fills with new water and the process repeats.

Some **geysers** gradually lose their power and stop erupting as the **volcanic heat** dies down. A **geyser** that experienced this is Iceland's **Great Geyser**, which gave its name to all other **geysers.**

Hot Springs

Hot springs are **gushes** of hot water that are found on the land surface. As molten materials **deep in the earth** cool down, they give off **water vapor** and **carbon dioxide.** This hot vapor then find its way upward through the **cracks** in the rocks, cooling as it goes, until it condenses to become water. Finally, it gushes from the ground as a **hot spring.** This water may be pure and clear, but it is **rich in mineral** salts dissolved from the rocks it has passed through on its way to the surface. As water cools around the edges of the spring, the minerals form **crystals** and **grains.** These build up into beautiful rock layers shaped like waves, basins and terraces. These **hot springs** can be found in Japan, New Zealand, Kenya and Iceland.

A geyser is a hot spring that periodically erupts, throwing water into the air. Though that sounds simple, geysers are extremely rare. As of August 2008, the total of active geyers on earth numbered approximately 1000.

Conditions must be just right present for geysers to exist: of heat, and unique come from volcanic activity, thrown into the air, geyser. Geyser scientists and observers particularly effective at can deposit a water-tight of the geyser fields in the rocks (like ignimbrite). The for geysers to occur. Three components must be an abundant supply of water, an intense source plumbing. Water is common in nature, heat can but the plumbing is critical. For water to be plumbing must be water- and pressure-tight have identified the volcanic rock rhyolite as being hosting geysers. Rhyolite is high in silica, which seal along the walls of the geyser plumbing. Most world are found in rhyolite or similar silica-laden mixture of water, volcanic heat, and plumbing is exceptional at Yellow National Park. Over one-half of the world's geysers are located within the park's boundaries.

It is increasingly apparent that geysers must possess a fourth characteristic to exist: remoteness. Within the last fifty years, volcanic heat and abundant water have been increasingly harnessed to turn turbines for electricity production. Geothermal energy can be produced at any site where volcanic heat and water are readily available. Unfortunately, geyser fields are ideal for this type of energy production. Geothermal energy production steals the geysers' water, and destroys geyser activity (for example, Wairakei, New Zealand). A growing threat to geysers stems from mineral extraction. Hot groundwater may precipitate gold or other valuable minerals, and extraction may require removing the geyser plumbing itself. For example, in May 2003, mineral exploration at South America's second largest geyser field (Puchuldiza, Chile), caused cessation in the field's geysers. Few realize the actual rarity of geysers. As a result, many geyser fields have been destroyed and many others are being threatened.

How do geysers work?

The following is an excerpt from Scott Bryan's **GEYSERS OF YELLOWSTONE, 3rd edition, copyright 2001.** It is reproduced here for educational purposes. Scott Bryan's book not only describes each Yellowstone geyser in detail, but also includes descriptions of geyser fields worldwide. It is probably the best book on geysers out there. Buy it or check it out!

The hot water, circulating up from great depth, flows into the plumbing system of a geyser. Because this water is many degrees above the boiling point, some of it turns to steam instead of forming liquid pools. Meanwhile, additional, cooler water is flowing into the geyser from the porous rocks nearer the surface. The two waters mix as the plumbing system fills.

The stream bubbles formed at depth rise and meet the cooler water. At first, they condense there, but as they do they gradually heat the water. Eventually, these steam bubbles rising from deep within the plumbing system manage to heat the surface water until it also reaches the boiling point. Now the geyser begins to function like a pressure cooker. The water within the plumbing system in hotter than boiling, but "stable" because of the pressure exerted by all the water lying above it. (Remember that the boiling point of liquid is dependent upon the pressure. The boiling point of pure water 212 degrees Farenheit (100 degrees Celsius) at sea level. In Yellowstone the elevation is about

7,500 feet, the pressure is lower, and the boiling point of water is only 199 degrees Farenheit (93 degrees Celsius).

The filling and heating process continues until the geyser is full or nearly full of water. A very small geyser may take but a few seconds to fill whereas some of the larger geysers take several days. Once the plumbing system is full the geyser is about ready for an eruption. Often forgotten but of extreme importance is the heating that must occur along with the filling. Only if there is an adequate store of heat within the rocks lining the plumbing system can an eruption last for more than a few seconds. Again, each geyser is different from every - other. Some are hot enough to erupt before they are completely full and do so without any preliminary indications of an eruption. Others may be completely full well before they are hot enough to erupt and so may overflow quietly for some time before an eruption occurs. But, eventually, the eruption will take place.

Because the water of the entire plumbing system has been heated to boiling, the rising steam bubbles no longer collapse near the surface. Instead, as more very hot water enters the geyser at great depth, even more and larger steam bubbles form and rise toward the surface. At first, they are able to make it all the way to the top of the plumbing system. But a time will come when there are so many steam bubbles that they can no longer simply float upwards. Somewhere they encounter some sort of constriction or bend in the plumbing. To get by they must squirt through the narrow spot. This forces some water ahead of them and up and out of the geyser. This initial loss of water reduces the pressure at depth, lowering the boiling point of water already hot enough to boil. More water boils, forming more steam. Soon there is a virtual explosion as the steam expands to over 1,500 times its original, liquid volume. The boiling rapidly becomes violent and water is ejected so rapidly that it is thrown into the air.

The eruption will continue until either the water is used up or the temperature drops below boiling. Once an eruption has ended. The entire process of filling, heating, and boiling will be repeated, leading to another eruption.

Environmental Club

The Environmental Club is a group of concerned students that are committed to raising awareness about environmental issues, and to reducing our environmental impact as individuals, a school community, and as citizens of the world. During the 2008/2009 school year we worked on refining our recycling system, reducing paper waste, conducting a power use audit of the building (as part of our EcoSchools renewal application), creating a visual display in the school, setting up a recycling depot for items such as printer cartridges and cell phones, organizing waste-free lunch days, and hosting an Earth Day assembly. Thanks to all our efforts from the student and staff, we achieved gold status as an EcoSchool! The EA also helps to maintain the plants in our Solarium/ school.

"You must be the change you wish to see in the world." This famous quote by Mahatma Gandhi precisely explains what the JYC Environment Club is all about. We are a club with the mission to bring about a change. A change in the habits, thought and actions of people around us. Constituted 3 years back, the club has grown from strength to strength and is progressively involved in activities which aim at spreading environmental awareness among masses and help in promoting social well being of the underprivileged. All our events, activities and endeavors are inspired by our vision of creating a more beautiful world and uplifting our mother earth from its current deteriorated state. Taking into account the current trends of our environment and the issues associated with it, we believe it is time we take a stand and contribute to the fullest in helping Mother Nature recoup and sustain. Our purpose is to grow an environmentalist into each student of the university, to bring a smile on the faces of underprivileged people and to work towards a brighter, cleaner and more environment friendly future. We do not claim to drastically alter the current climate change trends or bring global warming under control. But we do believe in doing our bit and hope that our contribution would help in making this world a better place to live.

Dear EarthTalk: I'm thinking about starting an environmental club in my middle school. Can you give me some ideas about how to start? Can you connect me with other school clubs?

– Rosemary, Andover Township, NJ

Starting an environmental club at school is a great way to get students energized about taking care of the Earth and helping their community while learning about some of the most important issues facing the world in the 21st century.

EarthTeam, a non-profit environmental network for teens, teachers and youth leaders, offers many tips on how to start an environmental club. First and foremost is to make sure there are at least a half dozen or so other students interested in forming such a club to begin with, and then also finding a teacher, community leader or parent who is willing to serve as an adult sponsor. The sponsor's role is to provide advice along the way and to help ensure the stability of the group from year-to-year given that all of the students, even the founders of the club, will eventually graduate, or move on to other interests or endeavors.

Once the core membership and adult sponsor have been established, EarthTeam suggests all sitting down together to decide on the club's vision ("Why are we here?") and to brainstorm about possible activities or projects to undertake ("What do we want to accomplish?"). Once these questions have been answered, it's time to hold the club's first official meeting, which should be advertised as widely as possible to other students who may be interested in finding out what the group is about and how they can get involved, too.

The next step, according to EarthTeam, is to forget an action plan that focuses on one group-oriented, year-long project that has measurable benefits to the school or community and that can keep the interest of the student members–who will no doubt be spending long hours volunteering. Whatever project(s) the group decides on, members should develop a timeline that clearly lists goals, dates and responsibilities.

In addition to undertaking the one major project, clubs can also host or sponsor special events for extra visibility. Earth Team suggests getting students outside for a river or beach clean -up, a tree planting day, or a field trip to a local wetland, zoo or nature reserve. Another popular idea is to hold an Environmental Awareness Day to educate the entire student body about relevant green issues.

EarthTeam is also networking platform so clubs can work together and share experiences with each other to help get a sense of the bigger picture beyond one individual school's locale, given the global nature of most environmental issues. Another great networking resource is the Greenspan website, which lists clubs in 21 different U.S. states as well as Australia, Canada, Japan Ghana an Malaysia.

Another great resource for those starting up new or managing existing school environmental clubs is the U.S. Environmental Protection Agency's (EPA's) Student Centre website, which offers dozens of ideas for projects that both stimulate and enlighten participants while helping community. The website also provides links to several partner non-profit groups with club-worthy activities.

Green Accounting : A Virtual Resource Center

The centre provides a searchable database of various materials and Internet links related to integrated environmental and economic accounting also known as green accounting. It supports the objectives and programme of work of the UN Committee of Experts on Environmental-Economic Accounting (UNCEEA).

The database is organised in five categories: It provides a summary of each document and, where possible, a, link to its full text. The Search Documents functionality can be used to search the database. In cases where the text is not available online, the author's contact information will be provided. Upon request, UNEP ETB may also provide information and documentation on publications not available online.

1. Background and Introduction

The main objectives of this paper are to argue the case for Green Accounting for India (i.e. a framework of national accounts and state accounts showing genuin net additlons to wealth) and to present a preferred methodology and models to reflect natural capital and human capital externalities in India's national accounts, measuring as depreciation the depletion of natural resources and the future costs of pollution, and rewarding education as an addition to human capital stock. Our overriding purpose is to show that Green Accounting for India is desirable, feasible, realistic and practicable and that a start can be made with avalaible primary data already being collected by various official sources of the Government of India.

There is a dearth of focused sustainability analysis and information provided to policy makers at the National and State levels in India. As a result, the processes of public debate, government planning, budgetary allocation, and the measurement of economic results are in effect being conducted without a sustainability framework. High GDP growth usually accompanies investment in physical infrastructure, which places mounting pressure on the country's environment and natural resources. However, there is an asymmetry between man made and natural capital in that depreciation in the former reflects in GDP accounts but the latter does not. Recognizing that GDP growth is too narrow a measure of economic growth and not a measure of national wealth, we propose a "Green Accounting" framwork for India and its States and Union Territories.

This paper is an outcome of the "Green Accounting for Indian States & Uniton Territories Project" ("GAISP")[1] which aims to set up top-down economic models tor annual estimates of adjusted Gross State Domestic Product ("GSDP") for all Indian States, thus capturing true "value addition" not just at a National level but at State level too. India has a federal political structure - State legislature and administration has considerable impact on local environmental policies and standards. Whilst states are governed by the some national laws on environment, forests, and wildlife, their environmental attitudes and policies differ, & the range and effectiveness of their environmental and forest management programmes differ quite considerably. Known anecdotally. this feature is not ceptured systemetically, There are no metrics to distinguish sound and unsound environmental performance by India's State governments. Sustainable development at a State level remains unmeasurable, and therefore un-manageable.

The Central Statistical Organisation (CSO) in India is working on a methoclology to systematically incorporate natural resources into national accounts in different states for land, water, air, & sub-soil assets. However, the CSO approach develops accounts for some states & for some sectors, and their studies are still in progress. In contrast to the CSO approach, we use a top-down or macroeconomic approach to model adjustments to GDP/GSDP accounts for two reasons.

Firstly, a top-down approach has the advantage of providing a consistent and impartial national framework to value hitherto unaccounted aspects of national end state wealth and production. Secondly, it optimises extensive existing research which is not yet tied together in a manner to be useful for policy analysis. Thus we hope to provide a much-improved toolkit based on international best practice for India's policy makers to evaluate the economics of their policy trade-offs, and will enable them and the public to engage in a better informed debates on the sustalnability of economic growth, using national as well as inter-state comparisons. The materiality of calculated externalities in sectors such as education, health and natural resources are particularly interesting as these sectors are essential contributors to both sustainable development and poverty eradication.

Rationable for a system of "Green Accounting" for India

India has spent the past decade building a growth dynamic that was missing in the earlier quasi-socialist regime. The cumulative impact of the reform process appears to be generating growth, however, it is also desirable to monitor and channel the forces of growth and investment in order to ensure that they truly improve the quality of life for current and future generations, and to manage the economy sustainably, must also measure it with the lens of sustainability. Furthermore, there is an asymmetry between man made an natural capital int hat depreciation in the former reflects in GDP accounts but the later does not. In this context, it should be recognized that GDP growth is too narrow a measure of economic growth and not a measure of national wealth, and this is why we propose "Green Accounting" framework for India and its States and Union Territories.

Key externalities in the form of creation and destruction of Human Capital and Natural Capital both need to be explicitly measured, because they have a significant impact on the long-term sustainability of India's growth.

HONEY & BEE KEEPING

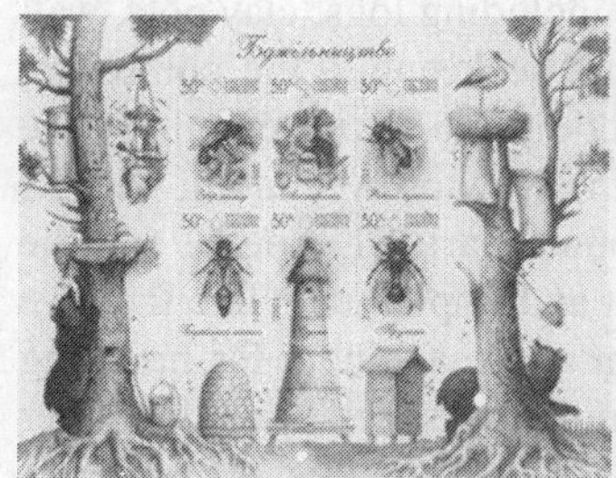

Honey and beekeeping have a long history in India. Honey was the first sweet food tasted by the ancient Indian inhabiting rock shelters and forests. He hunted bee hives for this gift of God. India has some of the oldest records of beekeeping in the form of paintings by prehistoric man in the rock shelters. With the development of civilization, honey required an unique status in the lives of the ancient Indians. They regarded honey as a magical substance that controlled the fertility of women, cattle, as also their lands and crops. The recent past has witnessed a revival of the industry in the rich forest regions along the sub-Himalayan mountain ranges and the Western Ghats, where it has been practiced in its simplest form.

In India beekeeping has been mainly forest based. Several natural plant species provide nectar and pollen to toney bees. Thus, the raw material for production of honey is available free from nature. Bee hives neither demand additional land space nor do they compete with agriculture or animal husbandry for any input. The beekeeper needs only to spare a few hours in a week to look after his bee colonies. Beekeeping is therefore ideally suited to him as a part-time occupation. Beekeeping constitutes a resource of sustainable income generation to the rural and tribal farmers. It provides them valuable nutrition in the form of honey, protein rich pollen and brood. Bee products also constitute important ingredients of folk and traditional medicine.

The establishment of Khadi and Village Industries Commission to revitalize the traditional village industries, hastened the development of beekeeping. During the 1980s, an estimated one million bee hives had been functioning under various schemes of the Khadi and Village Industries Commission. Production of apiary honey in the country reached 10,000 tons, valued at about Rs. 300 million.

Side by side with the development of apiculture using the indigenous bee, Apis cerana, apiculture using the European bee, Apis mellifera, gained popularity in Jammu & Kashmir, Punjab, Himachal Pradesh, Haryana, Uttar Pradesh, Bihar and West Bengal. Wild honey bee colonies of the giant honey bee and the oriental hive bee have also been exploited for collection of honey. Tribal populations and forest dwellers in several parts of India have honey collection from wild honey bee nests as their traditional profession. The methods of collection of honey and beeswax from these nests have changed only slightly over the millennia. The major regions for production of this honey are the crests and farms along the sub-Himalayan tracts and adjacent foothills, tropical forest and cultivated vegetation in Rajasthan, Uttar Pradesh, Madhya Pradesh, Maharashtra and Eastern Ghats in Odisha and Andhra Pradesh.

Resources and Potential

The raw materials for the beekeeping industry are mainly pollen and nectar that come from flowering plants. Both the natural and cultivated vegetation in India constitute an immense potential for development of beekeeping. About 500 flowering plant species, both wild and cultivated, are useful as major or minor sources of nectar and pollen. There are at least four species of true honey bees and three species of the stingless bees. Several sub-species and races of these are known to exist. In recent years the exotic honey bee has been introduced. Together these represent a wide variety of bee fauna that can be utilized for the development of honey industry in the country. There are several types of indigenous and traditional hives including logs, clay pots, wall niches, baskets and boxes of different sizes and shapes. In modern beekeeping, the combs are built on wooden frames that are moveable. This facilitates inspection and management of bee colonies. Three types of moveable frame hive are in common use: the Newton type along with its standardized version ISI Type A, the Jeolikote Villager, and its counterpart ISI Type B, and the Langstroth type. Besides the hives, the beekeepers need equipment and implements like the hive stand, nucleus box and smoker. The industry also needs equipment and machinery for handling and processing of honey, beeswax, for manufacture of comb foundation sheets, and for other operations.

India has a potential to keep about 120 million bee colonies, that can provide self-employment to over 6 million ural and tribal families. In terms of production, these bee colonies can produce over 1.2 million tons of honey and about 15,000 tons of beeswax. Organized collection of forest honey and beeswax using improved methods can result in an additional production of at least 120,000 tons of honey and 10,000 tons of beeswax. This can generate income to about 5 million tribal families.

Apiculture Technology

Production of honey has been the major aim of the industry. Modern beekeeping also includes production of beeswax, bee collected pollen, bee venom, royal jelly, propolis, as also of package bees, queen bees and nucleus colonies. All these are possible only with a proper management of bees, utilizing the local plant resources and adapting to the local climatic conditions.Modern beekeeping makes heavy use of beekeeping equipment and honey processing plant. This results in high efficiency and also ensures the quality of the processed honey.

Seasonal management of bee colonies varies in different parts of the country although the basic management methods are the same. Flow management, dearth management, provision of feeding, and control and cure of bee disordres, bee diseases, pests and enemies, are some of the routine measures to keep bee colonies healthy and strong. There are special management techniques like queen rearing, migration for honey production or for colony multiplication, which the beekeeper takes up after he gains sufficient knowledge and experiences in handling bee colonies.

Social Forestry

The term social forestry was used by the National Commission on Agriculture in 1976, to denote tree programmes to supply fire wood, small timber and minor forest products to rural population.

Rural income generation through massive plantation work & revenue earning from wood stock value, selling the medicinal plants and energy crops generated by inter crop management are the important task to improve socio-economic condition of rural masses.

Carbon credit earning through Clean Development Mechanism (C D M) will be an additional INCOME benefit by SOCIAL FORESTRY for Afforestation/Reforestation and Waste land development.

It's a community based work on massive plantation through PANCHAYAT/VILLAGE ASSEMBLY involving farmers, village workers, Govt & private bodies etc under JOINT VENTURE programme.

SOCIAL FORESTRY is a management and protection of forest and afforestation on the degraded land with the purpose of helping in the Environment,Social & Rural development.

Introduction & Objectives

According to the National Forest Policy 1988, one - third of the geographical area of the country should be maintained as forest. But Gujarat has only 9.69% of its geographical area declared as forest. More than 60% of the population resides in the rural area, which is directly dependention forest resources for their fuelwood, fodder, small timber for the agriculture and other requirements. Due to these reasons, regeneration, maintaining the productivity level and sustainability of the non-forest lands, particularly the common lands has become imperative for the state for a continuous supply of rural needs. Promotion of tree planting on non forest land, through Social Forestry programme has, therefore, been considered to be the only answer to improve the green cover in Gujarat state. With this background, as early as 1969-70, Gujarat Forest Department launched a "Social Forestry Programme" for planting trees on non-forest lands and became a pioneer and leading state in this field.

Objectives of Social Forestry

- Afforestation in lands outside forest areas
- Increasing the number of trees in Gujarat
- Promote the participation of institutions and people in the field of growing of trees.
- Increasing the yield of timber and other non-timber forest produce like fruit, firewood, fodder, etc to ensure easy supply to people.
- To put less fertile and unproductive land to productive use.
- Augment the income of people by tree planting.
- Increase the employment opportunities of rural poor.

Social Forestry Programme

The Social Forestry Programe which has got its impetus with funding from world bank US-AID and JBIC along with internal funding of Govt of Gujarat is at present implementing the following schemes.

- Strip Plantation
- Village Woodlot
- Rehabilitation of Degraded Farmlands
- Environmental Plantation

- o Decentralized School/Kisan Nursery
- o Van Mahotsava
- o Biogas/Solar Cooker Distribution
- o Improved Crematoria
- o Special Component Plan
- o Afforestation of Barren Hillocks
- o Planting of fruit trees on the sides of public roads with participation of farmers and landless labourers.
- o Different Prize Schemes

QUESTIONS

Short answer type questions :–

1. Differentiate exponential growth with geometric growth.
2. Discuss population growth in India.
3. How HIV/AIDS is spread ?
4. What do you mean by Indira Mahila Yojana and National Commission for women ?
5. Discuss the child development programmes in rural areas.
6. Write the causes of population explosion.
7. Explain the importance of family welfare programmes.
8. HIV and AIDS stands for.
9. Name any two organization concern with the women and child welfare.
10. What do you know about social factory?

Long answer type questions :–

1. Define population growth in various way. How it varies among nations ?
2. What is meant by "population explosion" ? Discuss it in Indian scenario.
3. What are family welfare programmes ? To what extent family planning is useful for population control.
4. Discuss the salient features of Human right movement.
5. What is value education ? How it can be described ? Give aims and methodology.
6. Discuss the role of information technology in environment and human health.
7. Write informative notes on :–

   (*i*) HIV/AIDS
   (*ii*) Zero population growth.
   (*iii*) Women and Child welfare.
   (*iv*) Natural Geysers
   (*v*) Environmental club
   (*vi*) Honey & Bee keeping
8. What is population explosion ? What steps are taken to check the population growth.
9. Write an essay on "Human Rights".
10. Describe the role of information technology in environmental studies.
11. Explain BOD. How it is determined?
12. Differentiate between DO, COD & BOD.
13. Write explanatory notes on (i) DO & (ii) COD.

BIBLIOGRAPHY

1. "Environmental Chemistry" by A.K. Dey Wiley Eastern limited - New Delhi 1989.
2. "Environmental Science & Engineering" by J. Gyn Henry et. al, Prentice Hall N.J. 1989.
3. "Environmental Engineering" by G.N. Pandey & G.C. Carney Tata Mc Graw Hill Publishing Co. Ltd. New Delhi.
4. "Nuclear & Radiochemistry" by J. Fried Landers and Kannedy.
5. "Air pollution" by H.C. Perkins Mc Graw Hills, Yew York.
6. "Air pollution" by A.C. Stern Academic Press, New York.
7. "Environmental Chemistry" by J.W. Moore & F.A. Moore Academic Press.
8. "Water and Waste Water Engineering" by Raju Tata Mc Graw Hill.
9. "Water Pollution" by V.P. Kudesia Pragati Prakashan Meerut.
10. "Water and Waste Water Technology" by M.J. Hammer, John Wiley & Sons Inc., New York.
11. "Solid Waste Management" by Hagerty, Pavoim, Heev & Jr. Van Van Nostrand Rein hold co., New York.
12. "Industrial Water Pollution, Origin Charasteristics and Treatment" by H.L. Nemerow, Addition - Wisley Publishing Co.
13. "Handbook of Pollution Control and Waste Minimization" Edited by Abbas Ghassemi Marcel Dekker Inc. 2002.
14. "Environment Regulations and Policy in India" by Ram Boojh. APH Publishing corporation, New Delhi
15. "A Text Book of Environmental Chemistry and Pollution Control" by S.S. Dara, S.Chand & Company Ltd., New Delhi.
16. Ministry of Water Resources & Management Reports Mammals.
17. "Environmental Science Working with Earth" by G. Tayler Milter Jr.
18. "Environmental Chemistry" by B.K. Sharma.
19. "Elements & Environmental Engineering" by Duggal.
20. "Water Resources Handbook" by Larry W. Mays, Mc Graw Hill, New York.
21. "Asian Tsunami Disaster" Published by Student Book Depo & Publisher, New Delhi.
22. "Natural Disaster & their Mitigation" by Indian Institute of Remote sensing, Dehradoon.
23. "The Violent Earth" by Subodh Mahant, Delhi.
24. "Earthquake Tips" by CVR Murty IIT, Kanpur.
25. "Earthquake Rebuilding in Gujrat India" Report by EEER Institute.
26. "Taemming the Flood" by Jeereny Parseglove, Oxford University Press, New York.
27. "Landslide" by Glissements Deterrain, New York.
28. "Indian Forest Law" by Wadhwa.
29. "Environmental Laws" by Babel.
30. IIT Publications Web Sites & bureau of Energy Efficiency.
31. Central Ground Water Board Reports.
32. Reports of Central Water Commission.
33. India's Mineral Resources by S. Krishnaswamy, Oxford & I.B.H. Publishing Co., New Delhi.
34. Hand Book of Agriculture I.C.A.R., New Delhi.
35. Energy Policy for India (Govt. of India Publication).
36. Applications of Non-Conventional & Renewable Energy sources (Urja Vikas Nigam).